UPDATED 6TH EDITION

ULTIMATE GUIDE
PLUMBING

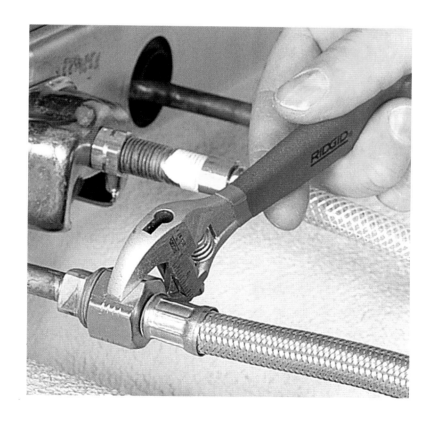

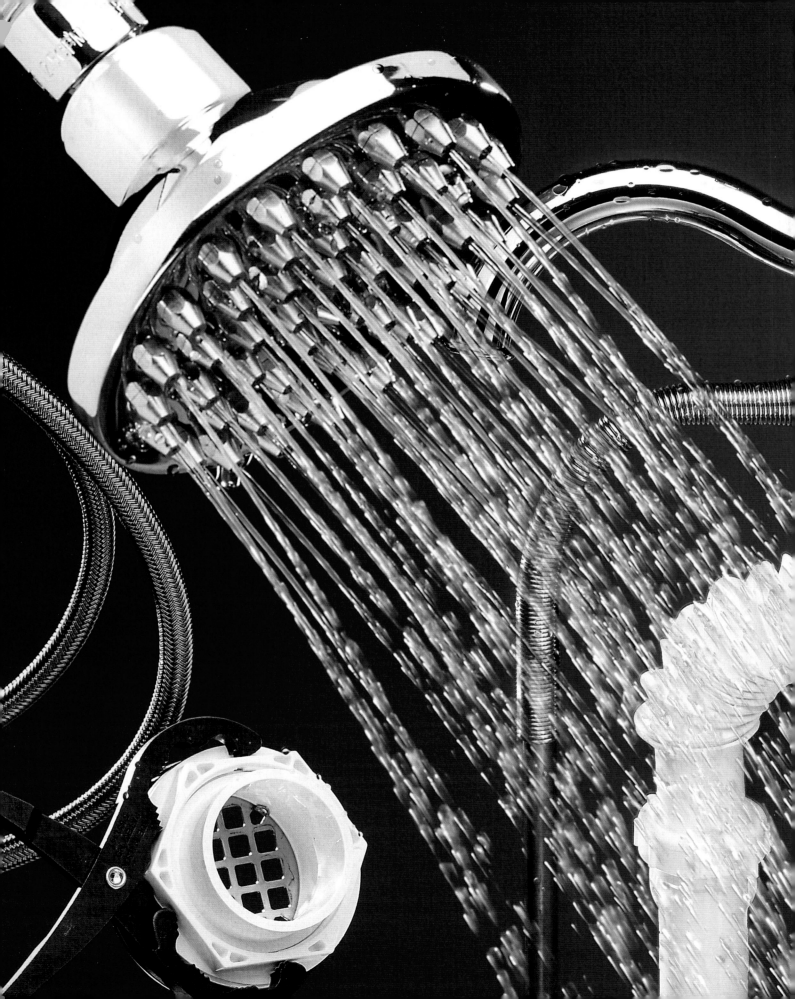

CREATIVE
HOMEOWNER®

UPDATED 6TH EDITION

ULTIMATE GUIDE

PLUMBING

MEETS 2024 **NATIONAL PLUMBING CODE** STANDARDS

Technical Editor for Updated Edition: Charles T. Byers
Associate Professor, Residential Remodeling Technology, AAS, AST,
Thaddeus Stevens College of Technology

CRE▲TIVE
HOMEOWNER®

ULTIMATE GUIDE: PLUMBING

PRINCIPAL AUTHOR:	Merle Henkenius
CONTRIBUTING AUTHOR:	Steve Willson
MANAGING EDITOR:	Fran Donegan
GRAPHIC DESIGNER:	Kathryn Wityk
PHOTO COORDINATOR:	Mary Dolan
JUNIOR EDITOR:	Angela Hanson
PROOFREADER:	Sara M. Markowitz
DIGITAL IMAGING SPECIALIST:	Frank Dyer
INDEXER:	Schroeder Indexing Services

UPDATED 6TH EDITION

TECHNICAL EDITOR:	Charles T. Byers, Associate Professor, Residential Remodeling Technology, AAS, AST, Thaddeus Stevens College of Technology
COVER DESIGNER:	David Fisk
ACQUISITIONS EDITOR:	Lauren Younker
MANAGING EDITOR:	Gretchen Bacon
EDITOR:	Sherry Vitolo
DESIGNER:	Matthew Hartsock
PROOFREADER AND INDEXER:	Kelly Umenhofer

ISBN 978-1-58011-602-2

Library of Congress Control Number: 2024939825

We are always looking for talented authors. To submit an idea, please send a brief inquiry to acquisitions@foxchapelpublishing.com.

Printed in China
First printing

Creative Homeowner®, *www.creativehomeowner.com*, is an imprint of New Design Originals Corporation and distributed in North America by Fox Chapel Publishing Company, Inc., 800-457-9112, 903 Square Street, Mount Joy, PA 17552.

Safety

Although the methods in this book have been reviewed for safety, it is not possible to overstate the importance of using the safest methods you can. What follows are reminders—some do's and don'ts of work safety—to use along with your common sense.

▌ Always use caution, care, and good judgment when following the procedures described in this book.

▌ Always obey local plumbing codes and laws, available from the building inspector. This book is based on the National Standard Plumbing Code, which despite its name, is one of several regional plumbing codes in force in the United States. As of the time of publication, there is no truly national plumbing code.

▌ Always use a flame shield to protect combustible materials when using a torch for soldering. And keep a fire extinguisher nearby whenever using a torch, just in case.

▌ Always be sure that the electrical setup is safe, that no circuit is overloaded, and that all power tools and outlets are properly grounded. Do not use power tools in wet locations. Use a battery powered flashlight when working near or with water.

▌ Always read labels on solvents and other products; provide ventilation; and observe all other warnings.

▌ Always read the manufacturer's instructions, especially the warnings, for using a tool or installing an appliance.

▌ Always remove the key from any drill chuck (portable or press) before starting the drill.

▌ Always use a drill with an auxiliary handle to control the torque when using large-size bits.

▌ Always pay deliberate attention to how a tool works so that you can avoid being injured.

▌ Always wear the appropriate rubber gloves or work gloves when handling chemicals, soldering, or doing heavy construction.

▌ Always wear a disposable face mask when you create dust by sawing or sanding. Use a special filtering respirator when working with toxic substances and solvents.

▌ Never try to light a gas appliance, like a water heater, if you smell gas. Do not touch any electrical switch or use any telephone in the same building. Go to a neighbor's house, and call the gas supplier. If you cannot reach your gas supplier, call the fire department.

▌ Always wear eye protection, especially when soldering, using a plunger or auger, using power tools, or striking metal on metal or concrete; a chip can fly off, for example, when chiseling concrete.

▌ Never work while wearing loose clothing, open cuffs, or jewelry; tie back long hair.

▌ Always be aware that there is seldom enough time for your body's reflexes to save you from injury from a power tool in a dangerous situation; everything happens too fast. Be alert!

▌ Always keep your hands away from the business ends of blades, cutters, and bits.

▌ Always hold a circular saw firmly, usually with both hands.

▌ Always check your local building codes when planning new construction. The codes are intended to protect public safety and should be observed to the letter.

▌ Never work with power tools when you are tired or when under the influence of alcohol or drugs.

▌ Never cut tiny pieces of pipe or wood using a power saw. When you need a small piece, saw it from a securely clamped longer piece.

▌ Never change a saw blade or a drill bit unless the power cord is unplugged. Do not depend on the switch being off. You might accidentally hit it.

▌ Always know the limitations of your tools. Do not try to force them to do what they were not designed to do.

▌ Never work in insufficient lighting.

▌ Never work with dull tools. Have them sharpened, or learn how to sharpen them yourself.

▌ Never use a power tool on a workpiece—large or small—that is not firmly supported.

▌ Never carry sharp or pointed tools, such as utility knives, awls, or chisels, in your pocket. If you want to carry any of these tools, use a special-purpose tool belt that has leather pockets and holders.

Contents

Introduction

THESE ARE GOOD TIMES for do-it-yourself plumbers. Plumbing materials are lighter and easier than ever to install, and the range of quality products sold to homeowners is unprecedented. Fifteen years ago, many of these products were sold only through wholesalers to plumbers.

And the materials are affordable. Many faucets and fixtures cost less at home centers than they do at wholesale houses. This may be bad news for plumbers, but it's good news for you. With these advantages, all you need is help with the installations. That is the purpose of this book.

Ultimate Guide: Plumbing is, of course, loaded with detailed projects, but it also provides context. Part I takes you step-by-step through tasks, projects, improvements, repairs, and solutions to specific problems. Chapters 1 and 2 introduce you to working with all types of water and waste piping, including soldering copper, solvent-welding plastic, and working with PEX tubing. Chapter 3 deals with toilet repairs and installations. Chapters 4 and 5 help you with sinks and faucets, as well as waste-disposal units, dishwashers, and hot-water dispensers. Chapter 6 teaches you how to clear all kinds of drainpipes; then it's on to repairing and installing tubs and showers in Chapter 7. The final three chapters deal with so-called mechanicals: water heaters in Chapter 8; sump pumps, filters, and water softeners in Chapter 9; and septic systems, wells, and lawn sprinklers in Chapter 10. Every major (and some minor) system is covered.

Part II provides the background needed to accomplish almost any plumbing task, from the materials and tools needed to the importance of vents and traps in a properly functioning plumbing system.

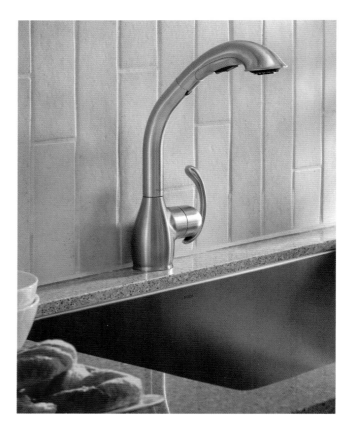

Faucets come in a variety of styles, such as the single-handle kitchen faucet, top, and the two-handle bath faucet, immediately above.

About Plumbing Codes

Many plumbing projects require permits, so **Ultimate Guide: Plumbing** discusses industry standards and code compliance, project by project. All projects in the book are based on the National Standard Plumbing Code, which despite its name is one of several regional plumbing codes, along with countless local codes, in force in the United States. As a practical matter, the only codes that really matter are those adopted by your local municipality. You'll learn enough about the fundamentals to work intelligently with your codes office. Because codes vary widely, it is best to consult with your local building department before beginning any large project. The officials there can tell you what is required in your area.

SIDEBARS

The numerous sidebars focus on a product, technique, or situation, providing additional detail or insight.

CONNECTING TO A DRAIN BENEATH CONCRETE

WHEN A TUB WILL BE SET ON CONCRETE, a 2-inch drain line is usually brought to a boxed out area and taped off. The box, usually made of framing lumber, is then partially filled with gravel. If a tub will not be installed immediately, concrete is poured over the top of the gravel and finished. This keeps insects, radon, and moisture from entering the living space, but leaves a weak spot for easy access. If the tub will be installed as part of the primary construction, the box is usually left open.

Only after the tub and drain assemblies are installed is this pipe connected, using a PVC P-trap. If the waste pipe is made of cast iron and ends in a hub, press a rubber gasket into the hub. Then dull the end of a short length of 2-inch PVC pipe; lubricate both the pipe and gasket; and push the pipe into the hub. Immediately out of the hub, install a 2-inch plastic coupling with a 2 x 1½-inch reducing bushing. Pipe a 1½-inch P-trap into this bushing. All other aspects are the same as those of an upper-story connection.

If a cast-iron line is without a final hub, join the plastic to the iron with a banded coupling. If the drain line is also plastic, make the connection with solvent-cemented fittings. When you've completed the connection and tested it with water, mix a small batch of concrete and seal the opening around the trap.

Connecting a Tub Drain beneath Concrete

Labels: 1½" Waste/Overflow Drain; Tub; Concrete Floor; Stud Wall; Mortar Patch; 1½" PVC Ground-Joint Adapter; 2" Cast Iron; Rubber Gasket; 2" PVC Coupling; 2" x 1½" PVC Bushing; 1½" Solvent-Welded PVC Trap

STEP-BY-STEP PHOTO SEQUENCES

A listing of skill level, tools, and materials accompanies each step-by-step photo sequence.

TOOLS & MATERIALS
- Dispenser, fittings ▪ Screwdrivers
- Open-end wrenches ▪ Groove-joint pliers
- Tubing cutter

Skill Level   Detailed Tools and Materials List

About the Projects and Text

Every numbered step-by-step project is keyed to the level of difficulty (below, right) and the average time required by a reasonably skilled homeowner to complete the project. Each project also has a listing of tools and materials as well as a useful tip for getting the job done right.

In addition, you'll find a host of **Smart Tips,** or insider information based on the author's many years of practical plumbing experience; **informative sidebars,** which provide additional information on points of interest, products, and techniques relating to plumbing; and **Green Solutions,** or tips, information, and projects relating to saving energy, conserving precious natural resources like water (by stopping leaks, for instance), and the like.

SMART TIPS

Insider information, shortcuts, new techniques, pitfalls to watch for—this (and more) is the stuff of Smart Tips.

smart tip
TO CAULK OR NOT TO CAULK?

WHEN IT COMES TO CAULKING AROUND THE BASE OF THE TOILET, OPINIONS VARY. SOME LOCAL CODES REQUIRE IT, SOME FORBID IT, AND OTHERS DON'T ADDRESS IT AT ALL. THE ADVANTAGES OF CAULKING AROUND THE BASE OF A TOILET ARE THAT THE CAULK 1) HAS GOOD ADHESIVE QUALITIES, 2) HELPS PRESERVE THE SEAL BY KEEPING THE TOILET FROM MOVING, 3) KEEPS THE JOINT FREE OF BACTERIA AND OTHER DIRT, AND 4) LOOKS BETTER THAN A DARK, SOILED PERIMETER CRACK (AS LONG AS THE JOINT IS NEATLY DONE). THE DISADVANTAGE IS THAT WHEN A LEAK DOES OCCUR, IT CAN'T BE SEEN, SO IT DOES MORE DAMAGE.

THERE ARE NO STRONG ARGUMENTS ONE WAY OR THE OTHER, EXCEPT IN TWO CIRCUMSTANCES:

▪ Caulking is usually recommended for toilets that are set on concrete. Concrete is never level around a toilet flange, and the added support from the caulked joint helps keep the bowl from rocking.

▪ Caulking is usually not recommended in a bathroom with wood floors (a poor flooring choice). These floors buckle with the slightest penetration of water, and you'll want to see a leak as soon as it starts.

TO CAULK AROUND THE BASE OF A TOILET, USE LATEX TUB-AND-TILE CAULK. LEAVE THE BACK OF THE TOILET BASE OPEN FOR WATER TO ESCAPE IN CASE A LEAK OCCURS. START BY WETTING THE FLOOR AND TOILET BASE; THEN APPLY A LIBERAL BEAD OF CAULK IN THE SEAM. DRAW A FINGER ALL AROUND THE JOINT SO THAT ANY VOIDS ARE FILLED, AND WIPE AWAY THE EXCESS USING A DAMP RAG OR SPONGE. KEEP IN MIND THAT LATEX CAULK WILL SHRINK A BIT WHEN IT CURES, SO LEAVE A LITTLE MORE IN THE JOINT THAN SEEMS NECESSARY. THE CAULK WILL DRY TO THE TOUCH WITHIN THE HOUR BUT WILL TAKE SEVERAL DAYS TO ACHIEVE ADHESIVE STRENGTH. IN THE MEANTIME, FEEL FREE TO USE THE TOILET.

If codes require caulking around the base of the toilet, use a good-quality mildew-resistant latex or silicone tub-and-tile caulk.

GUIDE TO SKILL LEVEL

 Easy. Even for beginners.

 Challenging. Can be done by beginners who have the patience and willingness to learn.

 Difficult. Can be handled by most experienced do-it-yourselfers who have mastered basic construction skills. Consider consulting a specialist.

PART I: Projects, Improvements, Repairs

working with water piping

THE TYPICAL HOME PLUMBING SYSTEM may appear to be a jumble of different size pipes running in all directions. But there is a logic to a well-designed system that almost anyone can learn to understand. The easiest way to approach the task is to separate the incoming water pipes from the outgoing drainage pipes. For an overview of how water piping and drainage systems work together, see Chapter 11 "Plumbing Basics," page 248. This chapter will cover working with water piping to help you get started with making repairs and adding new plumbing fixtures.

One drawback to working with water piping might be the water-pipe joining method. Most water pipes are made of copper, which must be soldered. Well-soldered joints require some skill, but with practice and the information in this chapter, good soldering technique is not beyond your reach.

Of course, you might worry about leaks, but leaks are fairly easy to repair. Just drain the system, and redo the offending fitting. It may be inconvenient and time-consuming, but it's not difficult.

Plastic water pipes rely on a solvent for attachment. This is a skill that is easy to master. The problem is that many codes restrict plastic to drainage systems.

Cross-linked polyethylene (PEX) is an alternative that has shown tremendous growth in the plumbing industry. These systems use crimp-ring and barbed fittings for joining pipe that are easy to master.

COPPER WATER PIPING

Copper piping for water supply is available in two forms—rigid, or hard tubing, and drawn, or soft tubing. Rigid copper comes in 10- or 20-foot lengths, while soft copper comes in 60-, 100-, and 120-foot coils. You typically use rigid copper for in-house, above-concrete water-piping installations, and soft copper for belowground applications and for connecting stub-outs with faucets. Along with black steel pipe, some codes allow the use of soft copper for both natural gas and propane piping.

Rigid copper is available in Type M and Type L wall thicknesses. Type M, thinner than Type L, is used predominantly in residential systems. Type L is more common in commercial installations. Soft copper comes in Type L and Type K wall thicknesses—Type K is heavier. You use Type L most often aboveground, as both water and gas piping, while you use Type K almost exclusively for underground water piping. Type K soft copper is also used to run water service lines between public mains and private homes.

You can join rigid copper with soldered—or sweat—fittings, compression fittings, and push-fit fittings. You can join soft copper with compression and flare fittings. Threaded adapters are available for joining copper to any other threaded material, including threaded steel and CPVC plastic. Only soldered and threaded fittings can be hidden in walls, however.

Cutting Water Pipes

The methods and equipment you need to cut water pipes depend on the piping material itself. Many people cut copper and galvanized steel with a hacksaw, but a tubing cutter leaves a more uniform edge. You can also cut plastic pipe with a tubing cutter, but most do-it-yourselfers reach for a hacksaw instead. The reason has less to do with the quality of the cut than with the availability of the tool. Tubing shears are probably the best cutting tool for plastic.

A clean, straight cut is also important. A tubing cutter can leave a compression ridge inside the pipe, while hacksaws leave coarse burrs. Ragged burrs protruding from a pipe's edge will eventually break off and make their way into control valves, appliances, and faucets. Severe edges also create friction in the water flow, called line friction, which can reduce pressure. And finally, raised edges generate turbulence, which can eventually erode the pipe wall. To prevent these problems, ream any severe edges left by a cutting tool before you install the pipe.

To ream a copper or plastic pipe, lift the triangular reaming attachment from the top of the cutter, insert it into the end of the pipe, and give it several sharp twists. When dealing with steel pipe, you'll need a more aggressive reaming tool—one with hardened-steel cutting blades. You can rent many of these tools. If you are making only a few cuts, use a rat-tail file.

USING HACKSAWS

CUTTING WITH A HACKSAW. Use as much of the blade as possible in long, easy strokes. If you work too fast, the blade will heat up and start binding. A hot blade also leaves a ragged pipe edge. The best approach is to steady the pipe on a solid surface and cut just to the left or right of the support. Some people like to use a miter box to ensure straight cuts.

Close-Quarters Hacksaws. When you need a smaller saw, you'll find that there are a variety of miniature hacksaws on the market. While it's not sensible to try plumbing an entire job with a tiny saw, they work wonders in cramped spaces. In fact, close-quarters hacksaws often work in situations too cramped for thumb cutters. The design shown in the photo below is usually preferable. In a pinch, some people remove the blade from a full-size hacksaw and use it alone. But hacksaw blades are fairly brittle, so remember to wear gloves.

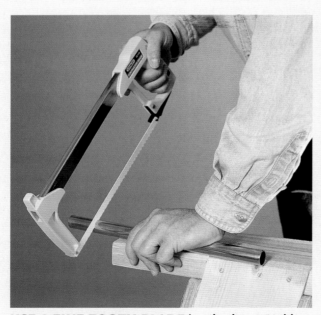

USE A FINE-TOOTH BLADE in a hacksaw. Hold the tubing steady, cutting near the support.

YOU CAN USE A MINIATURE HACKSAW as you would a full-size one, but it fits tight spaces.

USING TUBING CUTTERS

Use all tubing cutters in a similar fashion. First, mark the tube to length. Then clamp the cutter onto the tube, centering the cutting wheel on your mark. Rotate the tool's handle clockwise until the cutting wheel just bites into the tubing. Don't overdo it. If it's difficult to rotate the cutter around the tube, unscrew the handle, but just slightly. Then rotate the tool around the pipe several times. You'll feel slight resistance on the first one or two turns, but after that, the cutter will roll easily around the tube. This is your cue to tighten the wheel

TOOLS & MATERIALS

▌ Measuring tape
▌ Pencil/permanent marker
▌ Tubing cutter
▌ Deburring tool

against the pipe again. Rotate the cutter and repeat this procedure until you cut all the way through the pipe.

You use the same method for cutting plastic, steel, or copper. The only difference is that tubing cutters made for steel pipe are much heavier than those for copper or plastic. Because of this, there's a substantial cost difference. Buying a tubing cutter for plastic and copper does not make sense—one will work for both—but cutters made for steel pipe are strictly rental items.

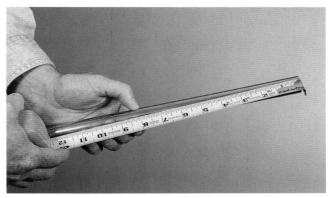

1 Hold the tubing in your left hand and the measuring tape in your right. Hold the location with your thumb; then mark it using a permanent marker or pencil.

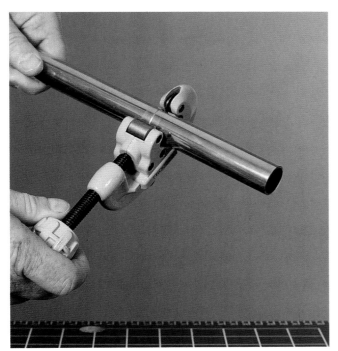

2 Tighten the wheel cutter to clamp it onto the tube at the mark, and rotate the cutter with the wheel following the rollers. Work slowly for the first couple of revolutions to make sure the cutter doesn't slip out of the cut.

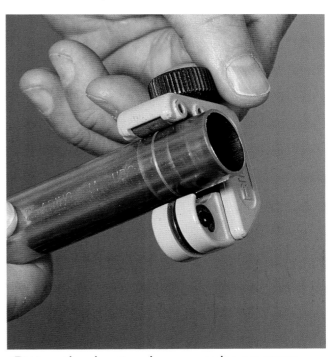

3 Use a thumb cutter when you need to cut copper tubing and you are confined by a tight working space. These tools don't provide the same mechanical advantage as longer cutters, so they are harder to tighten and to turn. Deburr the inside of the copper tube after cutting, using a deburring tool.

Working with Water Piping

Solder

Solder is metallic filler that bonds two metallic surfaces to itself. Flux helps this bonding to occur. Flux works by ridding the copper surfaces of oxidation and other contaminants. It pulls molten solder into the joint, even when the fitting is upside down. Where the flux goes, solder will follow. Without flux, molten solder will just bead up and fall away.

Until the 1980s, most of the solder used in residential plumbing was a 50-50 amalgamation of tin and lead. Lead also makes brass more easily machined, so most quality faucets contained lead until very recently.

The plumbing industry didn't realize that water, under fairly common conditions, could leach lead from soldered fittings. Even small amounts of lead ingested by a human being can cause brain damage. Today, the U.S. Environmental Protection Agency (EPA) bans the use of lead-based plumbing solder.

Soldering Copper

The three most common mistakes in soldering are using dirty fittings and using too much or too little heat. If fitting and pipe aren't clean down to shiny copper, the solder may not adhere well to the metal. With too much heat, you'll cook the flux from the fitting hubs, resulting in a weak bond, and with too little heat, the solder may not flow properly, also resulting in a weak joint. Other trouble sources are water left in fittings, which prevents the metal from heating up as the water absorbs the heat; heating the top end of a fitting first, resulting in uneven heat distribution; and not allowing for the greater mass and density of brass

EXISTING LEAD HAZARDS

WHAT IF YOUR HOME ALREADY HAS lead-based solder joints? Short of installing an expensive distiller or an equally expensive reverse-osmosis water filter, there's little you can do. However, you can take precautions, as discussed below.

Almost all soldered copper piping installed before 1988 had lead soldered joints. Does this mean that your home's plumbing puts you at risk today? It's hard to say with certainty, but probably not. There's a limit to the amount of lead available in a soldered fitting, and most of the exposed lead leaches out in the first 90 days. Water conditions also make a difference. Soft water is inherently more aggressive, so it dissolves some soft metals at a faster rate, depleting them sooner. With even slightly hard water, the joints scale over in a few years, sealing off the problem. The greatest likely hazard occurred when these systems were new.

Water must remain in contact with soldered joints six to eight hours before it can absorb much lead, so the greatest lead concentrations are present when the system hasn't been used for a while, such as overnight or while you're at work. If you flush a toilet, take a shower or let a little water run through the faucet before taking a drink, you pretty much avoid the lead risk. If you're concerned, let a faucet run for 15 to 20 seconds before drinking.

smart tip

START AT THE BOTTOM

ALWAYS START WITH THE LOWEST HUB ON A FITTING. THIS IS CRITICALLY IMPORTANT. WHEN YOU SOLDER THE LOWEST JOINT FIRST, THE SOLDER IN THAT HUB COOLS AND THICKENS SLIGHTLY BY THE TIME YOU MOVE TO THE UPPER JOINTS. IN THIS WAY, THE HOTTER, THINNER SOLDER FED INTO THE TOP OF THE FITTING DOES NOT DRAIN THROUGH THE BOTTOM. IF YOU REMEMBER THIS ONE PROCEDURE, YOU'LL CUT YOUR SOLDER LEAKS IN HALF.

valves and fittings when applying heat, which may result in the fitting not getting hot enough to melt the solder.

Bringing the Heat. Use a mapp (methylacetyline propane) gas torch, available at hardware stores for soldering. Mapp gas makes a hotter flame than propane. When you heat a joint for solder, always heat the fitting, not the pipe. Keep the torch moving, side to side, to avoid hot spots that can cook the flux. Heat just one fitting hub at a time, not the entire fitting.

Try to keep the torch tip about ¾ inch away from the fitting, and always keep it moving. As soon as the flux in the near side of the fitting begins to crackle and spit, move the flame to the far side of the fitting and heat it, again moving the torch in a side-to-side arc.

How much solder should you give each joint? As a rule a ½-inch fitting should get ½ inch of solder wire per hub, a ¾-inch fitting, ¾ inch of solder, and so on. When the fitting is uniformly hot, the flux will draw the solder completely around the joint quickly. When the joint cools a bit, look for the solder to draw into the rim slightly. When this happens, you'll know that the joint was a good "take." If the solder seems to just lie on the rim, add a little more heat until it gets drawn in. If that doesn't do it, brush new flux around the rim and add a little more heat and solder.

SOLDERING PROBLEMS AND SOLUTIONS

NOT ALL SOLDERING JOBS ARE SIMPLE.
Common problem situations include tight workspaces, fittings installed too closely to structural timbers. You can often pre-solder fittings that must rest against structural timbers or be installed deep inside cantilevers or walls. **A.**

When you can't avoid soldering against studs or joists, you can keep from scorching the wood by sliding a double thickness of sheet metal between the fitting and the structural member. **B.** It's handy to keep a 6-inch fold of sheet metal in your toolbox for this purpose. The sheet metal must have two layers. Plumbing outlets also sell squares of woven fireproof protective fabric, which also work well. **C.**

SOLDER DEEP-SET FITTINGS, like this freeze-proof sillcock, to their pipes before installing them.

USE A DOUBLE THICKNESS OF SHEET METAL to keep from scorching the wood.

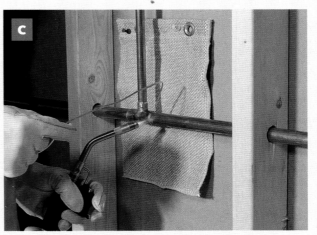

YOU CAN ALSO USE A FLAME SHIELD of fire-proof woven fabric to protect larger areas.

SOLDERING COPPER TUBING

project

At first glance, soldering together copper tubing and fittings seems like a messy job. You have dirty pipes, greasy flux, and dripping solder at nearly every turn. But the connecting points between tubing sections and fittings are anything but dirty. In fact, the most important step in sweating joints is to make sure all mating parts are clean before any flux, solder, or heat hits the joint. Specialized tools make this job easier. But don't forget, you can get good results with nothing but a pad of steel wool.

TOOLS & MATERIALS

- Cleaning tool or pad ▌ Flux & brush
- Tubing & fittings ▌ Solder ▌ Rag
- Gloves & goggles ▌ Soldering torch

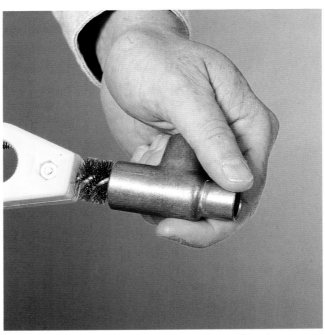

1 Use a combination tool, wire brush, or abrasive pad to clean the inside of each hub on the fitting. Combination tools come with one end for cleaning the inside of the fittings and a round recessed brush for cleaning the outside of the tubing. Tools are available for ½- and ¾-dia. tubing.

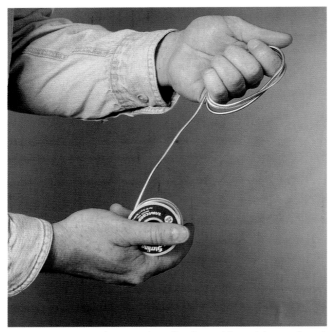

4 Because you have to handle the solder with one hand and the torch with the other, it's a good idea to pull about 24 in. of solder from the roll and wrap it around your hand. Holding solder this way and advancing it into the joint is much easier than struggling with a tightly wound roll of solder.

5 Shutoff valves are standard fittings used for regulating water flow throughout the system. Most feature some rubber and/or plastic parts. These will be damaged when the fitting is soldered unless you remove the valve stem first. Use an adjustable wrench, and turn the stem counterclockwise.

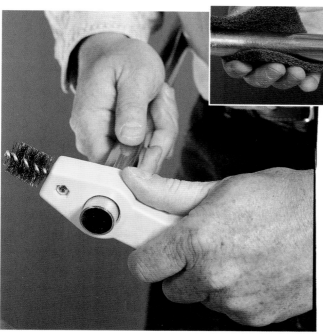

2 Use the tubing brush in the body of a combination tool to clean the ends on all tubing sections. Keep turning the tool until a 1-in.-long strip at the end of the tube is shiny. With a little more elbow grease, you can clean the tubing with an abrasive pad (inset photo) or steel wool.

3 Once the mating surfaces are clean, spread flux on the inside of the fittings and around the outside of the tubing using a small disposable brush. Then push the tubing into the fitting.

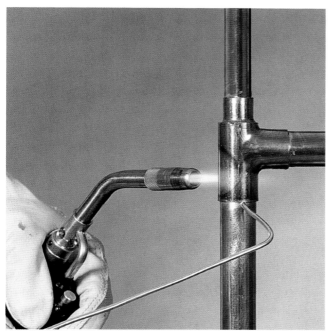

6 Start soldering at the lowest point. Adjust the torch flame so that the tip is blue colored, and direct this tip alternately to both sides of the fitting. Unroll a length of solder, and press it against the joint on the fitting and tubing. When the fitting is hot enough, the solder will melt and be drawn up into the joint.

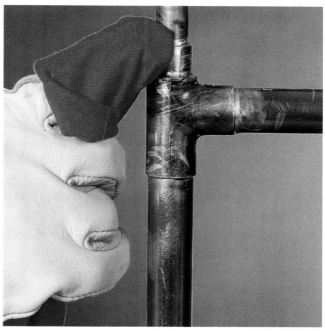

7 Joints are complete when they don't draw in more solder and any new solder starts to spill out. Once this happens, immediately wipe away any excess with a damp rag. If you wait for even 10 seconds, the solder can cool and harden, and you won't be able to wipe it away.

To stop a trickle of water, insert a special liquid-filled capsule before soldering.

A union installed at a low point makes a good drain-down fitting.

When Pipes Won't Dry Out

When old shutoff valves leak, the downstream pipes will continue to carry a trickle of water. It's difficult to make pipes and fittings that have even a small amount of water in them hot enough to accept solder. And even when you can make them hot enough, the escaping steam forces pinholes through the solder.

If that fails, you might try an old plumber's trick. Squeeze white bread into tight balls, and push it into the pipe using a pencil. If you work quickly, you can solder the fitting before the bread starts to disintegrate. When you have finished the job, detach the aerator from the nearest faucet, and flush the sodden bread from the line. Plumbing outlets also sell liquid-filled plastic capsules for this purpose, which you later dissolve with heat. **A.**

If all else fails and you simply can't keep water from trickling through the fitting to be soldered, cut the line at its lowest point so that it drains, and install a union or freeze-repair fitting when you have finished the work. **B.**

Other Methods of Joining Copper

While its best to assemble large piping projects with inexpensive soldered fittings, you can find other fittings for copper. These fittings fall into three categories, all of which are mechanical joints: compression fittings, flare fittings, and push-fit fittings.

Compression Fittings

A compression-type water fitting consists of a brass body—either an adapter body or valve body—with two or more pipe hubs. The fitting hubs have external threads and beveled rims. The nuts are open at the top so that you can insert pipes through them. A third component, a brass compression ring called a ferrule, makes the seal. The ferrule is also beveled, top and bottom.

You make the connection by sliding the nut and ferrule onto the end of a pipe and inserting the pipe into the fitting hub. As you tighten the nut, the beveled surfaces force the ring inward, cinching it around the pipe. Because the ring actually crushes the pipe a little, it locks the ring in place and makes the water seal.

You most frequently use compression fittings as conversion fittings under fixtures. Compression-type connectors normally come with shutoff valves, but they're also available as couplings and 90-degree L-fittings, in sizes ranging from 1/8 to 1 inch in diameter. You can also find valves and adapters with one threaded hub and one compression hub. Use these to join threaded brass water fittings to copper supply tubes.

COMPRESSION FITTINGS

Compression Valve

Ball Valve with Compression Fittings

Compression Fitting Adapter

INSTALLING COMPRESSION FITTINGS

project

Shutoff valves are the standard way to attach sinks and toilets to copper supply lines. They are located directly under the fixture and have a built-in shutoff valve. This valve lets you turn off the water to the fixture without going down into the utility room. These valves come in different configurations, as shown on the facing page. Often the valve has a pipe fitting on one end and a compression fitting on the other end that connects the tubing that goes to the fixtures. In most other cases the valve has compression fittings on both ends.

TOOLS & MATERIALS

▮ Stop valve and fittings ▮ Supply tubing
▮ Pencil ▮ Adjustable wrenches
▮ Tubing bender ▮ Tubing cutter

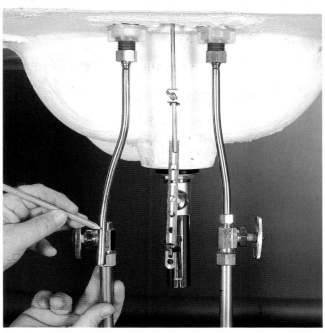

1 Gently bend the supply tube into shape using your hands or an inexpensive tubing bender. Test-fit the tube by attaching it to the bottom of the fixture and pushing it against the shutoff valve. Mark the tube to length; then remove it and cut it.

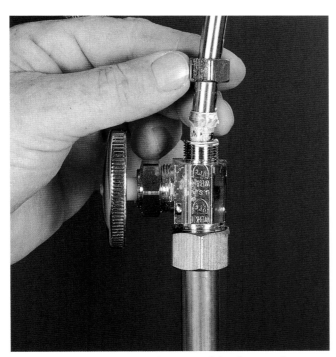

2 Attach the tube to the bottom of the fixture; then slide the compression nut and ferrule over the other end. Spread some pipe joint compound on the ferrule, and push the nut down onto the valve and tighten it with your fingers.

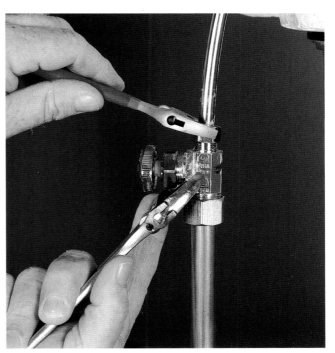

3 The best tools for tightening the nuts are a couple of small adjustable wrenches. Use one to hold the valve body and the other to tighten the compression nut. Do not over-tighten this nut. About 1½ turns should do the trick.

MAKING FLARE-FITTING CONNECTIONS

project

Compression fittings work on both rigid and soft copper pipe. But flared fittings work only on soft tubing, most often used for under-slab water piping and in-house gas piping. The goal is to expand—flare out—the end of a soft copper pipe to match the male end of a typical flared fitting. The most common type of flaring tool has a vise to hold the pipe and a separate flaring head that clamps to the vise and turns into the end of the tube. Once the end is flared the tubing nut is threaded onto the fitting.

TOOLS & MATERIALS

▮ Tubing and fittings ▮ Tubing cutter
▮ Pipe joint compound ▮ Tubing bender
▮ Groove-joint pliers ▮ Clamp flaring tool
▮ Adjustable wrenches

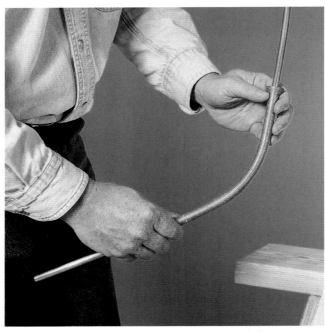

1 Bends in soft copper must be made carefully, or the tubing will kink. The best approach is to use a spring-type tubing bender. This tool just slides over the tubing, and you make the bend with your hands. The spring distributes the force over the entire length of the bender, so no kinks form.

4 Continue turning the flaring stem into the tubing. Generally, the flare is complete when a lip that is about ¹⁄₁₆ in. wide all around is formed. You don't have to measure this. Once the copper fills the tapered opening in the vise, the flare is done. This job usually takes a little trial-and-error to get right.

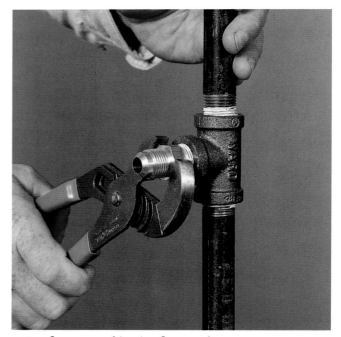

5 Soft copper tubing is often used to connect gas appliances to standard steel gas pipes. The flared fitting has pipe threads on both ends; install the standard (non-flared) end in the steel fitting. Apply a light coat of pipe joint compound to the threads before tightening the fitting in place.

Working with Water Piping

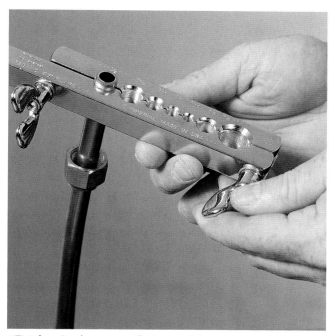

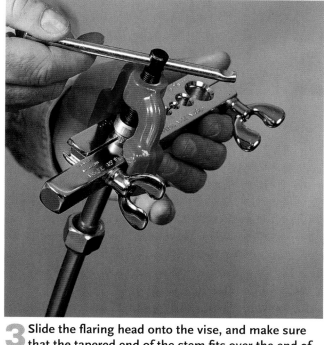

2 The vise base on a flaring tool has two parts. To use it, just slide the end of the tubing into its proper-sized hole and tighten both wing nuts securely. Be sure to slide the fitting nut onto the pipe before attaching the vise, and make sure the tubing extends ⅛ in. above the top of the vise.

3 Slide the flaring head onto the vise, and make sure that the tapered end of the stem fits over the end of the tubing. Once this stem is seated properly, turn the stem in a clockwise direction to start flaring the end of the tubing. Work slowly to ensure an even flare.

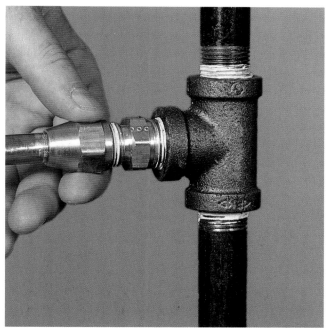

6 The free end of the flared fitting should be coated with pipe joint compound. Make sure to cover all the threads, but don't get the compound inside the fitting. Compound that spills into the fitting can move through the line when the gas is turned on and clog the orifices where the gas is burned.

7 Slide the nut to the end of the copper tubing, and press it against the fitting. Carefully thread the nut in place with your fingers to avoid crossing the threads. Don't switch to a wrench until the nut is at least halfway onto the fitting. Finish up by firmly tightening the nut with an adjustable wrench.

Working with Water Piping

HAMMER-TYPE FLARING TOOL

BEGIN BY SLIDING THE NUT onto the pipe. Because these tools can force the pipe slightly out of round below the flare, hold the nut less than an inch below the end of the pipe. If you do this, you won't have to worry that a slightly misshapen pipe will hold up the nut. Drive the tool into the end of the pipe with sharp, steady hammer blows. If you notice that one side of the pipe is being flared more than the other, even things out by striking the tool at a slight angle. Stop flaring when you see the outside diameter of the flare approaching the inside diameter of the fitting nut.

Push-Fit Fittings

Push-fit fittings, also known as crimp-ring fittings have gained wide acceptance because they are remarkably easy to use and almost never leak. They work on a variety of piping materials and can join dissimilar materials. They are available in a variety of configurations, the most ingenious of which is a braided stainless-steel freeze-repair coupling. To make an in-line repair, such as fixing a freeze rupture, you simply cut out the damaged section and splice in the repair piece, pushing each end fitting over its pipe. (See the photo on the opposite page.) Some codes don't allow push-fit fittings, however, and others allow them only when they remain exposed.

Another advantage to push-fit fittings is that you can rotate them on the pipe after you have installed them. This feature is handy when making retrofit installations.

Similar braided stainless-steel tubing is now offered in many forms, including toilet, sink, and clothes-washer supply tubes. You can fit these tubes with compression fittings, friction fittings, and crimp-ring fittings. While all

NATURAL- AND PROPANE-GAS PIPING

PROFESSIONAL PLUMBERS are usually the only ones to install in-house gas piping because, while there's little procedural difference between running water pipe and gas pipe, there certainly is a liability difference. A water leak can be costly, but a gas leak can be fatal.

With that in mind, there are aspects of the job that you should know about, if only to check the plumber's work. The piping between the gas meter and your appliances will be under either high pressure or low pressure. High-pressure systems allow smaller pipes, while low-pressure systems require larger pipes. High-pressure systems also require a pressure reduction regulator at each appliance, while low-pressure systems have a single regulator mounted on the meter.

Three piping materials are allowed for in-house gas piping: black steel, flexible stainless steel, or soft copper in Type L or Type K thickness. Low-pressure systems can use ¾-inch black steel or ¾-inch soft copper, in either thickness. As the ¾-inch feed line reaches each appliance, the branch line serving that appliance

will probably be reduced to ½ inch. When only one appliance remains, the feed line may also be reduced to ½ inch. With high-pressure systems, the entire run may be in ⅜- or ½-inch soft copper, with each branch line terminating in a regulator near the appliance. All flare fittings must have deep-shoulder nuts. Codes do not allow rigid copper. Some codes now disallow black steel, so make it a point to ask before any work is done on your system. It is being replaced by a new gas-only piping system made of flexible stainless steel, which is covered by a yellow plastic coating. Manufacturers use flare-type fittings, but each has its own design, and brands are not interchangeable. ***Note:*** *this new piping, called corrugated stainless-steel tubing (CSST), is not a DIY material.*

Codes require an approved gas shutoff valve within 36 inches of each appliance and a condensation-catching drip leg near each fixed appliance, including the furnace (or boiler) and water heater. All gas-pipe joints must be made with gas-compatible pipe joint compound, or in the case of black steel, gas-compatible

A good use of a push-fit fitting is this freeze-repair kit. Cut out the bad section of pipe, and bridge the gap.

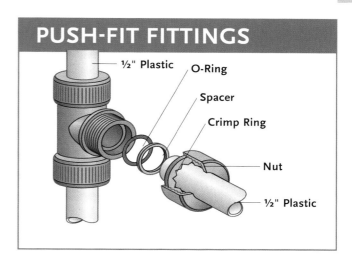

PUSH-FIT FITTINGS

½" Plastic — O-Ring — Spacer — Crimp Ring — Nut — ½" Plastic

codes allow stainless-steel-encased tubes, many do not allow nylon-reinforced versions.

How Push-Fit Fittings Work. Push-fit fittings come in two forms. You can remove the nuts from some of them, but others consist of one piece. If you take apart those with fastening nuts, you'll need to break off the crimping ring and install a new one.

Some push-fit fittings are brass, while others are plastic. The mechanism for all the fittings is similar. A push-

fit consists of a fitting body, neoprene O-ring, and metal crimping ring. You'll need to round the edge of both pipes with a file or grit cloth, lubricate the end of the pipes with plumber's grease, and push the fitting onto the pipes. When the pipe bottoms out in the fitting hub, the O-ring makes the seal and the crimping ring grips the pipe.

pipe-thread sealing tape. And finally, most codes now require braided stainless-steel connectors on movable appliances such as dryers and ranges. Some are available with built-in safety valves. Should the connector ever break, the valve would close immediately.

Testing Gas Lines. After charging the system and bleeding the air from the line, through a union or drip-leg cap, the plumber will test all the joints using an electronic gas detector. You can also test joints if

you ever suspect a leak. Use premixed testing soap (available at plumbing supply stores) or a mixture of dish detergent and warm water. Create a thick, soapy mixture by squirting about a tablespoon of liquid soap into a cup and mixing it with warm water. Then, using an inexpensive brush, coat each fitting connection with the mixture. A leaky fitting will produce bubbles. Have a plumber remake any joint where bubbles appear, and then retest it.

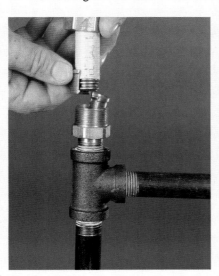

Plastic-coated CSST (corrugated stainless-steel tubing) for use with gas comes with a variety of proprietary fittings. This piping is not a do-it-yourself material.

Periodically test all gas fittings using liquid test soap or a mixture of household dish detergent and water. If you find a leak, turn off the gas and call in a plumber.

STEEL PIPING

Galvanized and black steel are the two types of steel pipe used in residential plumbing. You can install galvanized steel as water or gas pipe. However, use black steel, where allowed, only as gas pipe. Don't be tempted to use black steel in your plumbing system. The use of black steel for water is prohibited by many codes because water causes black steel to rust quickly.

Aside from their separate uses, there's little difference in how the two are cut and fitted. You can cut both types of steel with a heavy wheel cutter or a hacksaw and use threaded fittings on both. You can purchase short, threaded nipples at any hardware store, but you'll need to rent threading dies to cut threads on custom lengths. (See the "Rental Tools" photograph, page 263.) While few people use either type today, there is plenty of it in place, and you may need to know something about it to make repairs and additions.

Threading pipe is heavy work even when you use the specialized tools shown here. And assembling the parts can be even harder. But the result is strong, durable pathway for water or gas.

project

CUTTING, THREADING, AND FITTING STEEL PIPE

You can use a bench vise to work with steel piping, but a tripod-mounted pipe vise is a much better tool because it provides a more secure grip and greater mobility. You are likely to rent a cutter and threading tools anyway, so you may as well rent a pipe vise, too.

Your threading chores are limited to the ends of pipe, not to the fittings that join them. The fittings are sold as finishing units with the threads, whether male or female, already cut. They are sized to match standard pipe sizes.

TOOLS & MATERIALS

▌ Steel pipe and fitting ▌ Pipe vise
▌ Threading die ▌ Pipe cutter and reamer
▌ Pipe joint compound ▌ Cutting oil
▌ Pipe wrenches

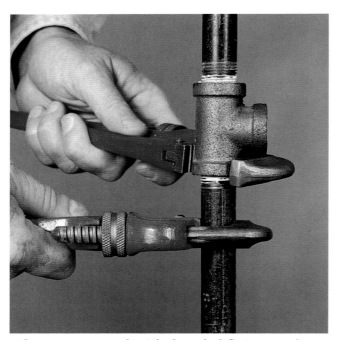

Whenever you work with threaded fittings, make sure you backhold the pipe with a second wrench.

3 Cut threads on the end of the pipe using a threading tool with interchangeable dies. First apply oil to the end of the pipe (inset); then press the die onto the pipe; and turn the handle clockwise until the die starts cutting threads. Keep turning the handle until the threads are complete.

1 Begin by marking the correct length on the pipe with a soft pencil or felt marker. Then clamp the pipe in the vise; slide the wheel cutter onto the pipe; and start tightening the handle by turning it clockwise. Once the cutter wheel hits the pipe, turn it around the pipe, tightening the wheel as you go.

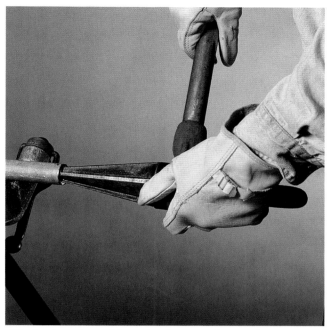

2 Cutting pipe with a wheel cutter almost always leaves steel burrs on the inside of the pipe. These can be removed with a round file, but a reaming tool works much faster. Just press it into the end of the pipe and turn. One or two turns should do the trick.

4 Remove any steel filings from the inside of the pipe; then spread pipe joint compound over the newly cut threads, using the applicator that comes with the container. Don't spread compound on the inside of the pipe because it will foul the water or gas that the pipe will carry.

5 After applying the joint compound, thread the appropriate fitting onto the end of the pipe, and tighten it by turning a pipe wrench in a clockwise direction. It's a good idea to support the fitting with a scrap pipe nipple loosely threaded into the open end of the fitting.

Working with Water Piping

ELECTROLYSIS

WHEN YOU JOIN COPPER OR BRASS with galvanized-steel pipes and fittings, you can cause a serious problem—electrolytic corrosion, or electrolysis. This corrosion breaks down the steel side of the joint. It stems from a disparity between each metal's inherent electrical charge. Copper and brass have more positively charged ions than the iron in steel, so the iron continues to give up electrons, and over time the steel side of the joint corrodes.

Electrolysis occurs when dissimilar metals come in contact, but the rate of corrosion varies locally. The damage may not reveal itself for decades, or it may take only a year or two.

When joining dissimilar metals in plumbing, dielec-

A dielectric union allows you to join copper and iron without electrolytic corrosion.

tric unions are normally required. A dielectric union differs from a conventional union in that a plastic spacer separates the two halves of the fitting. The threaded adapters are also different. In most cases, one is brass and the other is steel. The plastic spacer prevents direct contact, which prevents corrosion. You install these unions just as you install other unions. Check with your building department for requirements.

Alternative Methods of Joining Steel Pipe. If the thought of all this threading and wrenching wears you down, take heart. You'll often be able to repair or extend steel lines with newer, more friendly materials. In the case of black-steel gas piping, you might be able to cut the line and splice in soft copper tubing, using couplings that have a flared fitting on one end and male or female threads on the other.

If local codes allow plastic water piping, you might also splice in a length of CPVC plastic pipe, using easy-to-install threaded adapters. In this case, you'd remove the old pipe, thread male adapters in the female steel-pipe threads, and cement a length of plastic pipe into these fittings. But be sure you don't interrupt your home's electrical grounding system. (See "Electrical Grounding and Plastic," opposite.)

EXISTING LEAD HAZARDS

IN THE EARLY 1900s, plumbers used lead water piping. They considered lead a real problem solver because it is malleable and can accommodate moderate seasonal soil movement without creating pipe failure. Plumbers generally installed it in underground water service lines between public mains and private homes. They did not use it for the entire system. Even today, millions of homes still have lead loops in their water service lines.

To determine whether your older home—built prior to 1945—has lead service loops, check the service pipe as it enters the house. If you see galvanized-steel joints with visible threads, lead is not present. But if you see a dull gray loop of pipe that joins galvanized

piping with bulging, seamless joints, that's lead. These joints are seamless because the plumber repeatedly wiped molten lead over the lead-steel transition. To confirm your suspicions, scrape the pipe surface with a knife. If the scraped area is soft and shiny, it's lead.

The presence of calcified mineral deposits inside the pipes reduces the lead hazard in older systems. Because lead loops are part of the underground service pipe, replacement is costly. And because there's no way of knowing how a buried-service-line repair will go, few plumbers will give you a binding estimate. It will almost always be a cost-plus repair, and it's not something you can do on your own. That said, lead pipes should be removed.

Working with Water Piping

smart tip

INSULATING WATER PIPES

YOU CAN DO LITTLE THAT IMPROVES A WATER-PIPING JOB AS MUCH AS PIPE INSULATION. INEXPENSIVE AND EASY TO INSTALL, PIPE INSULATION DAMPENS MUCH OF THE NOISE OF RUNNING WATER AND LIMITS FREEZING. IT ALSO ELIMINATES CONDENSATION IN SUMMER AND REDUCES ENERGY COSTS YEAR-ROUND IN TWO WAYS. FIRST, IT REDUCES THE AMOUNT OF HEAT LOST THROUGH THE PIPE WALLS. SECOND, BECAUSE HOT OR COLD WATER TEMPERATURES ARE MAINTAINED LONGER IN THEIR RESPECTIVE LINES, YOU'LL SPEND LESS TIME—AND WATER—RUNNING THE TAP, WAITING FOR IT TO WARM UP OR COOL DOWN.

PIPE INSULATION COMES IN SEVERAL FORMS. THE EASIEST ONE TO INSTALL IS MADE OF FOAM RUBBER. IT COMES IN A VARIETY OF LENGTHS AND DIAMETERS, AND IS PARTIALLY SPLIT ALONG ONE SIDE. WHEN YOU INSTALL IT ON NEW PIPING, YOU SLIDE IT DIRECTLY ONTO EACH NEW LENGTH, LEAVING THE SPLIT INTACT. ON EXISTING INSTALLATIONS, CUT THE INSULATION TO LENGTH USING A UTILITY KNIFE, OPEN THE SEAM, SLIP THE INSULATION OVER THE PIPE, AND SEAL THE JOINT WITH DUCT TAPE OR PLASTIC SEALING TAPE.

To insulate existing pipes using foam insulation, open the seam and tape the insulation in place.

PLASTIC WATER PIPING

Plastic water pipe, made of chlorinated polyvinyl chloride (CPVC), has been around for years, and when properly installed, has proved to be durable. Its appeal, of course, is its ease of installation. CPVC piping can be installed with the most common of household tools and by people with almost no previous experience. The problem is that some plumbing codes have not come up to speed on CPVC as a potable water carrier.

Electrical Grounding and Plastic. Few homeowners understand the relationship between a home's plumbing and electrical systems. In many jurisdictions, the electrical panel is grounded through metallic water piping. Because the metallic piping inside the house connects to a metallic water service pipe that is buried underground, most codes require that the electrical system use this piping for all or part of its path to ground in order to have a safe installation.

If you cut out a section of cold-water trunk line and splice a length of plastic piping in its place, there's a good chance that you'll interrupt this path to ground. That's a dangerous situation. If you decide to splice plastic into a cold-water trunk line, install a heavy grounding conductor across the span. This jumper wire should be the same size as the service panel's existing grounding wire, usually 6 gauge. Attach the wire to the metallic pipes using code-approved grounding clamps, one on each side of the splice. You won't need a jumper where the grounding wire connects directly to the water service pipe on the street side of the meter.

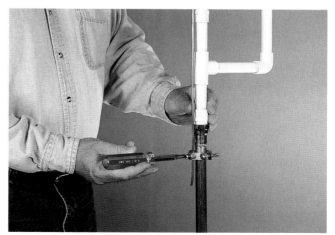

When splicing plastic into copper piping, you may need to install a copper jumper wire to maintain an unbroken ground for the home's wiring system.

CUTTING AND FITTING CPVC TUBING

CPVC tubing comes in a variety of diameters: ⅜, ½, ¾, and 1 inch. The fitting assortment made for other kinds of piping is also available in CPVC. The best way to cut it is using special shears. These cutters slice the plastic cleanly, leaving almost no ridges or ragged edges. Like the PVC made for drainpipes, CPVC comes with a surface glaze that you must remove before you cement it. The best remover is a solvent-primer, which comes in containers with lid-mounted applicators. Dab primer on the tubing area to be cemented. Do the same to the inner surfaces of the fitting hubs. Allow the primer to evaporate before making the joint. You can also cut the glaze by scuffing the final 1 inch of the tubing and the inner surfaces of each fitting hub.

After you remove the surface glaze, test-fit the joints you want to make. Make alignment marks on the fitting and the tubing using a pencil or marker. You'll use these marks later to line up the tubes and fittings in exactly the same positions when you rotate the parts as you put them together. Apply a thin but even coating of joint solvent cement inside the hub of the fitting and to the outside of the tubing, using the container's applicator. Insert the tubing into the fitting, and rotate it one-quarter turn, using the alignment marks you made previously as a guide. Rotating the tubing or fitting helps fill any voids in the cement. As with PVC drainpipe, once you have cemented the joint you'll have very little time to change your mind. If you find that you've made a mistake, pull the joint apart immediately; apply new solvent cement, and remake the joint.

TOOLS & MATERIALS

▌ Pipe-cutting shears ▌ CPVC solvent primer
▌ CPVC tubing and fittings ▌ Pencil
▌ CPVC solvent cement ▌ Measuring tape
▌ Deburring tool

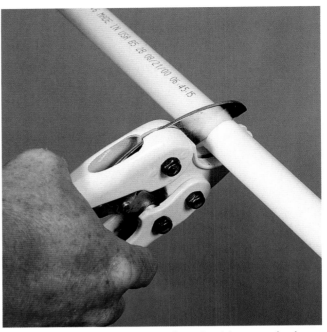

1 Plastic tubing can be cut to length with a standard tubing cutter or a common hacksaw. But the fastest and cleanest cut is achieved with plastic-pipe-cutting shears. To use this tool, just mark the tubing; hold the shears perpendicular to the tube; and squeeze until the tube is cut through.

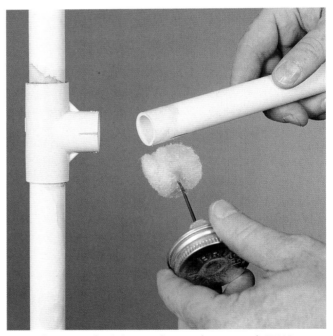

4 Take apart the joint, and spread approved CPVC plastic cement on the mating parts using the applicator that's mounted on the container's cap. Make sure the entire circumference of the tubing and the inside of the fitting are coated. Work quickly so the cement doesn't start to dry before the parts are assembled.

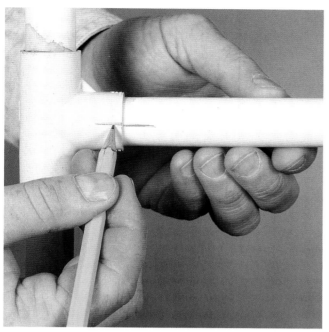

2 Before the mating parts are cleaned with primer, dry-assemble all the components, and test-fit the parts to make sure you are satisfied with their positions. Then make alignment marks at all the joints using a pencil or felt tip marker. Make heavy lines that won't rub off during assembly.

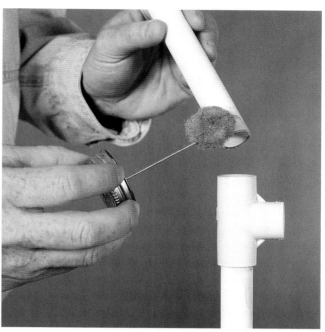

3 Once the tubing sections are cut, join them with a plastic fitting. Be sure to deburr the inside of the CPVC pipe, then clean the outside ends of the tubes and insides of the fitting with primer. This product removes the surface glaze and any oil and dirt. Use the applicator that comes with the primer container.

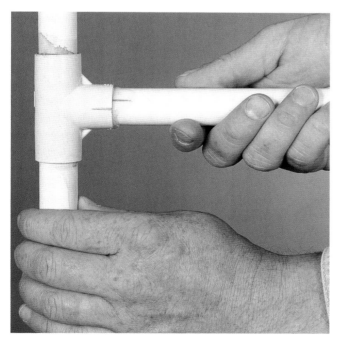

5 To assemble the parts, push the tubing into the fitting so that the marks are about ¼ turn out of alignment. Make sure the tubing is completely seated in the fitting; then turn the tube until the marks align. You only have a few seconds to do this job. Once the cement starts to set, you can't move the parts.

CPVC VERSUS COPPER

IS CPVC PLASTIC AS GOOD AS COPPER for common residential plumbing applications? In most cases, yes. But this is true only if it is not used underground or under concrete, if it has no chance to freeze, and when the installation is made according to manufacturer's specifications and is code-worthy. In some cases—for example, in cabins and second homes where water can stand long enough to corrode copper—plastic may actually be the best choice.

But in cold climate conditions, plastic can be a real problem. When water-filled plastic tubing freezes, the tubing tends to break before the fittings, usually along its entire length. Neither plastic tubing nor fittings can be repaired; they must be replaced. Copper plumbing is also damaged by freezing water, but the breaks tend to occur at the joints between tubing and fittings. These joints are easier and less expensive to fix.

Working with Water Piping

IN-LINE WATER-CONTROL VALVES

Shutoff valves, or stops, for water piping have a variety of control mechanisms. Each mechanism type has its advantages and disadvantages, and understanding those pluses and minuses will help you decide where to use each one. For example, some valve mechanisms restrict flow by as much as 50 percent. This is a real concern in certain situations. The main house valve, for instance, should not restrict flow, because a reduction there could affect the performance of the entire system. In contrast, the small-bore compression valves under sinks have little effect because the faucets they serve restrict flow in the first place.

Gate Valves

Gate valves, along with some ball valves, allow an unrestricted flow of water. As the name implies, the control mechanism is an internal gate. Turn the handle to the left, and the gate rises out of its seat. Turn it to the right, and the gate moves downward, slicing off the flow. The sliding gate makes gate valves larger than other valves.

Like most full-bore water valves, gate valves are not meant to be used frequently. They can't survive the wear and tear that a faucet endures on a daily basis. If the gate valve has a serious nemesis, it's hard water. Calcified minerals can encrust the mechanism, rendering it inoperable. Forcing the valve can break the stem from the gate. In most cases, replacement is the best option.

Because of their full-flow designs, gate valves are used close to the beginning of the water piping system, usually near the meter or pressure tank.

Ball Valves

Ball valves are available in flow-reducing and larger full-flow versions. The mechanism in these valves is a hollow nylon or metal ball that pivots in the valve body. Opposing sides of the ball are open. When these openings are oriented with the flow, water passes through the ball. When you rotate the valve against the flow, the closed sides of the ball stop the water. Ball valves are easy to use because it takes only one-quarter turn of the handle to open or close them. In addition, they are generally more durable than gate valves. They're certainly less vulnerable to calcification. Ball valves can be used anywhere a gate valve might be used.

Globe Valves

Globe valves have been the standard in the industry for years and are used most often in full-size, in-line piping, although they can restrict flow by nearly 50 percent. Some newer designs—with slightly larger bodies—are less restrictive.

Globe valves are popular for several reasons. To begin with, they are the least expensive types of valve. They are small and compact, and because they use a compression mechanism, you can service them easily and inexpensively. All it takes is a 5¢ washer. Unlike ball valves, however, globe valves do not withstand frequent operation. While the valve seat and mechanism hold up well enough, the graphite stem packing in many models leaks after each use. A half-turn of the packing nut stops the leak, but it's a nuisance. Some models now have O-ring packing and nylon bonnet-nut seals, which correct the problem.

Globe valves, in ½- and ¾-inch diameters, are common

IN-LINE VALVES

A gate valve has an internal wedge that controls water flow.

A ball valve uses a hollow ball at its core to control water flow.

A compression valve uses a threaded stem and rubber washer.

in residential plumbing. They do not work well as the first valves in the system, but they work in branch lines. Although not the best idea because they restrict flow, many people install them above water heaters in the cold-water inlet line. Check local codes, however.

Stop-and-Waste Valves

Stop-and-waste valves are usually globe valves that have a drain screw on the downstream side of the shutoff mechanism. This screw allows you to drain water from the downstream piping after you have shut off the valve. Stop-and-waste valves are useful for pipes that you need to shut down during winter, such as dedicated lines serving outdoor faucets, sprinklers, outbuildings, and the like. If you are draining an entire system, such as a summer home or cabin, however, a hose bibcock drain is the best choice. (See "Hose Bibcocks," page 32.)

Fixture Stops

Fixture stops, or shutoffs, come in two types: compression and ball-valve. Compression stops are small brass valves, usually chrome plated, that join permanent water piping to supply tubes. These stops have a compression-type shutoff mechanism, but they are called compression stops because compression fittings join them to piping. Valves that join copper pipes to copper supply tubes have two compression fittings, one large and one small. Those joining threaded iron pipe to supply tubes have ½-inch female threads at the lower end. The supply-tube connection has a compression fitting.

Ball-valve stops use a nylon ball mechanism instead of a compression-type one to control water.

The most common compression and ball-valve stops reduce in size from ⅝ inch outside diameter (O.D.) to ⅜ inch O.D., although other sizes are available. The valves come in straight (in-line) and angled configurations. You'll find straight stops when the water piping enters through the floor and angle stops when it enters through the wall.

Dual Stops. Dual stops are merely compression stops with two outlet ports. You'd use a dual stop to join a kitchen's hot-water supply line to the kitchen faucet and a dishwasher, for example. A two-stop valve allows you to work on the dishwasher without shutting off the faucet and vise versa.

CONFUSING TERMINOLOGY

THE TERM "COMPRESSION" has two meanings in plumbing. Any fitting that uses a tapered brass or nylon ring to make its seal is a compression fitting. These include water fittings and valves, as well as plastic tube traps and under-sink waste kits. Confusion arises when you talk about shutoff valves, however. People often call valves that have similar compression-style connectors compression valves. But the term compression also describes a shutoff mechanism in which a stem-fitted washer makes the seal when it is compressed against a raised seat. By this definition, globe valves, hose bibcocks, and many traditional sink faucets are compression valves, although they may or may not be joined to piping by compression fittings.

A stop-and-waste valve is drained by unscrewing a threaded cap.

A ball-valve fixture stop is the most reliable form of shutoff valve.

A dual stop valve splits one water line (usually hot) into two.

HOSE BIBCOCKS

A HOSE BIBCOCK IS A DRAIN VALVE with external hose threads on its spout. While it may be connected to its piping via male or female threads or through a compression fitting, a hose bibcock's identifying feature is its hose threads, which are larger and coarser than iron-pipe threads. You can use hose bibcocks to join permanent water piping to clothes-washer hoses and garden hoses. You'll also find hose bibcocks on boilers and water heaters.

Most hose bibcocks are brass and use a compression mechanism for controlling water. A hose bibcock used as an outdoor sillcock (a wall-mounted outdoor faucet) usually has female threads and a brass flange that is predrilled to accept screws. In this case, a threaded water pipe extends through the exterior wall about ½ inch, and the valve is threaded onto the pipe.

Male threads on some hose bibcocks allow you to attach the faucets to elbows and other appropriate fittings.

Female threads on other hose bibcocks allow you to thread the units onto supply pipes.

UTILITY FAUCET VALVES

Some valves are intended to control water for utility purposes (to a yard hose, for example) or for occasionally draining appliances like water heaters.

smart tip

PREVENTING LAUNDRY-ROOM WATER HAMMER

HOSE BIBCOCKS THAT HAVE MALE THREADS ARE USUALLY CONNECTED TO FEMALE FITTINGS MOUNTED ON A LAUNDRY ROOM WALL. WHEN THE SYSTEM PIPING IS COPPER, THE FEMALE FITTINGS SHOULD BE 90-DEGREE DROP-EARED ELBOWS THAT HAVE PREDRILLED SIDE TABS (EARS), WHICH ALLOW THE FITTING TO BE ANCHORED WITH SCREWS. BECAUSE SOLENOID VALVES IN WASHING MACHINES SHUT OFF ABRUPTLY, CAUSING THE BACKSHOCK KNOWN AS WATER HAMMER, IT'S IMPERATIVE TO HAVE ANCHORED FITTINGS HERE. (SEE "WATER HAMMER ARRESTORS," PAGE 216.)

Freeze-Proof Sillcocks

In areas with cold winter weather, sillcocks need some form of freeze protection. Traditionally, this has meant a two-valve assembly. You installed a hose bibcock valve outside, on the exterior wall, and an in-line valve just inside the house, in the joist space. When it began to get cold, the homeowner dutifully shut off the inside valve and drained the outside valve.

While this setup works well enough, it has two disadvantages. First, the faucet can't be used during the winter months. Second, homeowners are generally not good at routine maintenance like this.

First Freeze-Proofing Attempts. The freeze-proof sillcock was an attempt to correct the freezing problem, with generally good results. The trick was to move the stop mechanism well inside the house, where it could stay warm, while maintaining the handle and spout outside, where you could use them year-round. Those first freeze-proof sillcocks had a long stem that extended through an oversize tube, or drain chamber. Because the washer and seat operated in a warm environment and any water left in the chamber drained out when you shut off the faucet, it was a truly all-weather faucet.

These faucets would have been perfect but for one flaw: they only drained when the hose was removed. An

FREEZE-PROOF SILLCOCKS

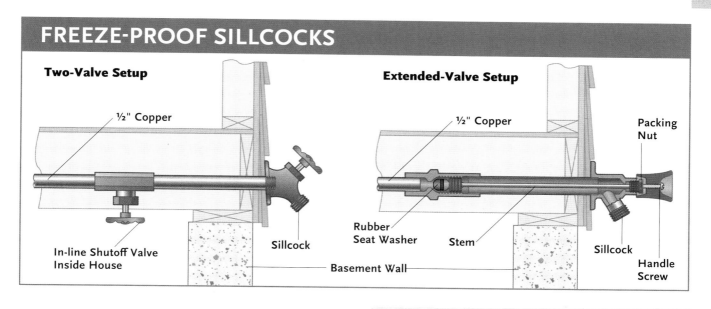

Two-Valve Setup

½" Copper

In-line Shutoff Valve Inside House

Sillcock

Extended-Valve Setup

½" Copper

Packing Nut

Rubber Seat Washer

Stem

Sillcock

Handle Screw

Basement Wall

attached hose created an air lock, which held water in the chamber. This water then froze and ruptured the sillcock. Because the freeze occurred in the chamber, on the downstream side of the stop mechanism, the leak only appeared when the sillcock was used again, usually the following spring. By then, of course, the hose had been stored away for months and the homeowner had forgotten last fall's early cold snap. Because an attached hose was about the only way to rupture a freeze-proof sillcock—and an attached hose voids the warranty—these sillcocks have been the source of a good many disputes between homeowner and plumber.

Modern Solution. Most new freeze-proof sillcocks have a vacuum breaker threaded onto the spout to solve the hose-related freeze problem and prevent back siphonage. The vacuum breaker releases the air lock created by the attached hose and drains the chamber. Some new so-called freeze-proof sillcocks include a top-mounted backflow preventer instead of a vacuum breaker. These prevent contaminating backflows, but they're not always freeze-proof with a hose attached. Nevertheless, codes now require sillcocks with either backflow preventers or vacuum breakers. When installing a freeze-proof sillcock, always prop up the rear of the chamber so that water can drain through the spout.

Freeze-proof sillcocks come in lengths ranging from 6 to 36 inches. Choose a length that places the stop a foot or so inside the house. If you are installing a sillcock in a cantilevered floor or wall, it will need to be longer, of course, as will those you install on a house with brick or stone facades.

FREEZE-PROOF SOLUTIONS

Aftermarket vacuum breakers can be installed on existing sillcocks.

Press plumber's putty around the freeze-proof sillcock mounting flange to seal the opening.

REPAIRING A FREEZE-PROOF SILLCOCK

Repairing freeze-proof sillcocks is not much more difficult than repairing a standard valve-stem-and-seat faucet. The only real difference is that any leftover rubber washer pieces that may have broken off the end of the valve stem are harder to fish out of the faucet body. If debris like this is left inside, the new replacement washer won't seat properly. The best tool for this job is long-handled needle-nose pliers. A good alternative is to duct tape the end of a shop vac hose to the end of the sillcock and vacuum out any loose material.

TOOLS & MATERIALS

▌ Replacement washer
▌ Screwdriver
▌ Adjustable wrench

1 Begin the repair by removing the handle and loosening the bonnet nut with an adjustable wrench. Look for an arrow that indicates the proper turning direction, either clockwise or counterclockwise. Once this nut is loose, pull out the valve stem.

2 Inspect the end of the valve stem to check whether the old washer is still in place. If it is, there shouldn't be any washer debris in the body of the sillcock. Replace the worn washer with a new one that's the same size. Coat it with heatproof grease to increase its lifespan.

3 Slide the handle onto the end of the valve stem, but don't attach it. The handle makes threading the stem into the body easier. When the valve is in, remove the handle; tighten the bonnet nut; and then reinstall the handle. Test the valve before turning on the water.

FREEZE-PROOF YARD HYDRANT

FREEZE-PROOF YARD HYDRANTS are the rural equivalents of freeze-proof sillcocks. Instead of being mounted through a house wall, they are buried underground, with only a short length of riser and the handle and spout showing aboveground. People typically use hydrants for water gardens, remote buildings, and livestock tanks.

Like freeze-proof sillcocks, these hydrants have a long stem and a drain chamber. But instead of draining unneeded water through the spout, hydrants drain downward, into a drainage pit at the base of the hydrant near the shutoff mechanism. When you raise the on-off lever, it lifts the long stem and the rubber stopper attached to it. This allows water to fill the riser and flow from the spout. When you lower the lever, the stem forces the stopper back into its seat, blocking the flow. The water left in the riser then drains through a small opening just above the stop mechanism; this keeps the riser from freezing. A gravel pit at the bottom of the trench stores the water until the surrounding soil can absorb it. Because you bury the stop mechanism and drain fitting below frost level—typically 3 to 6 feet deep, you can use the hydrant year-round.

Repairing a Hydrant. To repair a freeze-proof yard hydrant, shut off the water supply; thread the handle-and-spout assembly counterclockwise; and lift the assembly and stem from the riser. Remove the worn stopper from the stem, and install a new one. When you thread the handle-and-spout assembly off or on, be sure to backhold the riser with a second wrench. Backholding keeps you from damaging the piping connection at the bottom of the trench.

smart tip

FIND THE FROST LEVEL

YOUR LOCAL BUILDING DEPARTMENT CAN PROVIDE THE FROST DEPTH IN YOUR AREA. THIS IS VALUABLE INFORMATION FOR ANY BUILDING PROJECT AS WELL AS INSTALLING A FREEZE-PROOF YARD HYDRANT. ONCE YOU DIG EVEN A FEW FEET BEYOND THE FROST LEVEL, THE EARTH'S TEMPERATURE REMAINS RELATIVELY MILD.

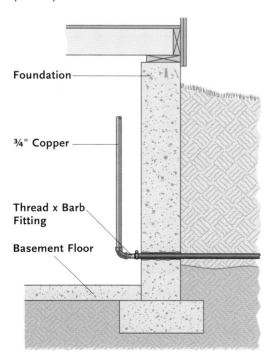

Foundation

¾" Copper

Thread x Barb Fitting

Basement Floor

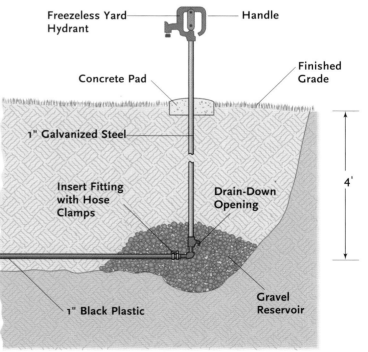

Freezeless Yard Hydrant

Handle

Concrete Pad

Finished Grade

1" Galvanized Steel

Insert Fitting with Hose Clamps

Drain-Down Opening

4'

1" Black Plastic

Gravel Reservoir

REGULATORS AND SAFETY VALVES

Sometimes you need to control the direction of water flow in a drainage or water-supply system so that contaminated water doesn't back up into the house or mix with potable water. That's where regulators like vacuum breakers and check valves come in.

Safety valves are important in controlling high-pressure systems like water heaters. If you didn't have a safety relief valve in such situations, dangerous explosions could easily occur.

Vacuum Breakers

As the name implies, vacuum breakers prevent air locks and back-siphoning of water in a piping system. As noted previously, modern sillcocks have a vacuum breaker that releases water held in the drain chamber. Vacuum breakers have an internal, spring-loaded diaphragm that is able to sense back-pressure. When back-pressure occurs, the diaphragm opens, venting the airlock through a series of perimeter openings.

While sillcocks and toilets have vacuum breakers or built-in backflow preventers, you can also get add-on devices. (See the photographs on page 33.) Some models fit the hose threads on outdoor faucets and laundry sinks, and others fit faucet aerator threads. When buying a toilet fill-valve or ballcock, make sure that it has built-in backflow protection. And if your current sillcock or hydrant doesn't have a vacuum breaker, buy an add-on breaker and install it. It's worthwhile protection for just a few dollars.

Check Valves

Check valves keep water from flowing backward through a line. They are available for both water and drainage piping. A check valve has a weighted flapper that allows water to flow in only one direction. Should the flow ever reverse, the flapper would fall against its seat and check the flow.

In drainage piping installations, full-size check valves can keep the city sewer main from backing into your home through your sewer service line.

In water-supply installations, check valves keep hot and cold water from cross-migrating through faucets. If you have a faucet that runs hot when you first turn the cold water on, cross-migration is the likely problem, and a check valve the probable solution.

Pressure-Reduction Valves

City water mains do not all have the same pressure. Those nearest water towers or pumping stations have greater pressure than those miles away, at the end of the line. It's not unusual for homes near a pumping station to have static pressures averaging 120 pounds per square inch (psi). This is more pressure than fixtures and appliances can manage. Too much pressure causes water-heater relief valves to leak, toilets to keep running, and faucets to pulse and pound when turned on and off.

The only solution to high pressure is a mechanical pressure-reduction valve. These devices come in two forms. Some have a set pressure, while others can be adjusted through a modest range of settings. Pressure-reduction valves have threaded female ports and are usually installed in-line, just before the water meter. If you suspect unreasonably high pressure, ask a local plumber or the utility company to run a pressure test. Forty to 60 psi is ideal. Anything above 70 to 80 psi probably deserves to be cut back.

CHECKING A SUMP PUMP

The most frequent use of check valves in drainage systems is in sump-pump installations. Without a check valve, the last bit of water pumped up the riser would fall back into the pit when the pump shuts off. This fallback water could be enough to activate the pump's float. Without a check valve to hold this quantity of water in the riser pipe, the pump would cycle on and off continuously, ruining the motor. Listen to the pump through a complete cycle, and replace faulty check valves.

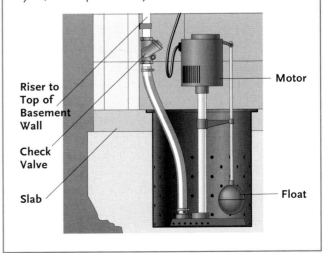

Riser to Top of Basement Wall

Motor

Check Valve

Slab

Float

INSTALLING A T&P VALVE

project

Temperature and pressure relief valves, usually called T&P valves, release water from a water heater when the temperature and pressure inside the tank become dangerously high. This usually happens when a thermostat sticks and won't shut off. People die from water heater explosions every year caused by defective (or missing) T&P valves.

TOOLS & MATERIALS

▌ New T&P valve
▌ Pipe wrenches
▌ Pipe-thread sealing tape

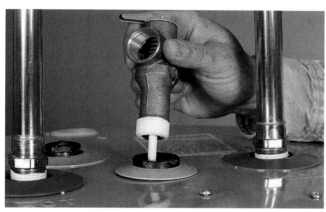

1 Wrap pipe-thread sealing tape around the threads of the new valve, and tighten the valve into the heater.

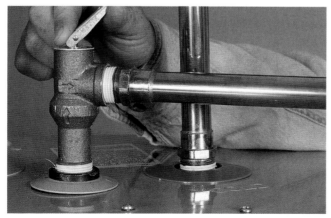

2 To test the T&P valve after installation, lift the release lever until water shoots through the overflow tube.

GREEN SOLUTION

smart tip

REDUCING HIGH PRESSURE SAVES WATER

ALTHOUGH THE MOST IMPORTANT CONSIDERATION IN INSTALLING A PRESSURE-REDUCTION VALVE IS TO REDUCE EXCESSIVELY HIGH, DAMAGING WATER PRESSURE, A HIDDEN BENEFIT IN INSTALLING ONE IS WATER SAVINGS. A FAUCET THAT RUNS FOR TEN MINUTES AT 100 PSI DELIVERS 45 GALLONS OF WATER. AT 50 PSI, THAT VOLUME IS CUT TO 30 GALLONS. YOU'LL STILL HAVE PLENTY OF WATER FOR YOUR NEEDS—WITHOUT WASTING IT—AT REASONABLY STRONG PRESSURE.

Pressure-reduction valves are adjustable. Thread the adjustment screw up or down, then tighten the locknut.

PEX WATER TUBING

Cross-linked polyethylene, or PEX, water tubing has been allowed in some locations for 20 years and pointedly disallowed in most others, but with the recent volatility in copper prices, resistance has begun to melt away.

PEX A and PEX B tubing is available in ½-in. or ¾-in. inside pipe sizes and three different lengths: 20-ft. straight pipes and 100-ft. or 300-ft. continuous coils.

PEX A and PEX B fittings are available in all common configurations that match current ½-in. and ¾-in. copper and CPVC fittings. Special adaptor fittings are also available to convert from copper or CPVC to PEX.

PEX A tubing has the best flexibility and the option of crimp or expanded fittings. PEX A if kinked can be heated with a standard heat gun and returned to its original unkinked state and size. A crimped fitting requires the use of the correct crimp tool for PEX A and a go/no-go gauge to ensure the proper compression of the fitting.

PEX B has less flexibility but has a higher resistance to chlorinated water over PEX A. PEX B will create a reduced

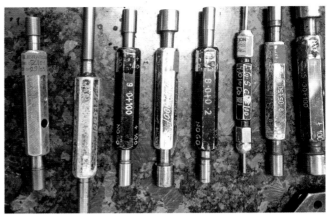

Various types and sizes of plug gauges (also known as go/no-go gauges).

inside pipe size as the fittings for PEX B are approved for crimp connections made with a limited expansion fitting connection. It is recommended to increase the inside pipe size tubing from ½-in. to ¾-in. to compensate for this issue. If PEX B tubing kinks, the section must be cut and removed as it can't be reshaped with a heat gun.

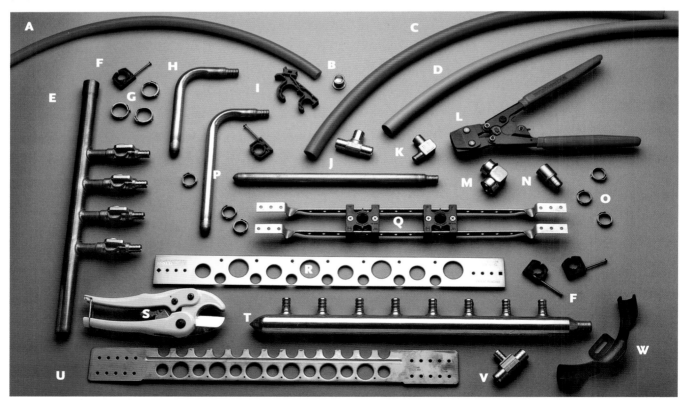

TOOLS AND EQUIPMENT FOR PEX TUBING: A—½-in. PEX hot, **B**—plug, **C**—¾-in. PEX hot, **D**—¾-in. PEX cold, **E**—copper manifold with shutoffs, **F**—tubing hangers, **G**—¾-in. crimp rings, **H**—copper stub-outs, **I**—dual tubing hanger, **J**—brass T-fitting, **K**—brass 90-deg. elbow, **L**—crimping tool, **M**—elbow with crimp collar, **N**—sweat x PEX coupling, **O**—½-in. crimp rings, **P**—straight stub-out, **Q**—carriage stub-out bracket, **R**—copper stub-out bracket, **S**—tubing cutter, **T**—manifold without shutoffs, **U**—alternate copper bracket, **V**—¾ x ¾ x ½-in. T-fitting, **W**—90-deg. tubing bracket.

CUTTING AND MAKING CONNECTIONS

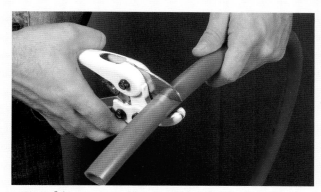

PEX tubing comes in a variety of colors and is easy to cut with a plastic tubing cutter. These cutters slice rather than rotate, making a square cut that is burr free.

PEX tubing comes with a variety of conversion fittings, which allow you to tap into existing plumbing materials. Here we thread a FIP x PEX fitting onto galvanized iron pipe. Be sure to use Teflon thread tape.

This brass plug just slides into the tubing and is secured by a crimp ring. The ring compresses the plastic tube into the barb recesses. Use plugs to water test the system.

Brass ball valves allow direct connection to PEX tubing. This one has barbs at both ends for an inline installation.

Here is a reducing T-fitting with barbed connections. The larger branch of the T is already secured with a crimp ring. Our crimping tool handles tubing sizes from ⅜ to 1 in. in diameter.

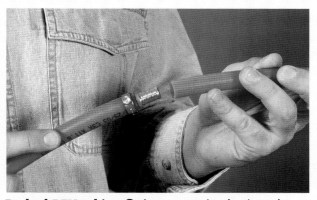

Barbed PEX tubing fittings come in plastic and brass, with brass having greater code acceptance. Insert couplings allow you to join tube lengths end-to-end. The tube fits the fitting loosely until banded and crimped.

INSTALLING A PEX MANIFOLD SYSTEM

You can install a PEX tubing system using a main branch and trunk lines as you would a copper system. But the material allows you to run lines from manifolds. In these systems, copper manifolds feed multiple PEX tubes, with each run of tubing servicing one fixture. So you can have one manifold to service the entire house, or many, depending on the size of the home and the number of plumbing fixtures it contains. Attach the PEX to the manifold outlets using crimp rings. Keep tubes organized with brackets and anchors.

TOOLS & MATERIALS

- Manifolds ▪ Copper tubing
- PEX tubing ▪ Crimp rings ▪ Crimping tool
- Hammer ▪ Tubing brackets and anchors

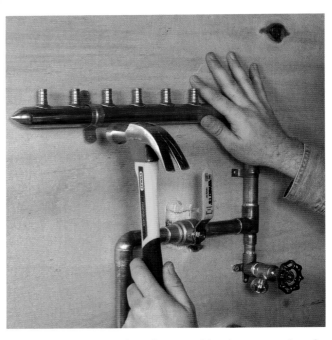

1 A popular approach with PEX tubing is to extend each line from a manufactured copper manifold that is fastened to a wall. Advantage: little pressure drop at one fixture when someone else uses another. Manifolds are available with shutoffs attached. Disadvantage: you waste more hot water waiting for it.

CONVERSION FITTING

THIS BRASS CONVERSION is designed to solder into a copper sweat fitting for a neat conversion to PEX. This fitting would work well when rigid copper is required early in the system. Interior tubing diameters are somewhat smaller than for copper.

4 Secure each feed line with a crimp ring. You will need a proprietary crimping tool for this function.

2 Connect a new PEX line to each manifold nipple. In this way, each fixture has its own line. To conserve tubing, skip the manifold, and use conventional trunk lines with fixture branches. Codes vary, but limit each ½ - in. branch to two fixtures.

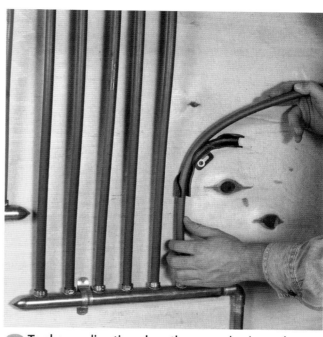

3 To change direction abruptly, use a plastic 90-deg. bracket. Screw the bracket in place, and snap the tubing into it.

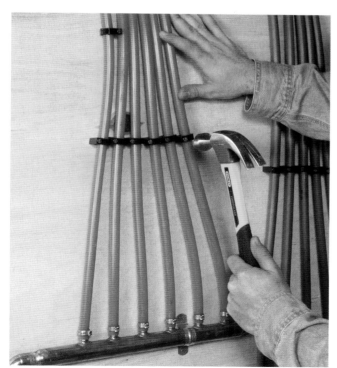

5 Use plastic tubing anchors to gather the various fixture feed lines. These anchors come with nails installed, so they're quick and easy to attach.

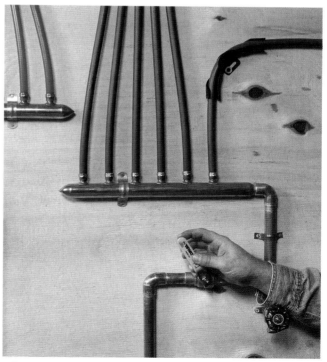

6 Install drain valves below the hot and cold manifolds, and test the system with water. Be sure every fixture stub-out is capped or plugged, and leave the system under pressure several hours for a complete test.

RUNNING PEX THROUGH WALLS

As with other types of pipe, PEX runs are hidden in walls, ceilings, and floors. If you must run PEX through structural lumber, be sure to install the material in the center one-third of the stud or joist to protect the tubing from nails and screws during the drywalling process. In general, PEX requires more anchoring than rigid copper, and there are a variety of anchors and brackets to help you do the job. One of the advantages of PEX is the ability to bend for a change of direction; use 90-degree brackets for that purpose.

TOOLS & MATERIALS

▌PEX tubing ▌Hammer
▌Drill ▌Screwdriver
▌Anchors and brackets

1 PEX is available in many colors, including clear, but many people are settling for blue for cold-water lines and red for hot. When running hot and cold water together, these dual tubing brackets are handy. Place one about every 2 ft. or at every stud or joist space.

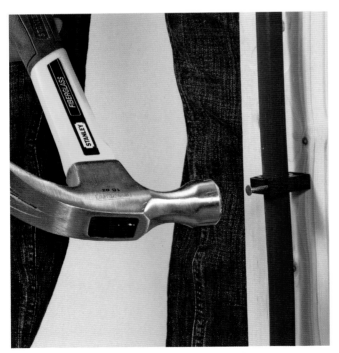

2 Don't forget to anchor PEX within wall spaces as well. PEX can take on a wavy look, especially on the hot water side. Plastic tubing brackets work well here and are inexpensive.

3 These 90-deg. brackets are used inside a wall. Center the holes in the stud width to prevent punctures by drywall nails, and nail the brackets in place. They have a bendable tab on one side, which allows you to lock the tubing into the bracket.

INSTALLING PEX STUB-OUTS

project

There are a variety of ways to terminate PEX tubing runs at plumbing fixtures, and a number are shown in the sequence that follows. For two stub-outs as you would use for a sink, there are brackets that allow you to set the stub-outs and the valves attached to them at precise locations. You can attach the valves directly to the PEX material or terminate the PEX run at copper stub-outs that are held in a copper bracket. In both cases, crimp rings seal the connection. For toilet stub-outs, use specially designed copper or galvanized brackets.

TOOLS & MATERIALS

▌ Stud-out brackets ▌ PEX tubing
▌ Tubing holders ▌ Hammer and screwdriver ▌ Crimping rings ▌ Crimping tool

1 Water stub-outs need to be rigid and properly spaced at the fixture, typically 4 or 8 in. apart. This carrier bracket provides ample support. Just nail it to the wall studs using galvanized nails, and slide the tubing fasteners for proper spacing.

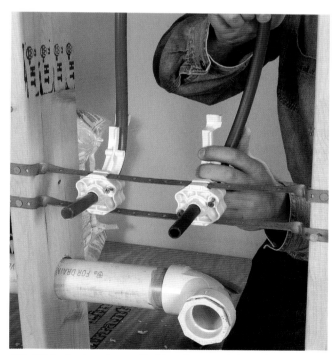

2 With the bracket nailed in place and the tubing holders in position, feed the tubing into the holders from the back. Bend the PEX to fit into the 90-deg. elbows. Cut each tube to the proper length, leaving the tubing stubbed away from the wall 3 to 4 in.

3 Use a screwdriver to secure the tubes in the fittings. Two Phillips-head screws compress the tubes in their holders.

Continued on next page.

Working with Water Piping

Continued from previous page.

4 With the wallboard in place, slide a chrome trim plate onto the tubing, and install a ½-in. PEX x ⅜-in. compression stop. Hold the shutoff valve upright, and clamp in. crimp ring over the tube. Leave about ¼ in. of tubing showing at the end.

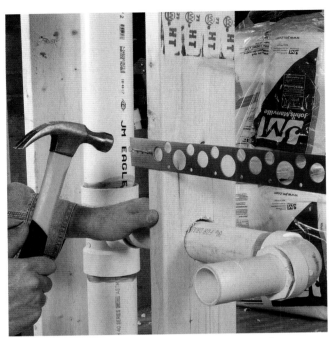

5 Another easy approach is to use copper stub-outs mounted in a copper bracket. Just nail the bracket across two studs using copper or galvanized nails.

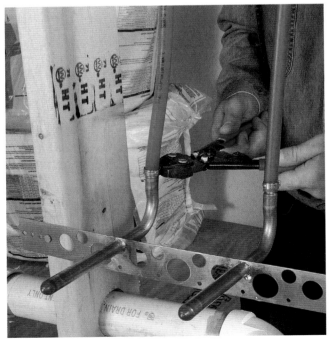

8 Secure each connection with a crimp ring; then anchor these risers every 3 ft. vertically. With PEX, more anchors are better than less.

9 When bringing water to a sink or toilet through the floor, a straight stub-out works best. This time, anchor the stub-out securely below the floor.

6 Slide two copper x PEX 90-deg. stub-outs into two of the bracket's holes; brush them with flux paste; and solder them in place. You won't need great soldering skills to master this approach. You can solder the stubs before or after nailing the brackets in place.

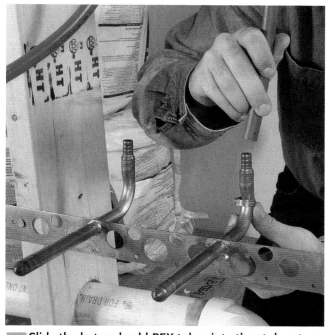

7 Slide the hot and cold PEX tubes into the stub-outs, and fit each with a crimp ring.

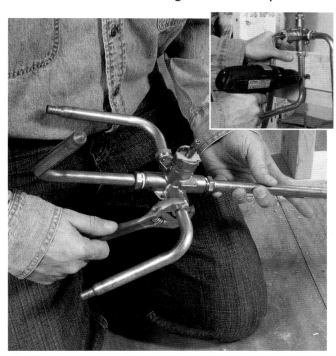

10 To create a rigid tub/shower faucet tree, just thread the ends of these shower-kit adapters into the faucet. Use thread-sealing tape or pipe joint compound to prevent a leak. To anchor the tub/shower tree in place, use two hole straps, and screw them to a backing board with drywall screws (inset).

STUB-OUT BRACKETS

Copper stub-outs with built in nailing flanges are another termination method. Here one is used to supply cold water to a toilet. As always, clamp the PEX to the stub-out with a crimp ring; then anchor the piping to the stud.

You can also use galvanized brackets to anchor stub-outs. This version has a nailing flange. Large-head galvanized roofing nails work best.

2

working with
waste &
vent pipes

BY LEARNING TO WORK WITH drain, waste, and vent (DWV) pipe and fittings, you'll be able to extend or replace some of the piping in your home rather than just make simple faucet and toilet repairs and improvements. Although having basic knowledge such as how to work with fixtures and faucets is useful and can save you money, making real upgrades involving new pipe and fittings will give you great satisfaction and can potentially save even more money.

Until about the 1960s and '70s, the traditional materials for waste and vent piping were cast-iron and copper. Modern plastic piping makes it easier to upgrade your plumbing, but most retrofits require splicing into those older materials, which in some cases requires fairly specialized skills. Only if you're plumbing a new home will you have the luxury of using new materials from start to finish.

As is so often the case, the things that seem the simplest need the most careful attention. Choosing drainage fittings that encourage gradual flow patterns, installing drain lines at just the right pitch, installing enough vents, and sizing pipes for the best efficiency—these are the things that matter most when working with the DWV system. In comparison, it's easy to learn to solder a T-fitting, solvent-weld plastic pipes, and cut cast iron. (For specific information on fitting selection, venting applications, and drainage installations, review Chapter 13, "Drains, Vents & Traps," pages 278–287, for information on designing vent systems.)

CAST-IRON DRAINAGE PIPING

Cast-iron pipe was once the universal material for under-floor, or basement, soil piping. It was also common in vertical stacks in houses built prior to the early 1960s. The above-floor branch lines serving these stacks may have been made of galvanized iron or copper, but cast iron was the backbone of every DWV system.

Some codes today still require cast iron below the basement floor, but the trend is clearly toward plastic drain and vent piping, from street to roof. Still, cast iron is present in most homes built prior to the 1980s, and builders in some areas continue to install it. If you hope to make any changes or additions to a cast-iron plumbing system, you'll need to know how to cut it. You'll also need to know how to fit it with either cast-iron or plastic fittings.

Cast-iron pipe comes in 3-, 5-, and 10-foot lengths, in no-hub, single-hub, or double-hub configurations. Two hubs make for less waste when you need custom-cut lengths, but you can't use them as they are because one of the hubs would always be backward. A double-hub pipe has to be cut in two.

Cast-iron pipes without hubs are available for use with no-hub fittings. Standard cast-iron pipe diameters for residential use are 2, 3, and 4 inches. Most existing homes have 4-inch soil and sewer-service pipes. However, with today's EPA-mandated low-volume 1.6-gallon toilets, it is becoming less common to install 4-inch pipes in new construction because of the reduced volume.

MAKING LEAD-AND-OAKUM JOINTS

Use a packing tool to tamp the joint two-thirds full of oakum.

Cap the oakum with lead wool, and tamp it into solid packing.

CONNECTIONS WITH NEOPRENE FITTINGS

Use banded couplings to splice plastic piping into a cast-iron drainage line.

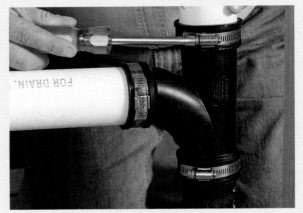

Install a no-hub flexible fitting for greater ease in retrofitting drainpipes.

CAST-IRON FITTINGS

CAST-IRON FITTINGS come in a variety of configurations, as shown in the illustration below. Several of the fittings also have side-inlet or heel-outlet openings. (For more on side inlets and heel outlets, see "Side-Inlet Ports" and "Heel-Outlet 90-Degree Elbows," pages 285–286.)

Traditional cast-iron pipes and fittings are formed with bells and spigots. A spigot, or male end, fits into a bell, or female hub. This is a loose fit, however, so you need a neoprene gasket or packing material to make the joint watertight. Prior to the 1970s, all cast-iron joints were packed with lead and an oily ropelike material called oakum, which expands when wet.

Today, however, plumbers join bell-and-spigot cast-iron fittings with neoprene rubber gaskets. You press the gasket into the cast-iron bell and lubricate it with a soapy gel before pushing the spigot end of the drainpipe through the gasket.

As mentioned earlier, you join no-hub pipes and fittings with banded (no-hub) couplings, which are remarkably easy to use. There are several banded coupling styles, but all work similarly. Each consists of a neoprene rubber sleeve banded by stainless-steel clamps, one at each end. Some brands use a thick but pliable neoprene sleeve and two stainless-steel clamps, while others use a thin neoprene sleeve backed by a wide stainless-steel band and two clamps. Both types are available as straight couplings; reducers, which allow you to join pipes of different diameters; and connectors such as sanitary T-fittings.

Because these couplings can also join dissimilar materials, such as cast iron and plastic or copper and plastic, they are perfect remodeling fittings, making permanent, corrosion-free joints. The thick-walled, heavy-duty all-neoprene type can be used underground, while the stainless-steel-collar type are confined to aboveground applications.

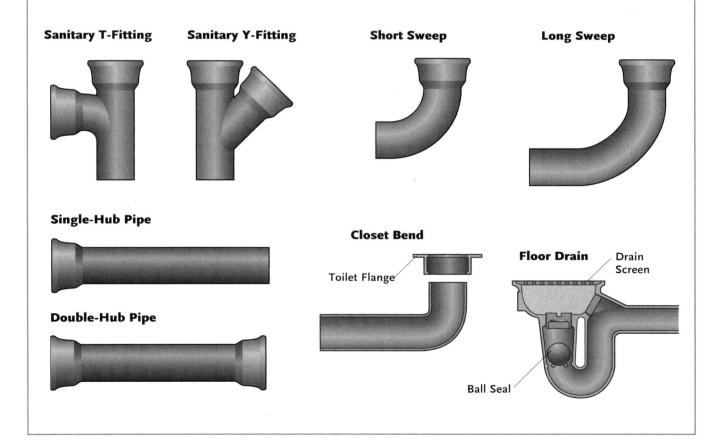

Sanitary T-Fitting **Sanitary Y-Fitting** **Short Sweep** **Long Sweep**

Single-Hub Pipe

Double-Hub Pipe

Closet Bend

Toilet Flange

Floor Drain Drain Screen

Ball Seal

CUTTING CAST IRON

project

You have a choice of methods when cutting cast-iron pipe. You can use a hacksaw, but it is slow work. You can also use a chisel and ball-peen hammer. But when professionals cut cast iron, they use a snap-cutter. As its name implies, this tool snaps—breaks—a pipe in two. It consists of a roller chain that has hardened steel wheels built into it. The chain is connected to a ratchet or scissor head. As you lever the head, the chain tightens, and the cutter wheels bite into the pipe with equal pressure. Eventually the pipe snaps in two.

TOOLS & MATERIALS

- Cast-iron pipe ▪ Ball-peen hammer
- Cold chisel or snap cutter
- Adjustable wrench ▪ Work gloves

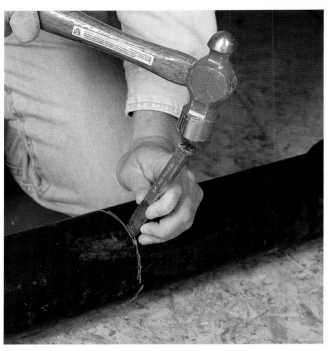

1 The low-tech approach to cutting cast iron pipe is to use a ball-peen hammer and a cold chisel. Begin by clearly marking the circumference of the pipe. Then score around the entire line with the chisel, until the pipe breaks in two.

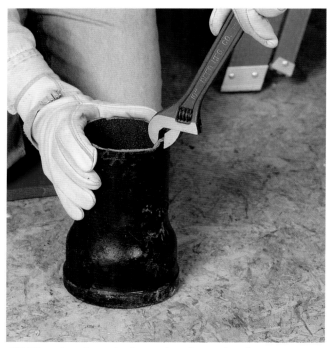

2 The pipe doesn't always break evenly when you use the cold chisel method. But any jagged section can be broken off using an adjustable wrench. Adjust the jaws so that they are slightly wider than the thickness of the pipe.

3 A snap cutter is the easiest way to cut cast iron pipe. This rental tool has a chain that's attached to a ratchet mechanism. To use it, position the chain around the pipe at the cut line; then tighten the ratchet until the pipe breaks.

JOINING CAST-IRON PIPE

project

These days, cast-iron pipes are joined with easy-to-use neoprene gaskets (shown on this page), or no-hub couplings. Both approaches replace the traditional method of packing the joints with oakum and lead wool. (See page 48.) While the lead-and-oakum system makes for a tight and durable joint, it takes skill and experience to execute properly. Even though neoprene gaskets are easier to use, they do have one drawback; they grow stiff in cold weather. So keep them in a warm location until just before you use them.

TOOLS & MATERIALS

- Neoprene gaskets
- Lubricating soap
- Small paintbrush
- File
- Hubbed piping
- Shovel
- Fittings

1 Before installing the gasket, smooth the inside of the fitting with a file or abrasive paper. Then push the gasket into the fitting. Roll up the bottom half and press in the top half until it's flush. Then push the bottom half down in place.

2 Before joining the pipe and fitting, lubricate the gasket with liquid dish soap (inset). Don't use a petroleum-based lubricant because it could damage the neoprene. Push the parts together using a shovel until the pipe bottoms out.

3 Fittings are installed just like pipe sections. First lubricate the gasket; then push the parts together until the fitting bottoms out in the end of the pipe. If the fitting binds a little, simply wiggle it from side-to-side as you push it in.

YOKE FITTING: BAD IDEA

A YOKE FITTING clamps onto a cast-iron stack or branch after you have cut a hole in the pipe. While this sounds reasonable, it's really not a good idea, and most code authorities don't allow yokes, even though you can buy them. The reasons are two-fold. First, it's tough to establish a good flow pattern. Second, cutting a hole in a cast-iron pipe with a cutting torch is messy business that burns off the protective coating around the hole. Without this coating, the stack deteriorates near the yoke. You also risk dropping the cut piece into the stack. You can find plastic yokes for plastic pipes, but again, they're usually not code compliant.

Working with Existing Cast Iron

Retrofit work differs from new installations in two ways. First, you don't usually have as much room to work. Second, in addition to extending the new, you'll need to adapt to the old. In the case of cast iron, that means finding the best place to splice into the line. In some cases, it means making the whole project fit current code requirements. Once you cut into grandpa's work, you lose the benefit of the grandfather clause. (See "Grandfather Clause to the Rescue," page 266.)

Where you choose to cut into an older cast-iron piping system will depend on what you hope to add or improve. If you've built a home addition that includes plumbing, you'll likely need to cut into the stack in the basement. If you're adding a bath in a basement, in an area without rough-in piping for a bath, you'll need to break the basement floor and cut into the underslab soil pipe. If you are simply re-modeling an upstairs bath, you may not need to cut into the stack at all. You may be able to simply cut into the existing horizontal toilet piping, just beyond the stack fit-ting. You can then extend all the fixtures from this full-size line. You may need to cut the stack in the attic to tie in a new vent, but that's usually easier than breaking the stack in a wall.

Hazards of Cutting Vertical Stacks. If you need to cut out a section of cast-iron stack on a lower floor, it's important that you make sure the stack above the cut can't come crashing down. Most stacks won't fall when cut, because upper-story vents or branch lines hold them in

place. Still, it pays to check, especially when you consider the weight of a 20-to-30-foot cast-iron pipe. The most sure-fire precautionary measure is to install a stack clamp that grips the stack and is supported by nearby wall studs. (See the photo below.)

If you can't locate a stack clamp or the local wholesaler won't sell one to you, you'll have to improvise. If neces-sary, you can make a clamp with strap iron and bolts. In most cases, you only need to support the stack until you can splice in the new fitting. Then you can remove the sup-ports. Dimensional lumber works well as blocking. If you're working in the basement, for example, and you plan to cut out a section of stack just below the toilet fitting, jam a 2×4 stud between the toilet T-fitting and the basement floor. If you've opened a wall to make your improvements, a short 2×4 jammed between the floor's soleplate and the sink's branch arm will work. You could also go into the attic and prop lumber under a re-vent just as it enters the stack. In this case, you'd use the lumber to bridge two ceiling joists or to block up from the plumbing wall's top plate. A final option is to suspend the pipe with hole strap or lightweight chain. This is the best way to support horizontal piping.

Install a stack clamp to secure the upper section of a cast-iron stack when cutting out a lower section.

SPLICING NO-HUB CAST-IRON FITTINGS

When you need to install a T- or Y-fitting into an existing cast iron pipe, local codes may let you install a plastic PVC fitting into the cast iron and then use PVC pipe and fittings for the rest of the job. But some codes require you to use only cast iron when you are adding to a cast-iron system. Fortunately, cutting into an existing cast pipe and installing a cast no-hub fitting isn't very difficult. Simply first cut an opening for the new fitting; then install it using neoprene sleeves and banded couplings. Common tools are all that's required.

TOOLS & MATERIALS

- Grease pencil ▪ Snap-cutter or hacksaw
- Cast-iron no-hub T- or Y-fitting
- Banded couplings ▪ Nut driver

1 Determine the best location for the fitting, and hold it against the pipe at this point. Mark the top and bottom of the fitting on the pipe, using chalk or a marking crayon. Leave about ½ in. of clearance on both marks for fitting room.

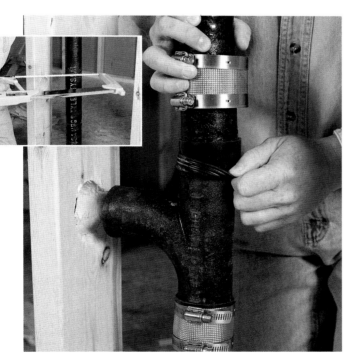

2 Cut out the marked section of the pipe using a hacksaw (inset) or a chain cutter. Then install the gasket sleeves onto both ends of the pipe, and slide the fitting and the banded couplings into the opening between the pipe sections.

3 While holding the fitting between the ends of the pipe, roll the neoprene sleeves over the joints at the top and bottom. Make sure the gaskets are flat; then slide the couplings into place, and tighten the bands.

Other Methods of Tapping into Cast Iron. Not all cast-iron fittings have bell-and-spigot inlets. Some older types accept threaded pipes or threaded adapters. You'll find these fittings in homes that have cast-iron stacks with copper or galvanized-steel branch lines. These threaded fittings offer yet another method of adapting plastic to cast iron, though not always as easily as with banded couplings. They also have the advantages of being extremely inexpensive and making neat, professional-looking retrofit connections.

In the case of copper or brass threaded adapters, just cut the old copper line near the cast-iron fitting, and back the adapter out with a large pipe wrench.

Galvanized-steel threads, in contrast, fuse with the cast iron over time. To break these threads free, you often need to heat the cast iron with a torch. Heat causes the female half of the fitting to expand slightly, loosening its grip. In any case, when you've removed the old threaded piece, clean the rust from the cast-iron threads using a wire brush. Then wrap plumber's pipe thread-sealing tape—only two full rounds—onto the threads of the new plastic adapter, and screw the adapter into the cast-iron fitting.

New Piping into Old Hubs

In some situations, the best approach is to start your new piping inside an old hub. One example is when a stack hub rests at basement floor level and you'd like to extend a branch line horizontally just above the floor. There are times when using an old hub can also save you from having to cut into a stack. Removing an old iron pipe from a lead and oakum joint is well within a homeowner's abilities. The job requires digging both the lead and oakum from the old joint so that you can lift out the pipe.

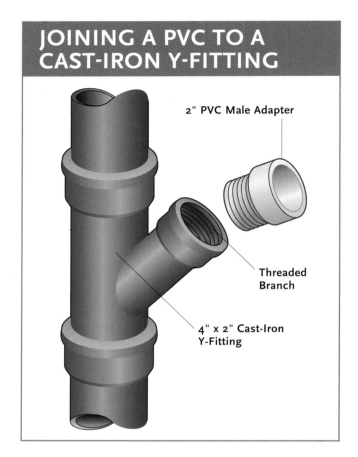

JOINING A PVC TO A CAST-IRON Y-FITTING

2" PVC Male Adapter

Threaded Branch

4" x 2" Cast-Iron Y-Fitting

Special tools for this job are available, but an old flat-blade screwdriver will work just as well as any specialty tool. Begin by driving the blade of the screwdriver diagonally through the ring of lead, prying it up. Grip the pried-out piece of lead with pliers, and peel the rest of it out. You can dig out the oakum in a similar fashion, but you'll have to remove it bit by bit. Needle-nose pliers work well. When you have cleaned out the joint and removed the old pipe, install the new pipe.

REMOVING GALVANIZED-STEEL WASTE PIPES

PLUMBERS NO LONGER INSTALL galvanized-steel waste piping at the residential level, so you will only be replacing or repairing this material. In either case, you must always cut galvanized steel into two pieces to work on it. This is because galvanized-steel piping sections were always installed sequentially, with threaded fittings. That mean when you try to loosen a length of pipe at one end, you're tightening it at the other—a no-win situation. You must cut the pipe so that you can back each end from its threads.

This is always easier said than done because steel threads rust almost from the moment they're installed. Coming along 40 years later, you're going to have to work to break this oxidation seal. When you've removed both ends of the pipe, thread plastic male adapters into the fittings and install PVC plastic piping. Or you can just cut out the steel and splice it with banded couplings.

JOINING PVC TO A CAST-IRON HUB

project

It's not a surprise to anyone that PVC plumbing pipe and fittings are easier to install and repair than cast-iron versions. In areas where local codes allow you to convert from cast to PVC, it's almost always worth the time to make the change. Fortunately this job is pretty easy because Schedule 40 PVC pipe is very close to the same size as cast iron in all the standard waste-pipe sizes. So you can use standard neoprene gaskets in the cast hubs to hold PVC pipe. Be sure to clean out the hub, and remove any metal burrs before installing the gasket.

TOOLS & MATERIALS
▌ Hacksaw ▌ File
▌ Gasket, lubricant, and brush
▌ Hammer ▌ PVC piping

1 Press a standard-sized neoprene gasket into the hub, and push it until the gasket lip is flush with the hub lip. If the gasket doesn't sit flat, remove it and clean out any debris that is preventing the gasket from seating completely.

2 Once the gasket is seated, spread liquid dish soap on the top and inside of the gasket to act as a lubricant (inset). Then cut the PVC pipe to length, and remove any plastic burrs from the inside and outside of the pipe with a file.

3 Place the de-burred end of the pipe against the top of the gasket, and drive it into the fitting until you feel and/or hear the pipe hitting the bottom of the hub. Hold the pipe securely when you strike it.

INSTALLING IN-GROUND DRAINPIPES

project

Every plumbing system has some of its drainage piping underground, even if it's only the sewer service line. Many houses with basements and all slab-on-grade homes will have soil pipes trenched in place before the concrete goes down. This piping must be able to support the substantial weight of the soil and concrete above it. This means that the trench you dig must be uniform, have proper slope, and avoid high spots and voids. Highs can squeeze (or break) the pipe while lows can cause sags that restrict consistent flow.

TOOLS & MATERIALS

- Cast-iron pipe ▮ Hubs ▮ Shovel
- Duct tape ▮ 4-foot spirit level
- Hammer, stakes ▮ Plastic cap

1 Dig a basement soil-pipe trench about 12- to 16-in. wide and as deep as necessary to meet the sewer line outside. In most cases, this means excavating under the foundation footing using a shovel, pickax, and a long pry bar. In soft soil, a posthole digger can also be useful.

3 In-floor drain risers for sinks, tubs, and toilets, such as the one shown here, can be knocked out of proper alignment when the concrete is poured for the floor. To maintain alignment, drive a stake clamp next to the riser so it's tight against the top of the pipe. Use a torpedo level to ensure the riser pipe is plumb.

4 You should protect all the drain openings, particularly floor drains like the one shown here, from debris falling inside. Covering these openings with a cap made of duct tape is the easiest way to protect them. Before applying tape, clean off the pipe or fitting to remove the dirt and oil.

2 Place the first length of pipe in the trench, and slide it under the footing. Check the slope with a 4-ft. level to make sure it drops ¼ in. per foot of run. If the pipe is high, remove it, and excavate deeper. If it's too low, remove it, and add sand to the bottom of the trench.

5 Toilet riser openings can be protected with duct tape or with an inexpensive plastic closet cap, like this one. Just push the cover onto the top of the riser; once the concrete is poured and cured, pull off the cap, and plumb the toilet to the riser.

REMOVING CAST IRON

TO DISMANTLE an antiquated galvanized-steel or copper plumbing system, you can cut it apart with a hacksaw or reciprocating saw. You can also cut out cast iron, but an easier method is to shatter the hubs with hammers. This may seem extreme, but it works well and gets the job done quickly. It also works when galvanized steel is joined using cast-iron elbows and T- and Y-fittings.

You'll need to wear face and eye protection, of course, but the method is simple. Starting near the top of the stack, strike the first hub simultaneously on both sides with two hammers of equal weight. Three-pound sledgehammers work well. The hammers should strike opposite sides of the cast-iron hub, hitting at roughly the same time.

When you reach the lowest section of piping on the stack, stop. From the basement ceiling on down, use a snap-cutter or a ball-peen hammer and cold chisel to make clean cuts. At this point, either dig the lead and oakum from the lowest hub, or use a snap-cutter to cut the stack a foot or so above the floor. Using a neoprene gasket or banded coupling, you can then extend the new piping upward.

PROVIDING CLEANOUT ACCESS

CODES STIPULATE that each drain stack must have a permanently accessible cleanout fitting at its base. (Dry-vent stacks don't need cleanouts.) Where you have reduced a 4-inch soil pipe to a 3-inch stack, make the cleanout T-fitting the size of the larger pipe. If the stack is more than 10 feet from the wall, codes require an additional cleanout fitting. You can place this cleanout just inside the wall, in the basement floor, or just outside the house, brought to grade. If you'll be finishing that area of the basement, an exterior cleanout fitting is practical. And, finally, most codes require additional cleanout fittings in above-grade kitchen lines. It makes the most sense to install one after a change of direction in the line.

Working with Waste & Vent Pipes

PLASTIC DRAINAGE PIPING

The plastic piping materials allowed by codes for use in drain and vent systems are Schedule 40 PVC, and to a lesser extent, Schedule 40 ABS. PVC is white; ABS plastic is black.

PVC has gained almost universal acceptance, so this discussion focuses on it; however, little difference exists between the two plastics in terms of installation. If you have ABS-plastic piping in place, you can make repairs and additions using ABS or PVC fittings, but use the more aggressive PVC joint cement or a universal solvent cement. ABS cement does not bond well enough to PVC.

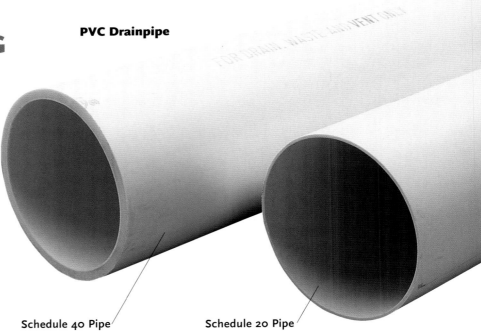

PVC Drainpipe

Schedule 40 Pipe   Schedule 20 Pipe

CONVERTING A LEAD TOILET RISER TO PLASTIC

YEARS AGO, plumbers used lead drain lines to join cast-iron and galvanized-steel pipes to traps and fixtures. The two most common places to find lead are on outdated drum traps and old toilet risers. No matter where you find these fittings, replace them. Lead is easy to recognize, as it shows a dull gray with age and is shiny bright when scratched. It is also soft. You can cut through lead easily using a hacksaw or even a sharp utility knife.

To replace a lead riser, cut through the lead with a utility knife. Remove the riser and flange, and cut the lead again, this time just above the hub. Install a banded coupling on the insert, and finish with a PVC pipe and flange.

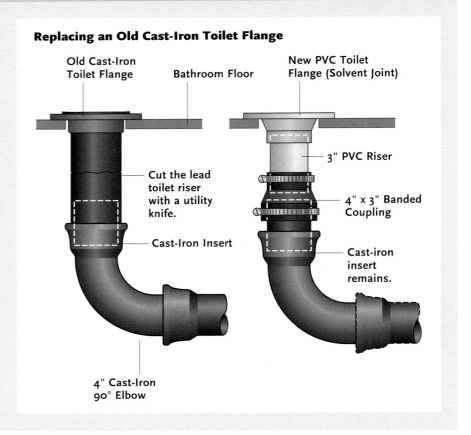

Replacing an Old Cast-Iron Toilet Flange

Old Cast-Iron Toilet Flange

Bathroom Floor

New PVC Toilet Flange (Solvent Joint)

Cut the lead toilet riser with a utility knife.

Cast-Iron Insert

3" PVC Riser

4" x 3" Banded Coupling

Cast-iron insert remains.

4" Cast-Iron 90° Elbow

Advantages and Disadvantages. Over the years, plastic has had to overcome an image problem as nothing more than a cheap substitute for metal. Actually, it's hard to imagine a better material for residential waste and vent systems. Almost every conceivable fitting is available in PVC or ABS. Plastic is remarkably easy to install with the simplest of household tools. Once installed, it's easy to alter. It fits in tight spaces and is universally available at reasonable prices. Plastic never rots or corrodes, and it stands up well to caustic drain-cleaning chemicals.

What are its disadvantages? It's noisy. You can hear water run through it, which some homeowners find annoying. It expands and contracts more than cast iron or copper with changes in water temperature, and if it's wedged against structural timbers, it makes a persistent ticking sound with the expansion and contraction. It also tends to have more abrupt flow patterns, which can lead to clogs. But for all of that, the advantages outweigh the disadvantages for most people.

Licensed plumbers often cut PVC pipe with a power miter saw (or cutoff saw), but a PVC saw, hacksaw or tubing cutter works about as well. If you use a hacksaw, be sure to clear the resulting ridges or burrs from the pipe edge before installing it.

Before cementing PVC joints, be sure to deglaze, or prime, both the pipe end and the fitting to roughen the surfaces and make it easier for the solvent to achieve a good bond. You can do this most easily using a primer-solvent, but sanding lightly or scuffing with an abrasive pad will also work. Many primers have a bright color additive, called an indicator. This additive has nothing to do with how the primer works. Rather, it tells the plumbing inspector whether or not a primer has been used.

Because primers are thin and tend to run all over the pipe and fitting, indicator colors tend to make a job (as well as hands and clothes) look messy. Understandably, many people prefer primers without indicator colors. When color indicators are required by code, ugly wins.

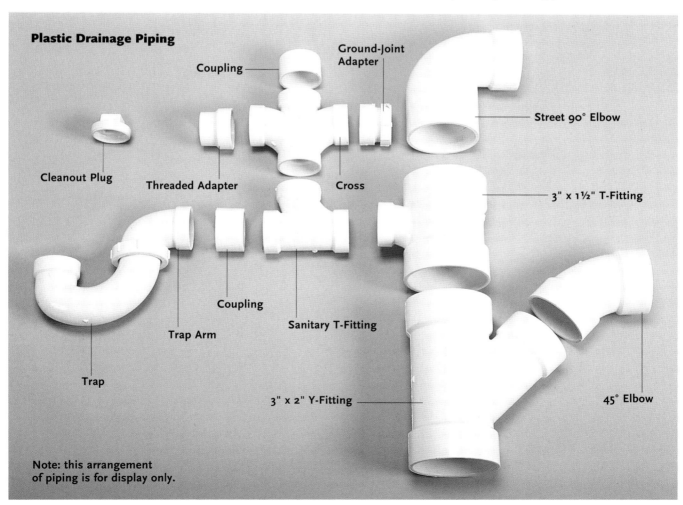

Plastic Drainage Piping

Coupling

Ground-Joint Adapter

Street 90° Elbow

Cleanout Plug

Threaded Adapter

Cross

3" x 1½" T-Fitting

Coupling

Trap Arm

Sanitary T-Fitting

Trap

3" x 2" Y-Fitting

45° Elbow

Note: this arrangement of piping is for display only.

Working with Waste & Vent Pipes

WORKING WITH PLASTIC DRAINPIPE

PVC (and ABS) pipes and fittings, once cemented together, stay that way. Unlike wood glues, which bind each piece of wood to itself, plastic pipe solvents actually melt one surface into the other, creating a chemical weld. With a 1½-inch pipe and fitting, you'll have about 30 seconds to change your mind about the joint. After that, it's permanent. So test-fit and mark each group of fittings with a pencil or felt-tip marker before gluing them in place.

When measuring for a pipe cut, be sure to include the depth of the fitting hubs in your total. You can make this calculation in your head, but holding an actual fitting in place helps to eliminate errors. When you've determined the exact pipe length needed, mark the pipe and cut it. You can use a hacksaw, handsaw (with miter box), PVC saw, wheel cutter, or power miter saw (cutoff saw).

Smooth the inside of the pipe end using a knife, sandpaper, or a deburring tool before cementing the joint together. Any rough edges will attract hair and strands of fabric sent though the system, causing clogs.

With the end of the pipe cleared of burrs and rough spots, apply primer-solvent to the outer edge of the pipe and to the hub of the fitting. Both primer-solvent and joint-cement containers come with applicators. When the primer evaporates, test-fit the joints, making sure that the pipe bottoms out in the fittings. When you're sure the joint is right, mark the pipe to show where the fitting should land on it in final assembly. Next, coat the first 1 inch of the pipe and the entire inside of the fitting hub with cement. Immediately insert the pipe

TOOLS & MATERIALS
▮ Measuring tape ▮ PVC pipe
▮ Hacksaw ▮ Deburring tool
▮ Pencil ▮ Primer and cement

and fitting. As soon as the pipe bottoms out in the hub, rotate the fitting about one-quarter turn. This fills any voids in the joint by breaking up the insertion lines. Of course, if you've test-fitted your joints first, you'll need to push the pipe into the fitting with the alignment marks about one-quarter turn out of sync, then rotate the fitting until the marks line up. Hold the parts together for about 10 seconds. Wipe off excess cement.

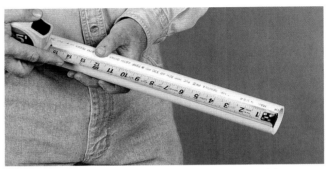

1 The easiest way to measure plastic pipe is with a tape measure. Just hook the end of the tape over the end of the pipe, and mark the proper spot. A soft lead pencil or a felt tip marker works best. Also, be sure to include depth of the fitting hubs in the overall length of the pipe.

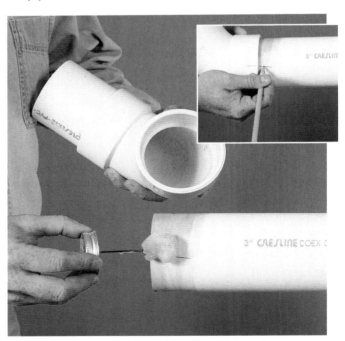

4 Test-fit the assembly, and mark each joint for alignment with a pencil (inset). You could also number the pieces. Apply PVC solvent cement with the applicator contained in the can. Cover both the pipe ends and the inside of each hub. PVC cements are available for use in a multitude of working temperatures. All weather medium clear PVC cement, for example, has an approved operating temperature of 15°F to 110°F.

60

PVC (and ABS) pipes and fittings are not interchangeable. PVC (which is usually white) and ABS (which is usually black) require different types of plastic cement to join and weld the mating parts together. Do not intermix these materials unless using a no-hub coupling as shown in steps 2 and 3 on page 53.

2 Cut the pipe with a PVC pipe saw. Make a square cut that's perpendicular to the circumference of the pipe.

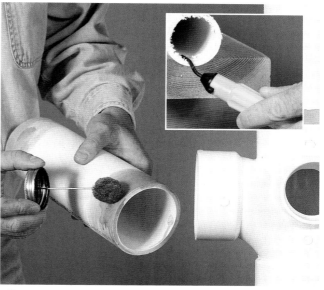

3 Use an inexpensive deburring tool or the rounded side of a file to smooth the ragged edge left by the saw (inset). Apply primer to the end of the pipe to cut the glaze. Also apply primer to the inside of the fitting socket. Many primers have an added colorant. Colored primer is often required by code.

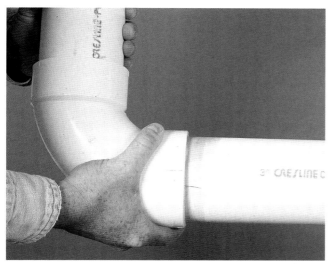

5 Once the mating surfaces are covered with cement, push the parts together so they are slightly out of alignment, and rotate them until the marks line up. This technique ensures that the cement is spread over the entire joint and fills any possible voids.

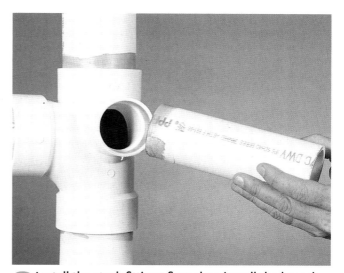

6 Install the stack fittings first; then install the branch lines. A side-inlet T-fitting, such as the one shown here, can drain both toilet and shower drains from opposite sides. Local building codes will stipulate the pipe size that's required for all the appliances and fixtures in the system.

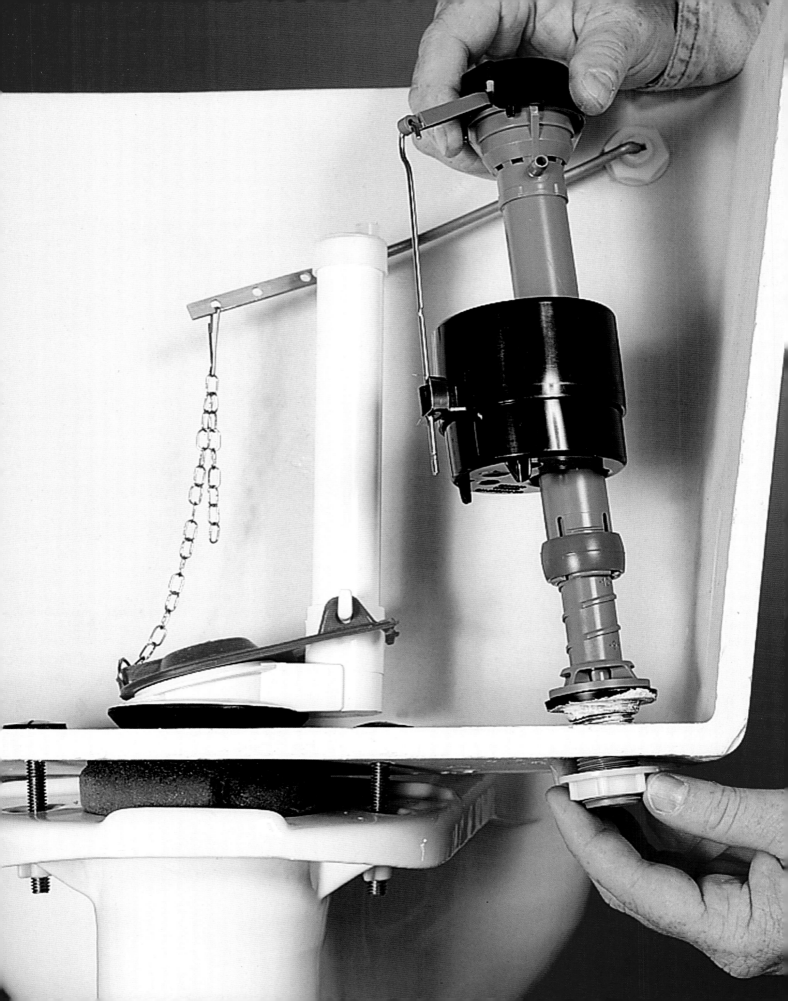

3

toilet repairs & installation

WATER-FLUSH TOILETS are miracles of simplicity. In fact, few devices accomplish as much with so few parts. They're so simply built that you can completely overhaul one in about an hour, for about $20 in parts. The latest generation of water-savers may be a little more complicated, but repairing them is work you can still do.

TOILET FUNDAMENTALS

Water-powered toilets come in two forms, gravity-flow and pressure-assisted models. Gravity-flow toilets, as the name implies, use only the force of gravity to flush wastewater. Pressure-assisted models use the extra power of compressed air to push water more forcefully. The basic waste-removal system is similar in both types. Water flows into the tank via the fill valve. When you flush the toilet, the water flows through the flush valve, into and through the bowl, and through the trap, taking waste with it. Knowing this sequence will later help you match symptoms with solutions.

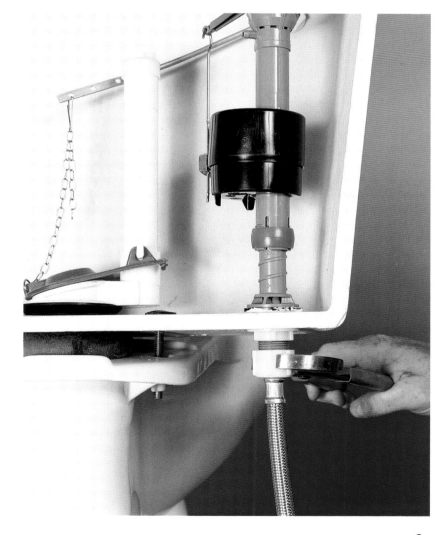

Toilet Repairs & Installation

HOW GRAVITY-FLOW TOILETS WORK

When you press down on the flush handle, its lever lifts a chain or lift wire attached to a flapper or tank ball. (Flappers have chains; tank balls have lift wires.) The chain or wire lifts the flapper or ball from its flush-valve seat, which allows water to escape through the valve.

Both flappers and tank balls are hollow. Some are open at the bottom, and some are closed, but both drain water and trap air, making them temporarily buoyant when you push down on the handle. Without this buoyancy, the flapper or ball would sink immediately, so you'd need to hold the flush lever down until the tank was empty.

As the flushing water recedes, the flapper or ball floats downward until it rests in the flush-valve seat and stops up the opening.

After passing through the flush valve, much of the water shoots through the siphon jet, the small opening across from the trap outlet. The rest spills through slanted holes in the rim.

The water rushing in through the siphon jet overflows the trap, priming it. Once primed, the trap siphons all the water it can over its weir, or crown, stopping only when there's not enough water in the bowl to sustain the siphon. At this point, all water on the house side of the trap slides back into the bowl.

Because the rim holes are slanted, the water entering the bowl through the rim travels diagonally around the bowl. This diagonal pattern scours the sides of the bowl, but it also sends the water over the trap in a spiraling motion, which improves the efficiency of the flush.

As the bowl empties, new water enters the tank through the ballcock or fill valve. This flow begins the moment you flush the toilet. Most of the water entering the tank does so through a tube that terminates near the bottom of the tank. Delivering the water to the bottom of the tank provides a measure of noise control. The sound is muffled as soon as the water in the tank rises above the end of the tube. At the same time, a small stream of water is diverted into the flush valve overflow tube—via a ⅛-inch-diameter fill tube—and falls directly into the bowl's rim. This stream restores the water in the bowl to its maximum level. As soon as the tank fills, the float shuts off the fill valve, and the toilet is ready for another flush.

GRAVITY-FLOW TOILET ANATOMY

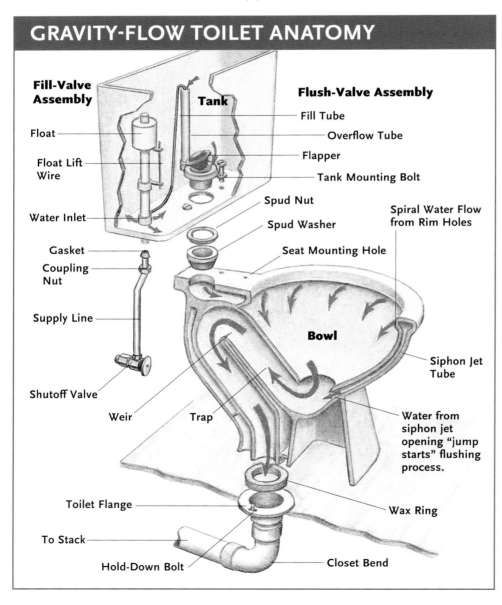

64

GREEN SOLUTION
WATER-SAVING TOILETS

IN THE LATE 1970s, a number of Scandinavian countries began using—and mandating—super-low-flow toilets, which flushed with an amazingly skimpy 1.6 gallons of water. Before long, these toilets appeared at trade shows here in the United States, and manufacturers began experimenting with low-flow toilets, trying to improve performance. Eventually, the U.S. enacted a national standard that limits to 1.6 gallons per flush (gpf) the water used by residential toilets made in this country after January 1, 1994.

Do low-flow toilets work as well as 3.5-gpf ones? The early models certainly didn't. From the start, manufacturers offered two distinctly different low-flow toilets; a gravity-flow model and a pressure-assisted model. The gravity-flow models were like traditional toilets but with minor changes all around, including new fill valves and flush valves. Engineers reduced trap geometry, along with water spots—the surface area of the bowl water—and outlet diameters to cut down the flow of water. These early water misers were so sluggish that they often needed additional flushes to clear and clean the bowl, and clogs were common. With steady engineering refinements, however, gravity-flow toilets now work reasonably well and are a good choice for most homes.

Pressure-Assisted Toilets. In general, pressure-assisted toilets work better than gravity models. These toilets use incoming water to compress air in a chamber inside the tank. (A water pressure of at least 20 pounds per square inch is required.) Flushing releases this compressed air in a burst, forcing water to prime the trap almost instantly. Air assist allows the tank to operate with less water, making more water available for the bowl. More water in the bowl means a larger water spot and a cleaner bowl. And finally, the tank-within-a-tank construction completely eliminates tank sweat caused by condensation during hot, humid weather.

With these advantages, you'd think everyone would want a pressure-assisted toilet, but that hasn't been the case. The most common complaints are that they're too noisy and too complicated. Starting each flush with a burst of compressed air does make them noisy, and they're certainly less familiar. Most people would recognize the tank components in a traditional toilet, but lift the lid on a pressure-assisted unit, and all you'll see is a sealed plastic drum, a water-inlet mechanism, a hose, and a flush cartridge. Most manufacturers use almost identical tank components.

1.6-Gallon Gravity-Flow Toilet

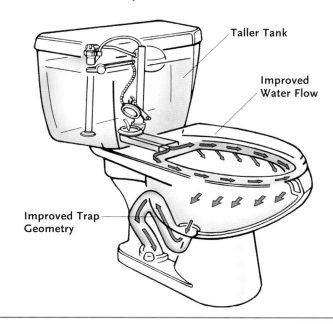

Taller Tank

Improved Water Flow

Improved Trap Geometry

Pressure-Assisted Toilet

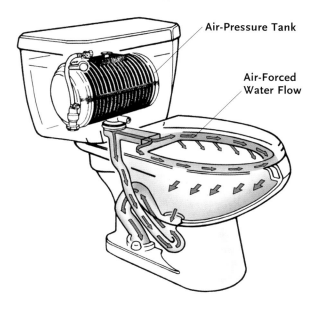

Air-Pressure Tank

Air-Forced Water Flow

TROUBLESHOOTING GRAVITY-FLOW TOILETS

Now that you know how traditional gravity-flow toilets are supposed to work, it's time to learn how and why they may not work and what to do when they don't. Keep in mind that poorly maintained toilets may display more than one symptom.

PROBLEM 1: Slow Toilet

Your toilet seems sluggish. It once flushed vigorously, but now the water seems to move slowly through its cycle, often rising high in the bowl before passing through the trap. You also notice large bubbles rising out of the trap during the flush. Sometimes the bowl even seems to double flush.

Fixing a Slow Toilet

These are classic symptoms of a partially blocked trap. An obstruction has made its way to the top of the trap and lodged in the opening. In many cases, enough of the trap remains open to keep the toilet working for a time. To clear a blockage, start with a toilet plunger, forcing the cup forward and pulling it back with equal pressure. If a plunger doesn't clear the clog, try a closet auger. If the closet auger fails, bail out the bowl with a paper cup or other small container, and place a pocket mirror in the outlet. Shine a flashlight onto the mirror, bouncing light to the top of the trap. The mirror should allow you to see the obstruction. When you know what and where it is, you should be able to pull it into the bowl by using a piece of wire.

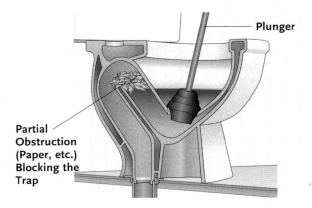

Plunger

Partial Obstruction (Paper, etc.) Blocking the Trap

PROBLEM 2: Bacteria- or

The toilet does not appear clogged, because water doesn't rise unusually high in the bowl, but it flushes sluggishly, and the bowl doesn't stay clean for long.

These symptoms suggest that the toilet bowl's rim holes—and possibly the siphon jet hole—are clogged with calcified minerals from hard water or with bacteria. To make sure, watch the water as it passes through the bowl. Open rim holes should send lots of water coursing diagonally across the sides of the bowl. If the water slides straight down, that may be a sign that the rim holes are partially clogged, either by bacteria or calcification. Dark, vertical stains beneath some of the holes suggest bacteria. Clogged siphon jets are almost always caused by bacteria.

Bacteria accumulations are soft and dark, ranging from orange to black. Mineral deposits are hard, scaly, and often light in color.

Cleaning a Mineral-Clogged Toilet

To remove calcified minerals left by hard water, pour vinegar heated to about 104 F into the overflow tube, and let stand for at least 30 minutes.

Ream each hole thoroughly. On heavily clogged holes, use Allen wrenches as reaming tools. Start with a small wrench, and use larger ones as you gradually unclog the hole. Remember that porcelain chips easily, so work carefully and use a pocket mirror to check your work.

Use a pocket mirror and Allen wrench to ream clogged holes around the underside of the rim.

Mineral-Clogged Toilet

Cleaning a Bacteria-Clogged Toilet

To remove bacteria, first kill as much of it as possible, not just in the bowl but in the bowl's rim and rim holes. Pour a mixture of 1 part household bleach and 10 parts water directly into the tank's overflow tube. Just lift the tank lid, and direct a cup or more of the bleach solution into the overflow. Allow the bleach to work for a few minutes; then flush the toilet, and carefully ream the rim holes with a pen knife or a piece of wire. Use a pocket mirror to check your progress. Scour any bacteria stains from the underside of the rim, using bowl cleaner and an abrasive pad. Add a final dose of bleach through the overflow; wait a few minutes; and flush the toilet. To clear a clogged siphon jet, ream it thoroughly with a stiff wire. You'll probably have to do this all-out ream-and-scour cleaning only once or twice a year if you add one or two tablespoons of bleach to the overflow tube periodically.

TOOLS & MATERIALS

▍ Measuring cup ▍ Bleach solution
▍ Insulated wire ▍ Pocket mirror (if necessary)

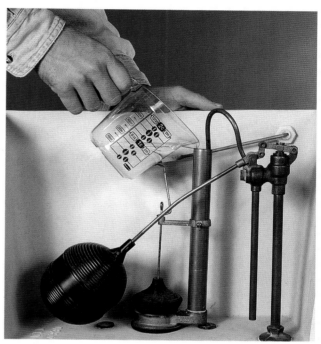

1 To kill a bacteria buildup in and under the toilet bowl's rim, pour a bleach solution directly into the overflow tube. A good mix has one part household bleach to 10 parts of water.

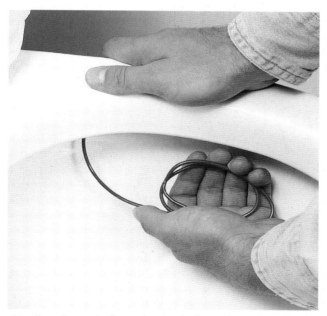

2 Clear bacteria from the rim holes using a length of insulated electrical wire. Approach each hole from several angles. The wire coil serves as a handle to manipulate the end of the wire.

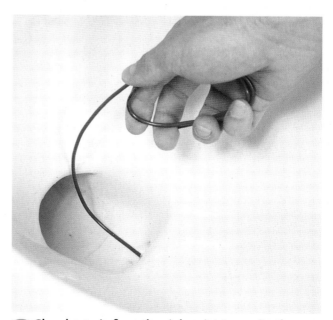

3 Clear bacteria from the siphon jet (opposite the trap hole) at the bottom of the bowl. Use a piece of insulated wire and turn it with a cranking motion. Clean the bowl with household cleanser.

PROBLEM 3: Slow-Filling Toilet

The toilet flushes well enough, but the tank takes 10 to 15 minutes to fill. You also hear a slight hissing sound when the house is quiet.

Removing Grit from the Diaphragm

This symptom indicates the presence of sediment in the fill-valve diaphragm. This problem can surface after the installation of a new toilet, after work is done on a nearby water main, or after a new well is put into service. The solution is to remove the diaphragm cover and pick the grit from the valve using standard first aid tweezers. This is usually a quick and easy fix. Just remember that sediment can move through your pipes at different rates. You may need to remove it several times before the system is clean.

TOOLS & MATERIALS

▌Screwdriver ▌Tweezers

1 To remove the ballcock's diaphragm cover, remove the three or four brass screws holding it in place, and lift it up and away. Avoid dropping the screws into the bottom of the tank because they can be difficult to retrieve.

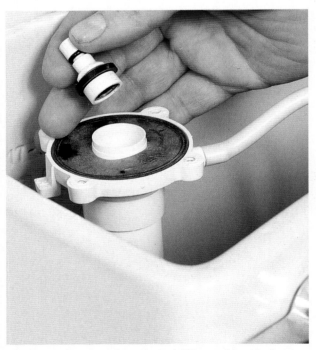

2 Once the cover is removed, lift out the valve's plunger and set it aside. Then carefully remove the diaphragm gasket and look for sand granules or other sediment underneath.

3 Use tweezers to lift out any sediment particles, rust flakes, or other debris. For very small particles, a cotton swab will do the trick. You may need to repeat this repair until all the debris is gone.

GREEN SOLUTION

PROBLEM 4: Running Toilet

The toilet often keeps running until you wiggle the flush handle.

Fixing Running Toilets

This symptom signals an adjustment problem, either in the flapper chain or tank-ball linkage. (It may also indicate a corroded flush lever, but in this case symptoms usually include the lever sticking in the up position.) If your toilet has a flapper, the lift chain may be too long. To correct the problem, remove the tank lid, and lift the chain from the wire hook that secures it to the flush lever. Reconnect it so that it has less slack. With the chain hooked, top and bottom, press it to one side. You should see roughly 1 inch of sideways deflection. (See the photo at right, top.)

If your toilet has a tank ball, expect the lift-wire guide to be out of alignment. You'll find this guide clamped around the flush valve's overflow tube, secured by a setscrew. (See the photo at right, middle.) Begin by removing the tank lid; then flush the toilet several times. Keep flushing until you see the tank ball fall off-center, showing you which way to move the guide. If the ball falls to the left, for example, move the guide to the right.

Shut off the water, and loosen the setscrew about two full turns. Rotate the guide about 1/16 inch, and then reset the screw. Turn the water back on, and watch the tank ball fall through several more flushes. If you've under- or over-corrected, you should be able to see it in the way the ball hits the flush-valve seat. It should hit dead center when it lands.

When making these adjustments, try not to apply too much pressure to the brass overflow tube. Brass gets brittle with age, and an old overflow tube might break off. If this should happen, simply pry the remaining bit of tube from its valve threads, and screw a new brass tube in place.

Adjust the flapper chain so that it has about 1 in. of slack.

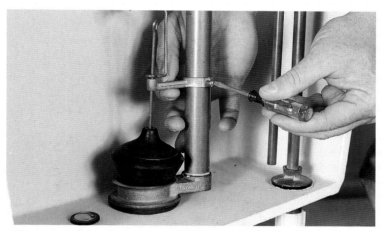

Adjust the lift-wire support if the tank ball falls off center.

When you replace an overflow tube that breaks off, coat the threads with pipe joint compound before screwing it down.

Toilet Repairs & Installation

GREEN SOLUTION

PROBLEM 5: Rippling Water

The toilet comes on by itself, runs for a few seconds, and then shuts off. You may also hear a steady trickle and see tiny ripples in the bowl water.

This is most likely a sign that your toilet's flapper or tank ball is worn out. It may also signal a bad flush valve, but check the flapper/ball first. In any case, a small stream of water is leaking through the valve. As the tank level drops, the float activates the fill valve, which replenishes the lost water and then shuts off.

When a flapper or tank ball fails, it's usually because the rubber has broken down. In the worst cases, the flapper will literally fall apart in your hands. An early warning sign of deterioration is a stubborn black slime that covers the surface of the rubber.

Replacing the parts is cheap and easy to do. Flappers and tank balls are universal, so just about any brand will work. Chlorine degrades rubber quickly, so if you use chlorine toilet-tank treatments, it's a good idea to install a flapper or tank ball made to resist chlorine. (Most of these are clear or translucent.)

Replacing a Tank Ball

To replace a tank ball, first shut off the water, and grip the top of the lift wire with pliers. With your remaining hand, thread the ball from the lift wire. If the ball crumbles, leaving only the brass insert attached to the wire, use a second pair of pliers to grip this fitting. Brass is soft, so you won't have any trouble backing the ball fitting from the lift wire with a good grip. Again, dress the flush-valve seat with an abrasive and a paper towel; then insert the wire through its guide and into the new tank ball. Finally, make any needed adjustments in the lift-wire guide. (See "Fixing Running Toilets," on page 69.)

If the flapper (or tank ball) seems to be in good shape and creates a good seal, the problem lies with the flush valve, and you'll have to replace it. (See "Replacing a Flush Valve," pages 76–77.)

TOOLS & MATERIALS
▌ Slip-joint pliers (if necessary)
▌ Tank ball replacement

Replacing a Flapper

With a flapper, shut off the water; unhook the chain; and lift the flapper from its pegs at the base of the flush valve or from around the overflow tube if the flapper has a collar. Before installing the new flapper, clear the valve seat of any old rubber (slime) or mineral deposits. Wipe the seat rim with a paper towel; then sand it lightly with fine sandpaper or steel wool. An abrasive pad from the kitchen will also work. You may need to use a little vinegar to dissolve mineral deposits.

The new flapper will likely have two types of flush-valve attachments for a universal fit: side tabs that hook over the valve's side pegs and a rubber collar for use on valves without side pegs.

TOOLS & MATERIALS
▌ Flapper replacement ▌ Vinegar ▌ Scissors
▌ Abrasive pad, steel wool, or scouring pad

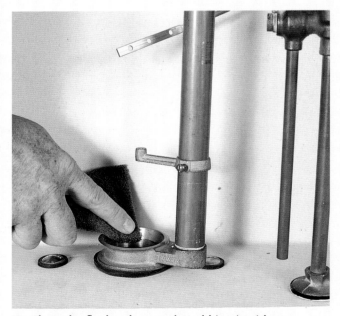

1 Clean the flush-valve seat by rubbing it with an abrasive material such as steel wool or a scouring pad until it is smooth. Check for imperfections by wiping your finger over the seat.

1 To replace the tank ball, first drain the water from the tank; then hold the ball with one hand and the lift wire with the other. Unthread the ball from the wire. If the two won't separate, hold each end with pliers and turn.

2 To install the new ball, slide it into place above the flush valve, and thread the lift wire down through the hole in the support arm and into the top of the ball. This ball has a weighted bottom to help it seat in the valve more easily.

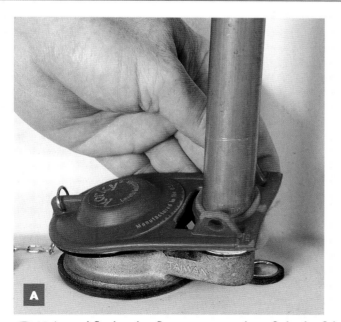

A

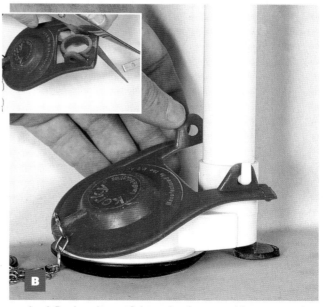

B

2 Universal flush-valve flappers are made to fit both of the standard flush valves. If the valve has no side pegs at the bottom of the overflow tube, slide the round collar at the back of the flapper over the tube, and push it down onto the valve seat (photo A). If the flush valve has side pegs, cut off the collar using scissors and discard it (inset). Then hook the eyelets on the flapper over the pegs on the overflow tube (photo B).

Toilet Repairs & Installation

PROBLEM 6: Hissing Toilet

The toilet doesn't shut off completely. You hear a hissing noise and see ripples in the bowl. This behavior starts inter-mittently but over time becomes constant.

Fixing Fill Valves to Cure Hissing

These symptoms suggest a problem with the fill valve or ballcock. (Remember that all ballcocks are fill valves, but not all fill valves are ballcocks. The term "ballcock" applies only to traditional fill valves, which have ball floats on the end of a pivoting arm.) It may be that you need to remove sediment from the fill-valve diaphragm. (See "Removing Grit from a Diaphragm," page 68.) Or you may be able to solve the problem by making a simple float adjustment. In most cases, however, the valve needs to be repaired or replaced. Don't put it off. A toilet that won't shut off com-pletely wastes lots of water.

Do the simple things first. If the water level is so high that it spills into the overflow tube before the float ball or cup can shut off the fill valve, adjust the water level. Aim for a level about 1 inch below the top of the overflow. With a ballcock assembly, tighten the adjustment screw on top of the fill-valve riser. If that doesn't work, bend the float-ball rod down slightly. (See the photo top right.) Use both hands, and work carefully. Newer fill valves have other float adjustments. For example, a common type has a stainless-steel clip that you use to adjust the height of a float cup. (See "Replacing a Fill Valve," on pages 80–81.) If adjustment doesn't solve the problem, a tiny amount of sediment may be in the diaphragm. See "Removing Grit from the Diaphragm," page 68, for how to clean the as-sembly.

Assuming that the float setting is fine and there is no sediment in the diaphragm, your next option is to replace the diaphragm and float-plunger seals. Begin by shutting off the water and removing the two or three screws that secure the combination diaphragm cover and float-arm assembly to the top of the riser. Lift the cover-and-float-arm assembly from the valve, and pull out the rubber seals. Expect a large rubber disk or stopper. Take these parts with you to a well-stocked plumbing outlet. If you can find replacement seals, install them in reverse order of removal, and test your work. You may be better off replac-ing the entire fill valve, especially if it's old. (See "When to Replace a Ballcock," page 74.)

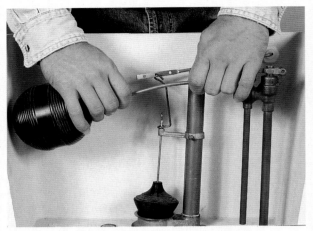

Adjust the float rod by bending it downward carefully.

Fine-tune the float by turning its adjustment screw.

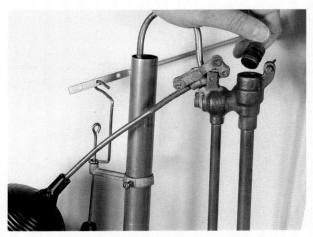

Replace the rubber seals on the plunger of a brass ballcock.

EASY-FIX FLUSH VALVE

HERE'S A FLUSH-VALVE ASSEMBLY unlike any other. It consists of an inverted cone that rides up and down on a plastic tower. A simple rubber gasket seals the valve. The design holds up well, but when it fails, all you'll need to do is replace the gasket. Thread the cap from the tower, lift the cone, and press a new gasket in place. (Mansfield Company, 150 First St., Perrysville, OH 44864; toll-free phone number: 1-877-850-3060.)

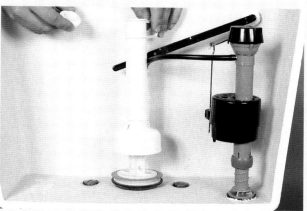

1 When it is time to replace the gasket, thread the plastic cap from the flush-valve tower.

2 Lift the tower to expose the sealing gasket. Remove the old gasket.

3 Replace the gasket by pressing it into place. When seated, reinstall the tower.

smart tip

WATER TANK CONDENSATION

IF YOU SEE WATER ON THE FLOOR NEAR THE TOILET, IT MAY BE NOTHING MORE THAN MOISTURE DRIPPING FROM THE SIDES OF THE WATER TANK. CONDENSATION ON THE SIDES OF A TANK OCCURS WHEN AIR IN THE ROOM IS WARM AND THE HUMIDITY IS HIGH. THE WARM AIR CONDENSES WHEN IT MEETS THE COOL SIDES OF THE TANK. DON'T TAKE THIS DRIPPING MOISTURE LIGHTLY. IF YOU DON'T WIPE UP THE PUDDLES (OR STOP THE DRIPPING),

WATER CAN SEEP BENEATH THE FLOORING AND CAUSE THE SUBFLOOR TO ROT. TO PREVENT CONDENSATION PROBLEMS, BUY A TOILET-TANK INSULATION KIT, WHICH CONSISTS OF POLYSTYRENE (USUALLY) TANK LINERS, FROM A HOME CENTER. SHUT OFF THE WATER TO THE TOILET, AND DRAIN THE TANK BY FLUSHING. CUT THE INSULATING LINERS TO SIZE; APPLY THE SUPPLIED ADHESIVE; AND INSTALL THE LINERS TO THE INSIDE OF THE TANK. THE POLYSTYRENE SHOULD PREVENT THE TANK SIDES FROM BECOMING SO COOL THAT THEY CAUSE DRIPPING CONDENSATION.

PROBLEM 7: Sticking Flush Handle

When you press down on the flush lever handle, it sticks—and you have to pull it back up.

Replacing a Flush Handle

Stiff flush levers are usually heavily corroded, while loose ones are probably broken. You may be able to free a sticking lever using a few drops of penetrating oil, but replacement is a good idea, especially considering how little a new one costs. Old lever handles also begin shedding their chrome or brass plating and look ugly.

There's nothing difficult about this repair, but in this case knowledge works better than leverage: you should know that the hex nut holding a flush lever assembly to the tank uses left-hand threads. (That's right, the threads are backward.) To loosen this nut, turn it clockwise rather than the usual counterclockwise. If the nut is too corroded to break free, cut the assembly apart with a hacksaw.

Your new lever and handle will likely come in one piece, with the fastening nut the only other component. The lever may be metal or plastic, while the handle will probably be chrome- or brass-plated metal. Snake the lever through the tank hole; then slide the nut over the lever, and screw it onto its threads. (See the photos below.) Finally, connect the flapper chain or lift wire.

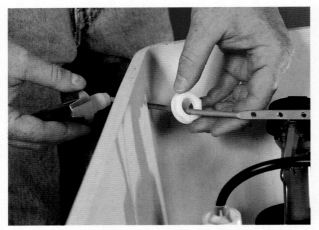

Remove the nut from the old lever, and pull the lever through the tank hole.

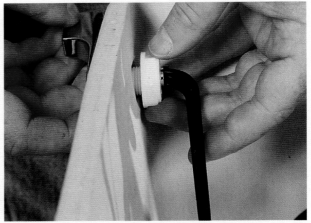

Slide the new lever in place, and tighten the nut. The nut has left-hand threads.

smart tip

WHEN TO REPLACE A BALLCOCK

WHILE REPLACING ONLY THE SEALS OF A BALL-COCK IS EASIER THAN REPLACING THE ENTIRE UNIT, IT'S NOT ALWAYS POSSIBLE. IT'S NOT ALWAYS EASY TO FIND REPLACEMENT PARTS, AND SOME WERE NEVER MEANT TO BE SERVICED. ALSO, BALL-COCK SEATS EVENTUALLY WEAR OUT. IN THESE CASES YOU'LL HAVE TO REPLACE THE BALLCOCK. REFERABLY WITH A MODERN FILL VALVE. (SEE "RE-PLACING A FILL VALVE," PAGES 80–81.)

PROBLEM 9: Leaking Base

Lately you've noticed water near the base of the toilet.

Fixing a Leaky Base

This is usually a sign that the bowl wax gasket on the toilet flange has failed. But that's not always the case, and because fixing a broken gasket requires taking up the toilet, it pays to investigate the quick fixes. Often, water that appears on the floor is not bowl but tank water. If the tank has been bumped, it's not uncommon for water to leak past the tank bolts, dripping

PROBLEM 8: Stuck Seat

The toilet's seat is broken or worn. You've bought a new seat but can't get the old one off.

Fixing a Stuck Seat

This is a common problem when an old toilet seat has brass bolts molded into the seat hinge. When you attempt to loosen the corroded fastening nuts, they stick tight, causing the bolt heads to break loose within their molded sockets. No matter what you do, the bolts just spin. The only way to deal with this situation is to saw through the bolts, just under the seat. To keep from marring the toilet's china surface, place a double thickness of duct tape on the bowl, all around the bolts. Remove the blade from a hacksaw, lay it flat against the bottom of the seat, and cut straight through the bolts.

With the old seat removed, position the new one, and insert the bolts through the seat and deck holes. Tighten the nuts. The good news is that you'll only have to do it once. The plastic bolts on new seats will never corrode.

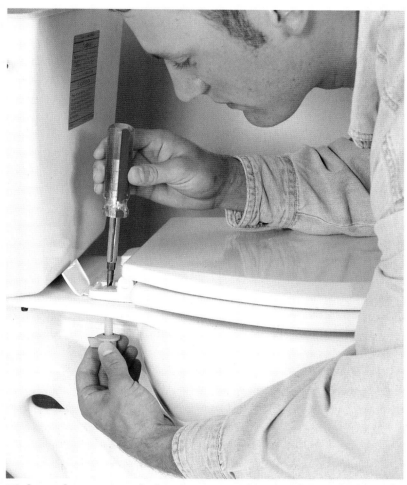

Tighten the new seat bolt using a screwdriver, and snap the hinged cover in place.

to the floor. Reach under the tank, and feel for moisture clinging to the tank bolts. If the bolts are wet, dry them using a paper towel. (Although a disturbed tank bolt may leak persistently, sometimes one bump makes one drip, and that's it.)

After an hour or so, check the bolts again. If they're wet, you have a tank leak. Carefully tighten the nuts on the tank bolts a half-turn to a full turn to stop the leak. If the leak persists, then tighten the nuts one or two turns. If the leak still persists, remove the tank bolts and install new bolt washers. In this case, you usually will not need to disturb the tank-to-bowl seal or the water connection. Just turn off the water; flush the toilet; and sponge all the remaining water from the tank. Then remove the bolts; replace the washers; coat them liberally with pipe joint compound; and reinstall the bolts.

If the tank is not leaking, the problem is with the wax gasket in the toilet flange. You'll have to take up the toilet and replace the seal. (See "Taking Up and Resetting a Toilet," pages 91–93.)

REPAIRING TOILETS

Replacing a Flush Valve

A flush valve fails when its valve seat can no longer form a seal to hold water. As water leaks past a defective flapper or tank ball, it can cut channels through the valve seat. When these voids appear, you have two repair options: in-

stall a retrofit flush-valve seat right over the damaged seat, or separate the tank from the bowl and replace the flush valve. Retrofit kits are easier, but a new flush valve makes a longer-lasting repair. (See "Retrofit Kits to the Rescue," page 78.)

To replace a flush valve, start by shutting off the water, either at the main valve or at the shutoff valve beneath the toilet. Lift the lid from the tank, and flush the toilet to

GREEN SOLUTION

REPLACING A FLUSH VALVE

project

You can replace a flush valve by yourself, but it's much easier if you have help. Removing the tank bolts is quite a stretch for arms that are attached to the same torso, and lifting the tank off the bowl can quickly remind you of just how old your back is. Both these jobs only take a couple of minutes with someone else.

TOOLS & MATERIALS

- Socket wrench ▪ Flat-blade screwdriver
- Hacksaw blade ▪ Adjustable wrench
- Groove-joint pliers ▪ New flush valve
- Pipe joint compound

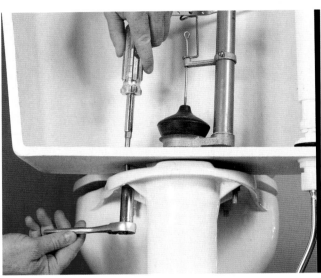

1 Loosen each tank bolt with a socket wrench on the bottom of the tank to turn the nut and a flat-blade screwdriver inside the tank to back hold the slotted screw. Corrosion can make separating the bolts and nuts almost impossible.

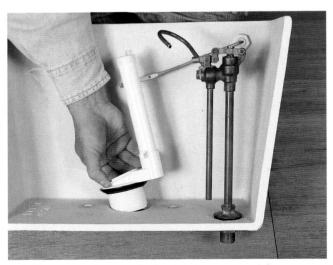

4 Clean any dirt or old pipe joint compound from the flush valve opening. Then install the new valve through the tank hole. Position the overflow tube on the new valve so that it will rest next to the fill valve on the fill valve.

5 Thread the new spud nut onto the bottom of the new flush valve, and make sure it sits flush on the bottom of the tank. Finger-tighten the nut, then securely tighten it with groove-joint pliers. When the nut squeaks against the tank it's tight enough.

drain as much water as possible. Sponge out any remaining water.

Remove the Tank. With the tank empty, loosen the coupling nut that secures the water-supply tube to the fill-valve shank, and remove the tube. Reach under the tank, and using a socket wrench, remove the nuts from the two tank bolts. If the bolts spin, backhold them with a large screwdriver. In most cases, the nuts will turn free, but if

your toilet is old and the nuts haven't been disturbed in many years, it's reasonable to expect them to resist. If the nuts seem really stuck, forgo the wrench and reach for a hacksaw blade. Brass bolts are relatively soft, and you should be able to cut through them quickly. Use the blade from a standard hacksaw, and wrap one end with duct tape to serve as a makeshift handle. Use only the blade, because you can't fit a hacksaw between the tank and the bowl.

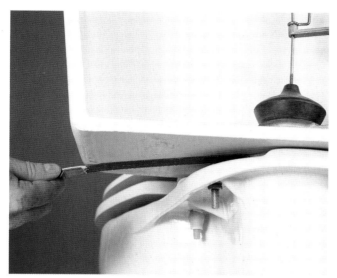

2 If the tank bolts won't loosen with a wrench and screwdriver, then cut them with a hacksaw blade. The bolts are usually brass, which is much softer than steel, so this job isn't too difficult. Wrap the end of the blade with tape to protect your hand.

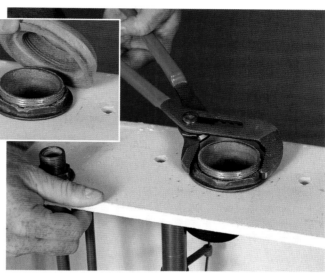

3 To remove the old flush valve, first remove the old spud washer from the bottom of the tank (inset). Then loosen the spud nut that holds the valve with large groove-joint pliers. Turn the nut in a counterclockwise direction.

6 Slide the new rubber spud washer over the plastic spud nut, and push it down so it's flush against the bottom of the tank. Then spread a uniform (and generous) layer of pipe joint compound over the top, but not the sides, of the washer.

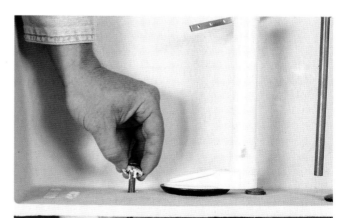

7 Slide rubber washers over the ends of the tank bolts, and coat the underside of these washers with pipe joint compound. Insert the bolts in the holes at the bottom of the tank and through the holes in the toilet bowl. Tighten the bolts.

RETROFIT KITS TO THE RESCUE

IF YOU'RE NOT UP TO SEPARATING THE TANK from the bowl to repair a damaged flush valve, or if the toilet has it's flush valve and overflow cast into the china, then a retrofit valve-seat replacement kit is a good choice. (Eljer and Crane are two manufacturers that may have toilets with these parts cast into the body of the tank.) The kit is an assembly that consists of a stainless-steel or plastic valve seat, an epoxy-putty ring, and a flapper. The super-tough epoxy putty ring adheres the entire unit onto the old seat. The kit can be used over brass, china, or plastic flush valves.

To install a retrofit valve-seat kit, start by shutting off the water and flushing the toilet to drain the tank. Sponge the remaining water from the tank. Remove the old flapper. If your toilet has a tank ball, remove the ball, the lift wire, and the lift-wire guide. Then sand the old seat to remove mineral deposits and to abrade the surface for better adhesion. (You may want to use vinegar if the mineral deposits are stubborn.) Wipe the seat clean, and allow the seat rim to dry.

With the old seat ready, peel one waxed-paper protector from the epoxy ring, and stick the ring to the bottom side of the retrofit seat. Peel the remaining waxed paper from the epoxy (photo at bottom left), and press the assembly firmly onto the old valve seat (photo at bottom middle). Finally, connect the flapper chain (photo at bottom right), and turn the water back on.

Epoxy kits are not your only valve-seat repair option. A simpler kit consists of a tube of silicone adhesive, which cures under water, and a plastic replacement valve seat. Sand the seat, and then apply a bead of silicone to the seat rim. Press the new seat in place; install a new flapper; and turn the water back on.

1 Peel the protective paper from the epoxy ring.

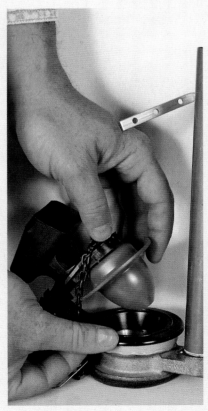

2 Press the replacement seat over the old seat.

3 Connect the chain with about ½ in. of slack.

ONE-PIECE SILENT-FLUSH TOILETS

ONE-PIECE SILENT-FLUSH TOILETS were the first alternatives to conventional two-piece toilets, hitting the market around the time Cadillacs grew fins and aimed at the same market. They were expensive, stylish, and discreetly quiet.

There are two basic types: One is a gravity-flow toilet with a conventional fill valve and flush valve. This type is easy to repair if you can find factory replacement parts. The other has a more complicated system of valves, which need to be calibrated to match a home's static water line pressure. One look at a repair kit, which resembles an automotive carburetor repair kit, and you'll get the picture: working on it is not easy. You should hire a good service plumber for this job. Ask up front whether the plumber is familiar with your particular make and model.

A silent-flush toilet is made in one piece. They are usually very quiet but may be difficult to work on.

A technologically advanced low-profile toilet does double duty as a bidet, incorporating a warm-water personal-cleansing system with its own handy control panel. The toilet also boasts dual-flush capability and an easy-to-clean concave rim.

REPLACING A FILL VALUE

project

If you're having trouble with a toilet's fill valve, it's usually best to replace the entire unit. You'll see several types on the market. The most familiar is the traditional ballcock, in brass or plastic. But you'll also see some that have floats that slide up and down on a vertical riser, and some low-profile valves that are activated by headwater pressure. Codes require fill valves to have built-in backflow protection to keep tank water from back siphoning into the water system. So check the product label and choose a valve with backflow protection.

TOOLS & MATERIALS

- New fill valve
- Pipe joint compound
- New supply tube (if necessary)
- Groove-joint pliers
- Adjustable wrench

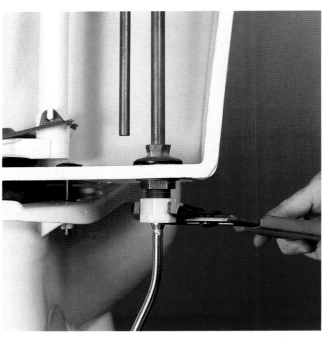

1 To replace a fill valve, first remove all the water from inside the toilet tank. Then loosen the coupling nut that connects the supply riser to the valve shank with large groove-joint pliers. These days, this nut is usually made of plastic, which doesn't corrode. So it's easy to remove.

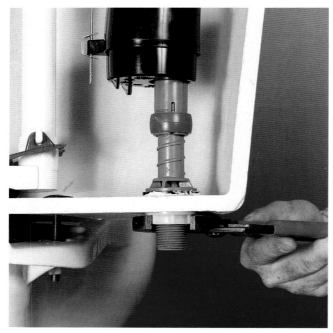

4 Most fill valves feature plastic jamb nuts, which are easy to install and never corrode. But be careful not to over-tighten them. As soon as the nut feels snug and starts to squeak, stop turning. If you are installing a new brass valve with a bronze jamb nut, just tighten the nut so it's snug.

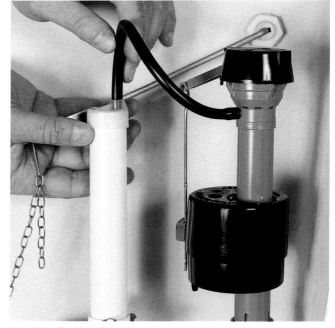

5 The flexible fill tube is the component that connects the fill valve to the overflow tube and transfers the water between the two. Cut this tube to length with scissors or a utility knife so that it bends gently from one to the other. Don't allow any kinks that would restrict the flow.

2 Once the supply line coupling nut is removed, loosen and unthread the fill valve jamb nut with an adjustable wrench. You, or a helper, should hold the top of the valve from inside the tank so that the whole assembly doesn't turn when you turn the wrench.

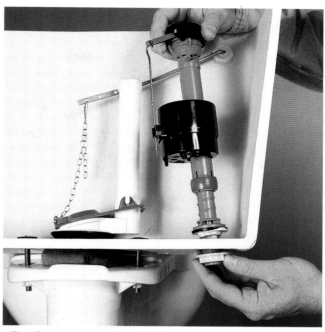

3 After removing the old fill valve, clean both sides of the tank hole, and apply pipe joint compound to the bottom of the valve's washer. Insert the new valve in the hole, and secure it with a jamb nut. Hold the valve from above and tighten the nut with an adjustable wrench from below.

6 Once the fill tube is installed, adjust the water level inside the tank by moving the fill valve float up and down. To do this, compress the clip that holds the lift wire to the float, and move the float to the desired height. Release the clip and the adjustment is complete.

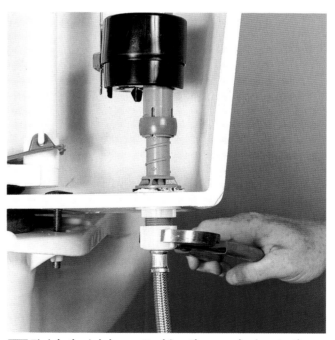

7 Finish the job by reattaching the supply riser to the bottom of the fill valve shank using groove-joint pliers. Then turn on the water, and check the new valve's performance. If the tank doesn't hold enough water to clean the bowl when flushed, readjust the float so the tank will hold more water. Also check for any water leaks.

TROUBLESHOOTING PRESSURE-ASSISTED TOILETS

Pressure-assisted toilets function differently from gravity-flow toilets. They are based on a different operating principle: use of air pressure. The toilets offer limited repair options, but you can attempt a few remedies.

TRADITIONAL TOILET SUPPLY LINE

THE TRADITIONAL supply tube is made of chromium-plated copper, with a rubber cone washer or a plastic flat washer, fitted to meet the fill valve's shank. You cut the other end to length and join it to the shutoff valve with a compression fitting. These tubes are supple, and if the offset is not too severe, you can bend them without a tubing bender.

Hold the tube between the fill valve's shank and the shutoff valve. This should give you a rough idea as to the degree of offset and the length you'll need. Make the offsets so that each end of the tube will enter its fitting dead straight. Trim the tube, and then slide on the coupling nut, followed by a compression nut and ferrule. Lubricate both ends with pipe joint compound.

To gain enough vertical clearance to fit the tube between the fittings, press down on the shutoff valve. Make the compression connection at the shutoff valve first. To keep from overtightening the compression nut, turn it finger-tight while wiggling the tube a little to keep it from binding. Then tighten it one full turn using a wrench. Finally, tighten the coupling nut at the tank, and turn on the water to test for leaks.

Traditional supply risers use compression fittings to make the seal at the shutoff valve.

PROBLEM 1: Running Toilet

The toilet keeps running between flushes or produces a loud buzzing noise after each flush.

Fixing a Running Pressure-Assisted Toilet

This behavior suggests that something is keeping the flush-valve cartridge—the central, top mounted fitting in the tank—from closing. Although an occasional trickle of water into the bowl between flushes is not unusual for these toilets, if the water actually seems to run while the toilet is idle, you'll need to investigate.

If the toilet has a flush button mounted in the tank lid, the button's trim collar may be riding on the activator. Remove the tank lid, and flush the toilet. If the toilet now flushes properly, the collar was the problem. Replace the tank lid, and remove the flush button. While sighting through the lid opening, move the collar until it is centered over the activator. When you replace the button, make sure it travels downward at least ⅛ inch before contacting the activator. If it doesn't, loosen the locking screw; adjust the activator; and reset the screw. If the flush button doesn't seem to be the problem or your toilet has a flush lever, shut off the water, drain the tank completely, and then turn the water back on. Hereafter, avoid pressing the flush button or lever before the tank fills completely. As with a flush button, the flush lever needs ⅛ inch of clearance.

If adjusting the activator doesn't help, suspect the pressure-regulating valve. (A faulty valve may also cause the toilet to take longer than normal to fill.) The regulating valve is sandwiched between the check valve and the relief valve in the water supply group. To replace it, you'll have to remove the entire supply group. Shut off the water, loosen the coupling nut beneath the tank, and remove the water supply tube. Then remove the jamb nut that holds the supply group in the tank. Lift the supply group out of the tank, and take it apart to expose the pressure-regulating valve.

TOOLS & MATERIALS

▌ Screwdriver ▌ Groove-joint pliers
▌ Adjustable wrench ▌ New pressure-regulated valve

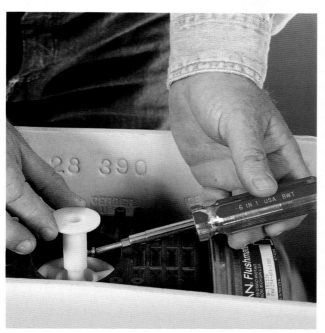

1 Adjust the flush activator setscrew so that the flush button (if you have one) has at least ⅛ in. of clearance before it contacts the activator. Avoid dropping the screw into the tank.

2 If there's no button, adjust the flush lever so there is ⅛ in. of clearance. If this doesn't help, drain the tank, remove the water supply group (inset) and replace the regulating valve.

3 Once the water supply group is removed from the tank hole, twist the assembly apart to expose the pressure-regulating valve. Lift out the old valve, and clean any sediment from inside the housing.

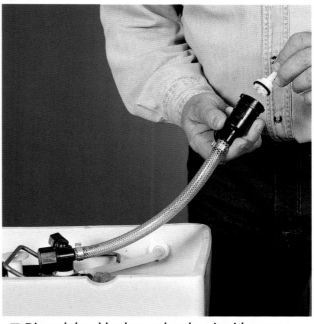

4 Discard the old valve, and replace it with a new one. Tighten all the connections with an adjustable wrench while back holding any nuts with a pair of groove-joint pliers.

FLUSH VALVE TEST

TURN OFF THE WATER, and flush the toilet. Hold the flush lever down a full 60 seconds to drain the tank completely. Then pour a little water into the hollow around the activator stem. (See the photo below.) Turn the water back on, and allow the tank to fill completely. Then check for air bubbles around the activator. If no bubbles appear, assume that you have a fouled air inducer. Remove the inducer and clean it. If you do see bubbles around the activator, you'll need to replace the entire cartridge.

To test for a leaky flush-valve cartridge, remove the water-tank top, pour water around the activator stem, and flush. The appearance of bubbles indicates a leak.

PROBLEM 2: Wasteful Flushes

The toilet flushes sluggishly and doesn't clear the bowl.

This symptom may simply mean a clogged trap, so use a plunger or closet auger on the bowl before tearing into the tank. If the trap is clear, suspect a faulty flush-valve cartridge. Lift the tank lid, and check for water on top of the cartridge. If you see any water at all, replace the cartridge. If you don't, you'll need to conduct a little flush-valve test. (See left.)

Cleaning a Fouled Air Inducer

To clean a fouled air inducer, shut off the water and drain the tank. Using pliers or an adjustable wrench, remove the inducer's plastic nut. Place a hand under the inducer to catch the internal components. Expect to find a small plastic or brass poppit fitted with a spring. Hold the poppit under running water to remove any calcification. If the poppit is really crusty, soak it in vinegar for about 30 minutes, and then rub off the scale. Carefully reassemble the components, and test your work.

TOOLS & MATERIALS

- Groove-joint pliers or adjustable wrench
- Vinegar (if necessary)

Replacing a Flush-Valve Cartridge

To begin, shut off the water, and drain the tank completely by holding down the flush lever for about a minute. Insert the ends of a large pair of needle-nose pliers into the fins of the cartridge's top nut. Carefully rotate the cartridge from the tank. If necessary, slide a large screwdriver between the pliers' handles for extra leverage. Buy an identical cartridge. Thread the cartridge into place. Tighten until it squeaks, then just a touch more. Turn the water back on to test your work.

TOOLS & MATERIALS

- Needle-nose pliers
- Plumbers grease
- Flush-valve cartridge

1 With the tank drained, remove the inducer's plastic nut, and catch the interior parts so they don't fall into the bottom of the tank.

2 Hold the tiny brass or plastic poppit fitting under running water, and roll it between your fingers to clean away and sediment or debris. You can also soak the fitting in vinegar.

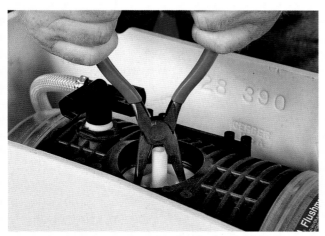

1 To remove the top nut that holds the flush-valve in place, insert the ends of a large pair of needle-nose pliers into the fins of the nut and turn.

2 Remove the old cartridge and discard it. Then coat the O-rings on the replacement unit with plumber's grease and install the cartridge.

TOILET FLANGES

A toilet flange, or closet flange, is a slotted ring, usually connected to a vertical collar. The collar fits through the floor, while the slotted ring, or flange, rests on top of the floor. Toilets are bolted directly to this fitting. In the case of a cast-iron flange, the collar slides over a riser pipe, which extends to floor level. The gap between the pipe and collar is packed with lead and oakum. If the pipe is cast iron and the collar is plastic, the gap is bridged by a rubber gasket. With plastic or copper piping, the flange is cemented or soldered to the riser. In all cases, the flange is anchored to the floor with screws.

Leaky Flange Gaskets

If you see water on the floor around the toilet, first try to determine whether it is from leaking tank bolts or condensation. (See "Problem 9: Leaking Base" and "Water Tank Condensation," pages 73–75.) If you can't find the source or if more water appears with each use, the water is probably coming from the floor flange. Correct the situation as soon as possible. Water can delaminate plywood, blister underlayment, and rot the subfloor.

Quick Fix. If the toilet was installed within the past few months, then merely tightening the closet bolts on the base of the toilet may reseal the bowl's wax gasket. New wax gaskets almost always compress a little after installation, which can leave the bolts loose. Continued use can then cause the toilet to rock in place, breaking the seal. There's often enough wax to create a new seal—but only if you can draw the toilet and floor flange together.

Start by popping the caps from the closet bolts at the base of the toilet. Pry under them with a screwdriver or putty knife. With the caps removed, use a small wrench to test the tightness of the nuts. If they turn easily, tighten them only until they feel snug, and then watch the base of the toilet carefully over the next few days. If the floor stays dry, you've solved the problem.

Replacing the Gasket. If water reappears or if the bolts were snug in the first place, you'll need to take up the toilet and install a new wax gasket and closet bolts. (See "Taking Up and Resetting a Toilet," page 91–93.) If your toilet has been in place for years, don't expect a quick fix to work. Replace the wax gasket at the first sign of trouble.

INSTALLING TOILET FLANGES ON CONCRETE

project

Installing a toilet flange on a concrete floor is a straightforward job. Just take a standard flange, center it over the drain opening, and mark the screw locations of the floor. Drill holes at these points for plastic anchors using a drill with a masonry bit. Then attach the flange by driving the screws into the anchors.

TOOLS & MATERIALS

- Hammer drill (or drill-driver) ■ Hammer
- Panhead screws (length as needed and anchors) ■ Plastic flange ring

1 Drill anchor holes and tap plastic anchors in place. Apply caulk to the bottom of the flange.

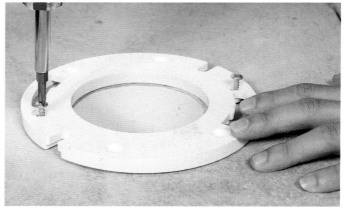

2 Attach the flange directly to the concrete using screws installed in the plastic anchors.

INSTALLING TOILET FLANGES ON WOOD FLOORS

project

Installing a toilet flange on the typical wood-framed floor is easier than doing the same job on a concrete floor, as long as the wood has no moisture damage. If it does, you should replace the finish flooring and any affected plywood subflooring before you install a new flange. One easy way to check for damage is to push the tip of a flat blade screwdriver into a dozen spots around the drain opening, moving out to a 16-inch-diameter circle. If the tip pushes easily into the wood, and especially through it, then you have damage that must be fixed so the toilet will be well-supported.

TOOLS & MATERIALS

- Drill-driver
- Panhead screws (length as needed)
- Plastic toilet flange Primer and cement

1 If you have cast-iron waste lines in your house, you can replace a cast toilet flange with a plastic one. Buy a two-part model that has the flange mounted on a sleeve. The sleeve has a rubber gasket to seal the joint with the cast iron riser.

2 To install the flange, first slide the sleeve into the riser, making sure that the rubber gasket is not damaged in the process. Then clean the sleeve and the flange with primer, and coat the mating surfaces with cement.

3 Push the flange onto the sleeve until the top lip hits the floor. Allow the cement to cure for a few minutes; then screw the flange to the floor using the holes provided. Don't over-tighten the screws because this can crack the flange.

TYPES OF FLANGE GASKETS

FIFTY YEARS AGO, the gasket material used to set toilets was plumber's putty. The plumber simply rolled out a quantity of putty and pressed it onto the flange. Putty, however, hardens with age, which can lead to leaks. The reason putty worked at all is because toilet bowls back then had four closet bolts instead of the two used today. Two were inverted, through the flange, as they are on modern toilets, but two more bolts located near the front of the flange anchored the bowl to the floor. Four bolts allowed very little flexing.

Wax. Beeswax rings, called bowl wax gaskets, eventually replaced putty as the preferred gasket material for toilets. (You'll also see the rings referred to as wax seals.) These wax rings were able to accommodate the slight flexing that occurs between a toilet and floor, so for generations bowl wax gaskets have been and still are the standard. You can buy gaskets in 3- or 4-inch-diameter sizes by about 1 inch thick. They are inexpensive and durable—a hard combination to beat.

Still, improvements are always in the works. One improvement was to incorporate a plastic, funnel-like insert in the traditional wax gasket. The insert was designed to deliver the water well past the flange surface, thereby eliminating leaks between floor flange and the wax. These special seep-proof gaskets are often used when a toilet is installed on a concrete slab.

Rubber. Next in the progression came flexible-rubber gaskets, which when compressed, block the lateral migration of water. Rubber gaskets can also re-seal themselves once disturbed, and they're reusable. They are sold in several thicknesses to accommodate a variety of flange heights relative to floor height, and you need to buy precisely the right thickness (unlike wax gaskets, which are more forgiving of small height differences). If a rubber gasket is even a little too thick, the toilet won't rest on the floor. If it's too thin, it won't seal.

The benefits of rubber and wax have recently been incorporated into a hybrid gasket, a neoprene rubber ring with a wax coating. (These gaskets are also available with seep-proof inserts.) The advantage is that the rubber will bounce back to reseal itself, while the layer of wax is more forgiving of sizing errors.

Measuring for Rubber. Wax gaskets come in standard thicknesses, and easily conform to slight job-site differences (flooring thickness and the like). In most cases, just knowing the horn length is good enough. To get the right thickness for a rubber flange gasket, you need to lay a straightedge across the base of the toilet and measure the length of the horn. Subtract the thickness of the toilet flange (minus any finish flooring such as tile), and add ⅛ to ¼ inch. The result is the ideal gasket thickness. Buy one as close to that thickness as possible. Measuring is not difficult, but it is bothersome. For that reason, you should use rubber gaskets only when the toilet is likely to be bumped frequently, as it might be when used by a physically handicapped person. A rubber gasket is better able to reseal itself after being disturbed.

Wax and Rubber Flange Gaskets

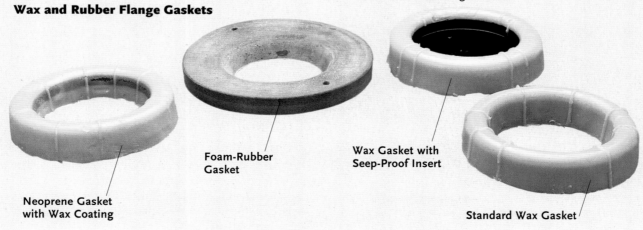

Foam-Rubber
Gasket

Wax Gasket with
Seep-Proof Insert

Neoprene Gasket
with Wax Coating

Standard Wax Gasket

Toilet Repairs & Installation

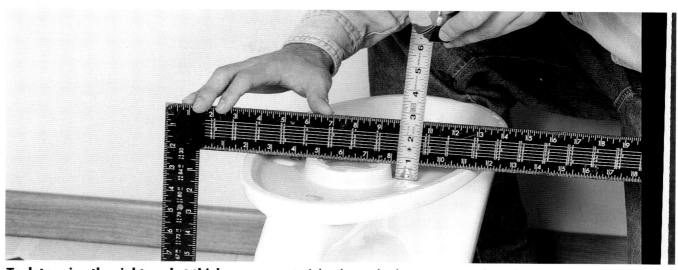

To determine the right gasket thickness use a straightedge and ruler to measure the toilet horn.

smart tip

INSTALLING A WAX-FREE GASKET

TRADITIONAL WAX GASKETS HAVE WORKED WELL FOR YEARS, BUT THEY HAVE SOME DRAWBACKS. THEY ARE NOT EASY TO USE IN TEMPERATURE EXTREMES, AND THEY ARE NOT FORGIVING WHEN YOU NEED TO DEAL WITH UNUSUAL FLOOR HEIGHTS DURING INSTALLATION: ALONE, THEY CAN BE TOO SMALL; STACKED, THEY MAY BE TOO BIG. FLUIDMASTER'S WAX-FREE BOWL GASKET MAKES LIFE A LITTLE EASIER. YOU CAN INSTALL IT IN ANY TEMPERATURE WITHOUT HAVING TO WORRY ABOUT CRACKED OR SOFT, STICKY WAX, AND IT IS ADJUSTABLE DURING INSTALLATION.

THE GASKET ASSEMBLY FITS INSIDE 3-INCH-DIAMETER DRAINPIPES (PLASTIC, COPPER, OR CAST IRON) AND COMES WITH A SLEEVE THAT FITS 4-INCH DRAINPIPES, SO YOU'RE READY FOR ANY SITUATION. O-RINGS ENSURE A TIGHT SEAL IN THE DRAINPIPES.

BEFORE YOU INSTALL THE GASKET ASSEMBLY, INSERT THE 3-INCH-DIAMETER GASKET INTO THE DRAINPIPE. IF IT IS TOO LOOSE, YOU'LL NEED TO USE THE 4-INCH SLEEVE. YOU WILL ALSO FIND TWO O-RINGS, A THICK ONE AND A THIN ONE, ONE OF WHICH GOES AT THE BOTTOM OF THE GASKET OR SLEEVE. USE WHICHEVER ONE GIVES YOU THE BEST FIT.

TO INSTALL THE GASKET, JUST BARELY INSERT IT INTO THE DRAINPIPE. THEN WRAP THE SQUARE CARDBOARD SPACER THAT COMES WITH THE UNIT (NOT SHOWN IN THE PHOTO) AROUND THE GASKET, AND SECURE IT AT THE TWO CUT CORNERS. PUSH THE GASKET DOWN INTO THE PIPE UNTIL IT TOUCHES THE SPACER OR IS 1 INCH ABOVE THE FLOOR, WHICHEVER COMES FIRST.

SLIP THE SUPPLIED FLOOR BOLTS INTO THE TOILET FLANGE'S MOUNTING ADJUSTMENT SLOTS, AND ATTACH THE SUPPLIED PLASTIC RETAINER NUTS TO HOLD THE BOLTS STEADY. LINE UP THE HOLES IN THE TOILET BASE WITH THE BOLTS, AND LOWER THE TOILET. APPLY WEIGHT TO THE TOILET TO ALLOW IT TO REST ON THE FLOOR. THE GASKET WILL ENGAGE THE TOILET'S HORN, AND THE CARDBOARD SPACER WILL COLLAPSE AND WILL NOT INTERFERE WITH THE GASKET. SECURE THE TOILET TO THE FLANGE, AND YOU'RE DONE.

REPAIRING A CAST-IRON FLANGE

project

Plastic toilet flanges are sturdy and seldom fail under normal conditions. Flanges made of cast iron and cast brass, on the other hand, are more vulnerable. The slotted portion of the flange is fairly narrow, so the slightest casting flaw will weaken the flange. As a result, it's easy to break a flange by overtightening the closet bolts. Once the flange is broken, the toilet bowl can rock from side-to-side and eventually damage the wax seal between the bowl and the flange. Fortunately, there's a way to repair one of these flanges without replacing it.

TOOLS & MATERIALS

▮ Tools and materials for removing a toilet
▮ Flange repair kit
▮ New wax gasket

1 To repair a flange, first remove the toilet tank and bowl and set them aside. Remove the wax-and-plastic flange insert from the drain opening, and scrap away any wax that's left behind. If the closet bolts are corroded, replace them.

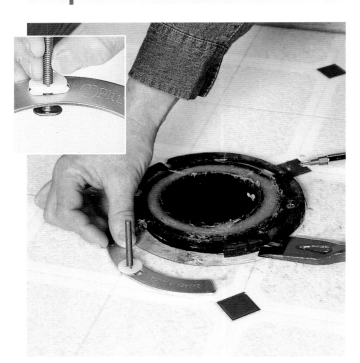

2 Buy a crescent-shaped repair strap at a plumbing supply store, and install a closet bolt into the hole provided. Loosen floor screws from the cast flange, and slide the strap underneath. Retighten the floor screws to hold the strap in place.

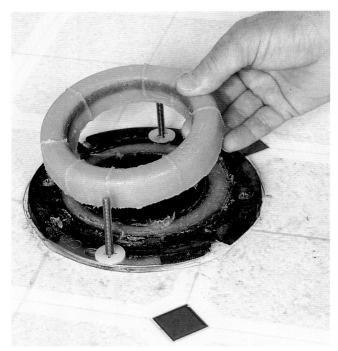

3 Align the closet bolt on the repair flange with the one on the other side of the cast flange. Install a new wax-and-plastic flange insert or a new wax ring over the toilet flange between the bolts (shown). Reinstall the toilet bowl and tank.

TAKING UP AND RESETTING A TOILET

project

There are lots of reasons why a toilet has to be removed. It may need to be replaced, repaired, or simply taken out of the way so it isn't damaged while other work is going on. And if a toilet is removed, eventually it will have to be reinstalled or replaced. While you can do both of these jobs yourself, some help is a big asset and makes moving the bowl and the tank much easier. All that's required are a few common home repair tools, a couple hours of time, and another toilet elsewhere in the house.

TOOLS & MATERIALS

- Adjustable wrenches ▌Needle-nose pliers
- Putty knife ▌Newspaper ▌Hacksaw
- New wax gasket kit ▌Water supply tube
- Plumber's putty ▌Groove-joint pliers

1 The first step in removing a toilet is to turn off the water supply at the shutoff valve under the tank. Then remove the compression nut that holds the toilet supply line to the shutoff valve using an adjustable wrench.

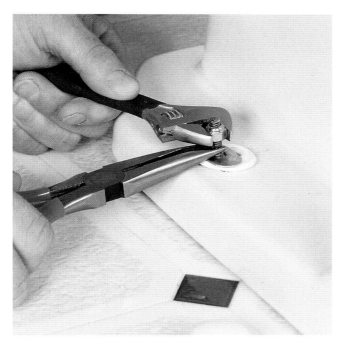

2 To remove the toilet, you have to remove the nuts on the toilet bolts. An adjustable wrench should do the trick. If the bolt starts to spin, this means that the flange underneath is broken. To remove a bolt like this, hold it with needle-nose pliers while turning the nut.

3 Once the toilet bolt nuts are removed, wiggle the bowl from side to side until the seal formed by the wax ring is broken. Then grip the toilet on both sides next to the seat hinges, and slowly lift the bowl using your legs and arms—not your back. Continued on next page.

Toilet Repairs & Installation

Continued from previous page.

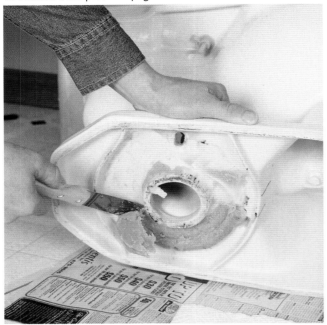

4 Place the toilet on several layers of newspaper, and tip it on its side. Work carefully so you don't damage the joint between the tank and the bowl. Remove the old wax from the bottom of the bowl with a putty knife. Then clean the opening with soap and water.

5 Remove the wax, and clean the surface of the toilet flange on the floor. Then insert new toilet bolts and a wax ring on the flange.

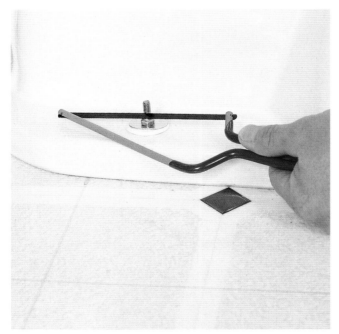

8 Depending on the length of the toilet bolts and the thickness of the bowl's base, the bolts may extend above the nuts so much that the decorative caps won't snap into place. If this happens, cut off the top of the bolts, just above the nut, using a hacksaw. Then retighten the nuts.

9 The toilet tank is joined to the house water supply through a supply tube. These days, this tube is usually made with flexible material protected with woven stainless-steel mesh. But older toilets often have a rigid supply tube made of chrome-plated copper.

6 Carefully lift the toilet and lower it onto the flange so the bolts go through the holes in the toilet's base. Push the bowl into the wax ring.

7 Wiggle the bowl from side to side while you press it down. This ensures that the toilet opening is completely sealed in the wax. Then thread the nuts onto the toilet bolts, and tighten them with an adjustable wrench. Stop turning when the nuts feel moderately snug.

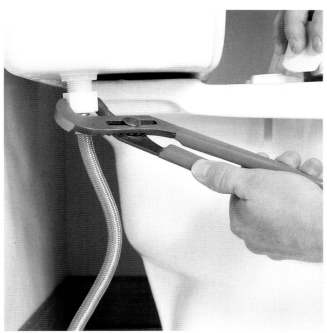

10 The top end of the supply tube is attached to the tank's fill valve at the bottom of the tank. Use groove-joint pliers to do this job, and once the nut is tight, turn on the water and check for any leaks. If everything is OK, flush the bowl, and check for leaks around the base.

smart tip

CHANGING FLOOR HEIGHT

IF YOU'VE TAKEN UP THE TOILET TO INSTALL NEW FLOORING, YOU MAY NEED TO ALTER YOUR APPROACH WHEN RESETTING THE TOILET. IF ALL YOU'VE DONE IS LAY NEW VINYL OVER OLD, YOU'LL BE ABLE TO RESET THE TOILET JUST AS DESCRIBED. IF YOU'VE INSTALLED ¼-INCH PLYWOOD OR CEMENT-BASED BACKER BOARD UNDERLAYMENT AND/OR GLAZED TILE, HOWEVER, YOU'LL HAVE ADDED TO THE HEIGHT OF THE FLOOR RELATIVE TO THE FLANGE. YOU'LL NEED TO COMPENSATE FOR THIS INCREASED DEPTH WITH A THICKER BOWL GASKET.

IF THE ADDED FLOOR HEIGHT IS LESS THAN ¼-INCH, YOU CAN SIMPLY KNEAD AND STRETCH A STANDARD WAX GASKET TO GIVE IT A TALLER PROFILE. WITH AN INCREASE OF ¼-INCH UP TO ABOUT 1 INCH, STACK ONE STANDARD WAX GASKET ON TOP OF ANOTHER.

Toilet Repairs & Installation

Setting Toilets On Concrete

Toilets set on concrete often ride high on the flange, with little of the base touching the floor. This is unworkable, so most toilets that are set on concrete need to be shimmed.

Set the toilet as usual, but draw the closet bolts down only three-quarters of the way. Then level the toilet by sliding cedar or plastic shims under the base. Plastic is best; never use metal shims. Only when you're satisfied that the base is well supported, without any obvious rocking, should you finish tightening the bolts. Trim the shims, and then caulk the base.

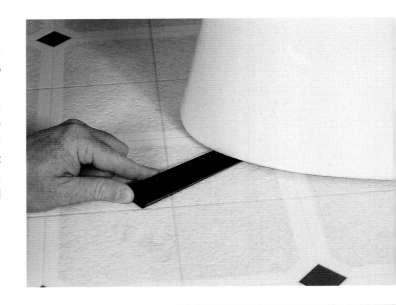

When setting a toilet on concrete, which is hardly ever perfectly level, you'll need to shim the bowl. Plastic shims work best. (Here the concrete is vinyl-covered.)

FIXING A ROTTED FLOOR

WHEN YOU ALLOW A TOILET TO LEAK long enough, the underlayment and possibly the subfloor around it swells and rots. The only option then, before replacing the flooring, is to cut out and replace the damaged underlayment area or cut the affected subfloor area back to the nearest joists to replace it. Remove the flooring, and probe the affected area using an awl or screwdriver to determine the extent of the damage and whether it extends past the underlayment to the subfloor.

Replacing Underlayment. When you replace just the underlayment around a toilet, you can leave the toilet flange in place and install the new plywood in two pieces. (The underlayment may be ½-inch or, more likely, ¼-inch plywood over ¾-inch subflooring.) Cut out the damaged area using a circular saw to cut just through the underlayment. Use a utility saw or reciprocating saw to cut through uncut areas in the corners or near walls. Cut the new plywood to size, and then cut it in half, with the cut intersecting the flange opening. Trim the plywood to accommodate the outside diameter of the

flange collar (not the flange itself), and slide each half under the flange. (See the illustration below.) Screw and glue the underlayment in place.

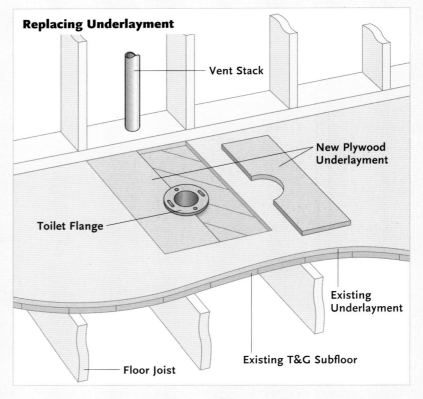

Replacing Underlayment

Vent Stack

New Plywood Underlayment

Toilet Flange

Existing Underlayment

Existing T&G Subfloor

Floor Joist

smart tip

VENT-RELATED PROBLEMS

1. POORLY VENTED TOILET. IF A TOILET FLUSHES SO SLOWLY THAT TWO OR MORE FLUSHES ARE NEEDED OR ADDITIONAL WATER MUST BE POURED INTO THE BOWL TO MAKE IT FLUSH CORRECTLY, SUSPECT A VENTING PROBLEM. YOU'LL NEED TO UP-GRADE THE PIPING SYSTEM. THIS IS MAINLY A PROB-LEM IN UNREGULATED AREAS, WHERE UNLICENSED PLUMBERS RULE. FOR INFORMATION ON VENT AND DRAIN DESIGN, SEE CHAPTER 13, "DRAINS, VENTS & TRAPS," STARTING ON PAGE 274.

2. FROST CLOSER. DURING WINTER, MOISTURE IN THE AIR RISING THROUGH A SMALL-DIAMETER STACK MAY FREEZE WHEN IT MEETS THE COLD AIR. ENOUGH FROST CAN ACCUMULATE TO FORM A CAP ON THE STACK, CAUSING SINK DRAINS TO GURGLE AND TOILETS TO FLUSH SLOWLY. MANY NORTHERN CODES REQUIRE STACKS TO BE 3 INCHES IN DIAM-ETER WHERE THEY PASS THROUGH THE ROOF, EVEN IF THE STACK BELOW THE ROOF IS ONLY 1½ INCHES. THE STACK SIZE IS TYPICALLY INCREASED ABOUT 1 FOOT BELOW THE ROOF LINE, AND NO STACK SHOULD EXTEND MORE THAN 1 FOOT ABOVE THE POINT OF EXIT FROM THE ROOF.

Replacing Subflooring. If the subfloor is damaged, you'll have to replace it. Cut the subfloor back to the nearest joist on each side where there is undamaged subflooring. A reciprocating saw is a good tool for this. Nail two-by blocking around the perimeter of the hole to provide nailing for the new plywood, sized and cut to match the existing subflooring. (See the illustration at right.)

There are two ways to proceed from here. One is to replace the old toilet flange and riser. If you have a cast-iron flange, convert the riser and flange to plastic. (See "Joining PVC to a Cast-Iron Hub," page 55, and "Toilet Flanges," page 86.) If you have plastic pipes, cut off the toilet assembly at the horizontal pipe, and rebuild it exactly the same way using a coupling. Leave the flange off for now. Measure and cut a hole in the new plywood for the riser before installing it. Nail or screw down the subflooring with fasteners around the perimeter every 6 inches on center. Lay down new underlayment and flooring, and then install the new flange.

The second way is to leave the

flange in place. Nail two-by blocking between the joists on each side of the toilet riser. Then cut the plywood down the middle the same way as for underlayment (left), and install it, sliding it under the toilet flange. Then lay new underlayment and flooring.

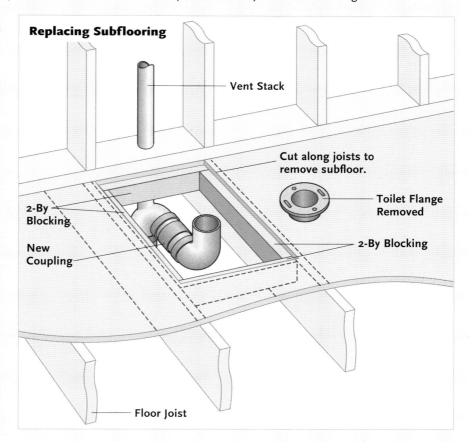

Replacing Subflooring

Vent Stack

Cut along joists to remove subfloor.

Toilet Flange Removed

2-By Blocking

New Coupling

2-By Blocking

Floor Joist

INSTALLING A NEW TOILET

A new two-piece toilet will come in two boxes, one containing only the bowl and another containing the tank, tank bolts, decorative bolt caps, retaining washers, and a large spud gasket, which forms the seal between the tank and the bowl. In addition (unless the model is provided with these items), you'll need to buy a toilet seat, a set of closet bolts, a bowl wax gasket, and a water supply tube.

Begin by emptying both cartons. Slide the new closet bolts into their flange slots so that they're centered across the opening and are the same distance from the back wall. Press a new bowl wax gasket onto the flange, and stretch the wax slightly so that the gasket holds the bolts upright.

Fasten the Toilet in Place. When carrying the new bowl, grip the seat hinge deck with one hand and the front rim with the other. Lift the bowl over the flange until you can see the bolts through the tank holes, and then settle the bowl onto the wax gasket. As with an assembled toilet, this is the time to step back and judge the straightness of the bowl. You won't be able to align the tank with the wall, so eyeball it as best you can. If you're not confident in sighting the unit, measure from the tank-bolt holes to the back wall. Each hole should be the same distance from the wall.

When you've aligned the toilet, unpack the cap-retaining washers and place one over each bolt with words stamped "This Side Up." Place a metal washer from the closet-bolt kit on top of each plastic retaining washer, and thread the nuts onto the bolts. Tighten the nuts, and cut the tops of the bolts off, as described in "Taking Up and Resetting a Toilet," pages 91–93.

REPLACING AN OLD DRUM-TRAP SYSTEM

AS NOTED EARLIER, many older plumbing systems (60 years old or more) have hopelessly outdated drum-trap piping configurations serving their bathrooms. In these setups, the sink and tub both flow into a central drum mounted in the floor. The drum trap then drains into the horizontal branch line serving the toilet. This piping cannot be properly vented, and after many years, clogs with frustrating regularity.

Unfortunately, drum-trap piping is impossible to reach without tearing out the floor from above or the ceiling from below. If you're planning a bathroom overhaul, make it a point to update this piping. It's not that difficult once you gain access.

Getting Started. Begin by taking up the floor and subfloor in the vicinity of the pipes. Cut out all the lead piping leading to and from the drum trap, as shown in the illustration at right. A reciprocating saw with a metal-cutting blade is ideal for this work, but a hacksaw will also do. Use a snap-cutter to cut the cast-iron horizontal branch pipe serving the toilet. Make this cut about 3 inches away from the hub of the stack T-fitting so that enough horizontal pipe remains to support new piping.

To rebuild the bathroom drains, first cut a 4-inch stub from 3-inch-diameter PVC drainpipe. Tighten a 4 x 3-inch banded coupling between the cast-iron stub

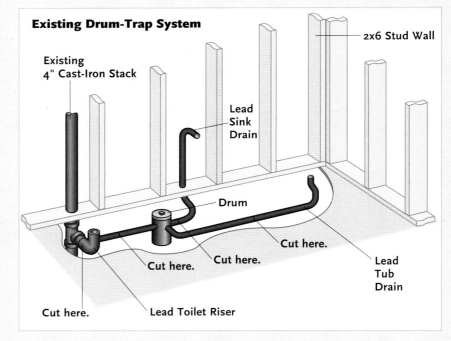

Existing Drum-Trap System

2x6 Stud Wall

Existing
4" Cast-Iron Stack

Lead
Sink
Drain

Drum

Cut here.

Cut here.

Cut here.

Lead
Tub
Drain

Cut here.

Lead Toilet Riser

Attach the Tank. With the bowl set, install the tank. The fill valve and flush valve are usually factory installed, but you should always check the tightness of the flush-valve spud nut. Use large groove-joint pliers or a pipe wrench to tighten the nut if necessary. If the nut moves easily, draw it down until it feels snug. If not, leave it alone.

Fit the large rubber spud gasket over the flush-valve threads and onto the tank bolts (along with the supplied rubber washers). Lubricate the tank side of the washers with pipe joint compound before installing them. Then lubricate the rest of the washers and the spud gasket with pipe joint compound. Insert the bolts through the tank holes, and set the tank on top of the bowl.

With the tank resting on top of the bowl, start a washer and nut on each tank bolt. Tighten these nuts carefully with

a small adjustable wrench, alternating from side to side. Focus on keeping the tank level, and stop when the nuts begin to feel snug.

Install the seat, and connect the supply tube as described in "Taking Up and Resetting a Toilet," pages 91–93. Turn on the water, and with the tank lid removed, watch the toilet closely through several flushes. If the water level is too high, bend the float rod down (or adjust the float cup) until the ballcock (or fill valve) shuts off with the water roughly 1 inch below the top of the overflow tube.

With the toilet working smoothly and the tank and flange seals holding without any leaks, use the toilet for several days. Then check to see whether the closet bolts have loosened. If they have, tighten them one last time, and snap the bolt caps over their retaining washers.

left by the snap-cutter and the plastic stub. Install a 3-inch 90-degree elbow with a 2-inch side inlet through which the sink and tub will drain. Check local codes regarding the capacity of a side inlet fitting, however. If this setup is forbidden, install the stub of PVC pipe in the hub of a 3 x 2-inch Y-fitting. Secure the Y-fitting in the banded coupling, with the branch of the Y canted slightly upward. Immediately out of the front of the Y, install a 3-inch 90-degree street elbow to serve as the toilet riser. The riser should be centered 12 inches from the finished wall or 12½ inches from a stud wall. If the fitting space is too tight and the riser falls an inch or two beyond 12 inches, the toilet will still work, though it will sit away from the wall slightly. In worst-case situations, the problem can be remedied by installing a toilet that has a 10-inch rough-in outlet instead of a 12-inch rough-in. Toilets are available to fit 10-, 12-, and 14-inch rough-ins.

Finishing Up. From the side inlet (or branch of the Y-fitting), run 2-inch PVC to a 2-inch Y-fitting. Then, from the branch of the Y, continue up into the plumbing wall to serve the sink. From the top of the sink T-fitting, continue into the attic with a 2-inch vent through the roof

or tied back into the main stack. Finally, from the front end of the Y-fitting in the floor, reduce to 1½ inches and continue to the tub trap.

If the floor joists run parallel with the plumbing wall, the piping will be easy to install. If the joists run perpendicular to the piping, you'll need to drill the center of each joist, splicing short sections of pipe together with couplings. *Caution: Don't notch the tops or bottoms of these joists.*

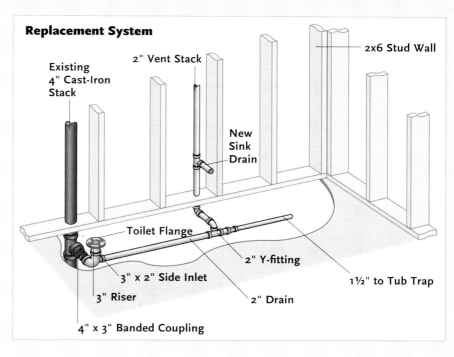

Replacement System

Existing 4" Cast-Iron Stack

2" Vent Stack

2x6 Stud Wall

New Sink Drain

Toilet Flange

2" Y-fitting

1½" to Tub Trap

3" x 2" Side Inlet

3" Riser

2" Drain

4" x 3" Banded Coupling

INSTALLING A NEW TOILET

project

There's not much difference between installing a new toilet and reinstalling an old one. The new toilet requires no cleaning, and all parts should be free of corrosion, which cuts installation time. But you do have to attach the tank to the bowl on the new unit, which takes more time than it does to clean up an old unit.

TOOLS & MATERIALS

- New toilet
- Wax gasket
- Closet bolts
- Water supply tube
- Pipe joint compound
- Adjustable wrenches
- Groove-joint pliers

1 Install new closet bolts in the toilet flange. If the bolts come with plastic washers, use the washers to hold the bolts in place.

smart tip

TO CAULK OR NOT TO CAULK?

WHEN IT COMES TO CAULKING AROUND THE BASE OF THE TOILET, OPINIONS VARY. SOME LOCAL CODES REQUIRE IT, SOME FORBID IT, AND OTHERS DON'T ADDRESS IT AT ALL. THE ADVANTAGES OF CAULKING AROUND THE BASE OF A TOILET ARE THAT THE CAULK 1) HAS GOOD ADHESIVE QUALITIES, 2) HELPS PRESERVE THE SEAL BY KEEPING THE TOILET FROM MOVING, 3) KEEPS THE JOINT FREE OF BACTERIA AND OTHER DIRT, AND 4) LOOKS BETTER THAN A DARK, SOILED PERIMETER CRACK (AS LONG AS THE JOINT IS NEATLY DONE). THE DISADVANTAGE IS THAT WHEN A LEAK DOES OCCUR, IT CAN'T BE SEEN, SO IT DOES MORE DAMAGE.

THERE ARE NO STRONG ARGUMENTS ONE WAY OR THE OTHER, EXCEPT IN TWO CIRCUMSTANCES:

- Caulking is usually recommended for toilets that are set on concrete. Concrete is never level around a toilet flange, and the added support from the caulked joint helps keep the bowl from rocking.

- Caulking is usually not recommended in a bathroom with wood floors (a poor flooring choice). These floors buckle with the slightest penetration of water, and you'll want to see a leak as soon as it starts.

TO CAULK AROUND THE BASE OF A TOILET, USE LATEX TUB-AND-TILE CAULK. LEAVE THE BACK OF THE TOILET BASE OPEN FOR WATER TO ESCAPE IN CASE A LEAK OCCURS. START BY WETTING THE FLOOR AND TOILET BASE; THEN APPLY A LIBERAL BEAD OF CAULK IN THE SEAM. DRAW A FINGER ALL AROUND THE JOINT SO THAT ANY VOIDS ARE FILLED, AND WIPE AWAY THE EXCESS USING A DAMP RAG OR SPONGE. KEEP IN MIND THAT LATEX CAULK WILL SHRINK A BIT WHEN IT CURES, SO LEAVE A LITTLE MORE IN THE JOINT THAN SEEMS NECESSARY. THE CAULK WILL DRY TO THE TOUCH WITHIN THE HOUR BUT WILL TAKE SEVERAL DAYS TO ACHIEVE ADHESIVE STRENGTH. IN THE MEANTIME, FEEL FREE TO USE THE TOILET.

If codes require caulking around the base of the toilet, use a good-quality mildew-resistant latex or silicone tub-and-tile caulk.

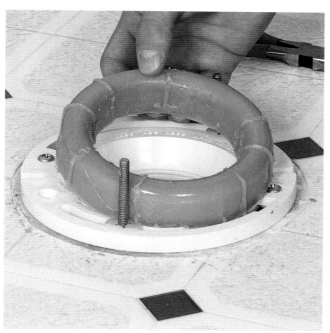

2 Once the toilet bolts are in place, lower a new wax seal onto the flange so it fits between the bolts. The wax ring is slightly wider than the distance between the bolts, so you can press the bolts into the sides of the wax. This holds the bolts upright so they don't tip when the bowl is installed.

3 One advantage of installing a new toilet is that you can place the bowl over the bolts and onto the flange without having the tank in place. This makes the job much easier for one person to do. Once the bowl is seated and attached to the floor flange, you can mount the tank to the bowl.

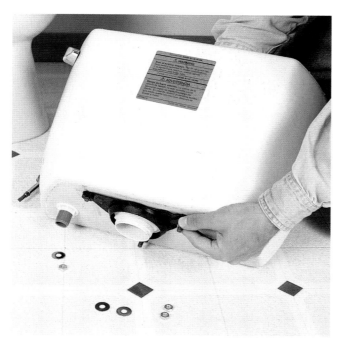

4 Toilets, with separate water tanks, are always bolted to the bowl and feature a gasket to keep the joint between the two from leaking. New toilets usually have spacer washers that fit between the tank and the bowl to evenly support the tank and stop any rocking movement.

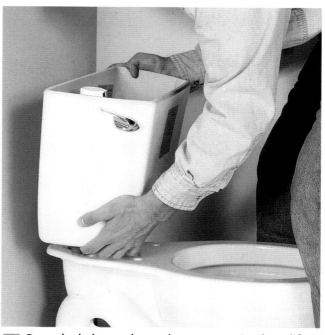

5 Once the bolts, gasket and spacers are in place, lift the tank onto the bowl. Tighten the bolts carefully, checking the top of the tank for level as you work. Make sure the bolts are tight enough to keep the tank from rocking. But don't overtighten the bolts because you can crack the tank.

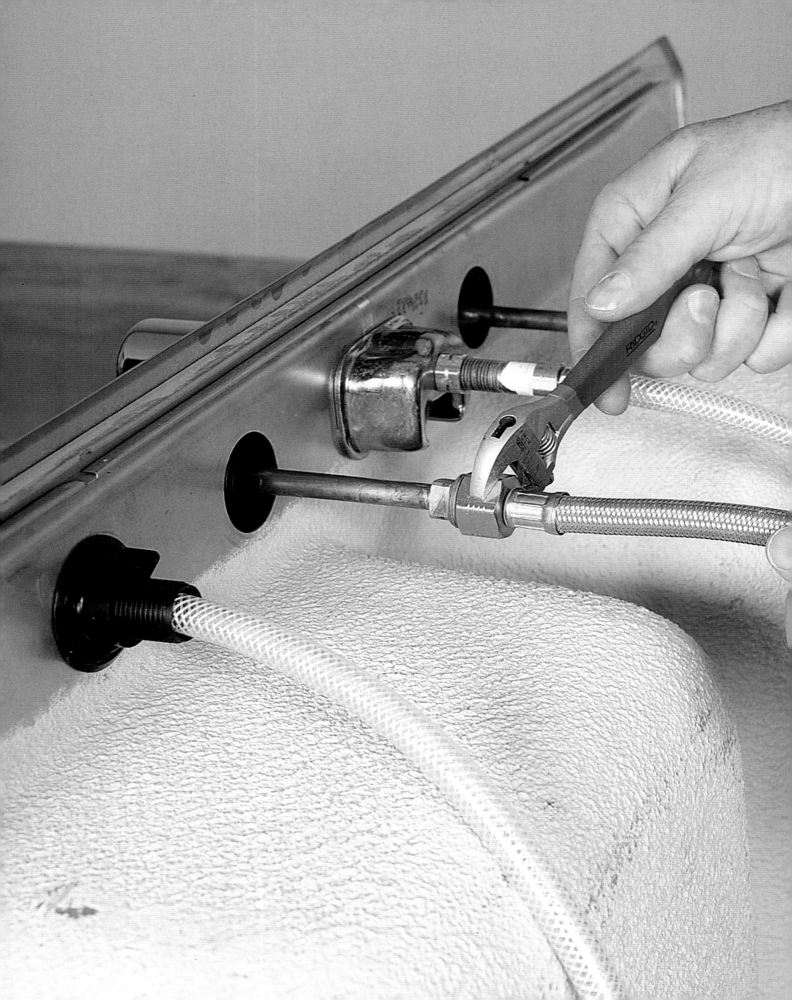

4

installing sinks & related equipment

THERE COMES A TIME when those old faucets and sinks just have to go. And given the range of products on home center shelves these days, you're sure to find a replacement you'd like to take home.

Product options range from simple faucet or drain replacements to complete sink and fixture upgrades. In the bathroom, you might also want to add a new vanity cabinet and top, which in many cases are almost of a piece with the sink. In the kitchen, you might want to add or replace an automatic dishwasher, a waste-disposal unit, or perhaps an instant hot-water dispenser. These days, you'll find dishwashers (and often, waste-disposal units) installed in nearly every new home as original equipment, and instant hot-water dispensers are becoming ever-more-popular kitchen add-ons. Besides covering the removal and installation of these fittings, fixtures, and appliances, this chapter explains how careful maintenance and judicious use can add to the efficiency and longevity of important household products such as dishwashers and waste-disposal units.

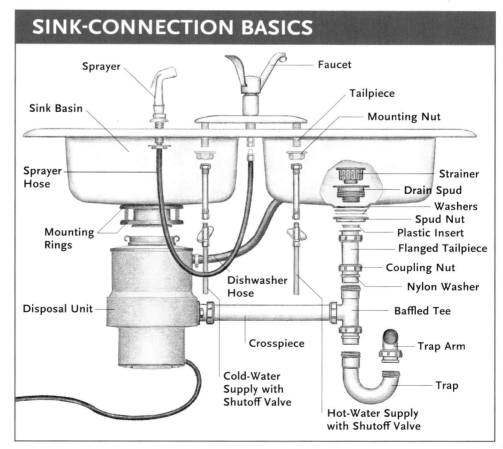

SINK-CONNECTION BASICS

Sprayer
Faucet
Tailpiece
Sink Basin
Mounting Nut
Sprayer Hose
Strainer
Drain Spud
Washers
Spud Nut
Plastic Insert
Flanged Tailpiece
Coupling Nut
Nylon Washer
Mounting Rings
Dishwasher Hose
Baffled Tee
Disposal Unit
Trap Arm
Crosspiece
Trap
Cold-Water Supply with Shutoff Valve
Hot-Water Supply with Shutoff Valve

WATER- & WASTE-CONNECTION BASICS

Almost all kitchen and bathroom sink faucets are connected to water piping through flexible supply tubes. In just about every case, you make the riser-to-supply tube conversion with a compression fitting. While some codes require shut-off valves only on the toilet supply, allowing simple compression adapters on other supply connections, the general trend is toward requiring shutoffs on all fixtures fed by supply tubes. Even when shutoffs are not required, they're a good idea. You'll thank yourself for installing them the first time a fixture or appliance needs servicing.

As for supply tubes, you can choose soft-copper tubes, which are extremely inexpensive, or stainless-steel enmeshed polymer tubes, which cost substantially more but are prefitted and almost foolproof. Stainless-steel mesh tubes are perfect for the poorly equipped and the mechanically timid. At this writing, many codes do not allow plastic supply tubes. (See "Types of Piping and Tubing," page 253.)

Sink Drains

When it comes to joining a kitchen sink to permanent drain piping, use 1½-inch tubular waste kits and P-traps. Waste kits for double sinks are sold with or without waste-disposal-unit connections. One is called a sink-waste kit and the other, a disposal-waste kit. These kits usually include all the pieces needed to drain both compartments of a sink into a single P-trap. Traps are usually sold separately. Use a 1½-inch flanged tailpiece and P-trap to drain a single-compartment sink.

Bathroom sink drains are always 1¼ inches in diameter, but you can fit them with either a 1¼- or 1½-inch trap. You'll find a special reducing washer packaged with each trap to make the conversion. All waste kits and traps are available in PVC plastic or chrome-plated brass. Plastic is much easier to cut and assemble. It's less expensive, and unlike brass, it doesn't corrode. This is one case where plastic beats metal, hands down.

This sink-drain connection is made using a ground joint adapter, which has a nylon compression washer.

Dishwashers

When hooking up a dishwasher, use ⅜-inch soft copper water piping and compression fittings. Run this line from the hot-water shutoff (or dual stop) under the sink, through the side of the sink cabinet, near the back, to the inlet on the dishwasher's solenoid valve.

Dishwashers usually come with more than enough discharge hose attached. The hose will be made of ribbed plastic or heat-resistant rubber. If you need to extend a rubber discharge line, automotive heater hose is a good choice. You can make all connections using conventional hose clamps. If the sink has no waste-disposal unit attached, connect the free end of the hose to a disposal unit T-fitting under the sink, just above the P-trap. If the sink is equipped with a waste-disposal unit, connect the hose to the discharge nipple cast into the metal body of the unit.

VANITY SINK ANATOMY

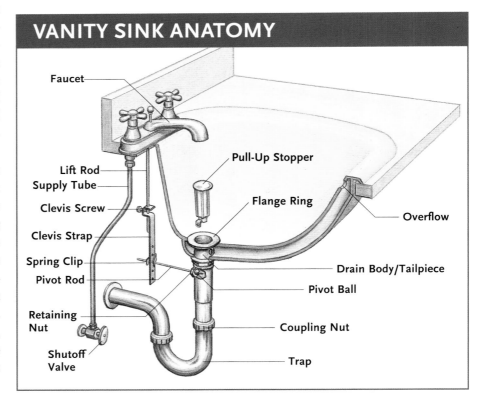

Faucet
Pull-Up Stopper
Lift Rod
Supply Tube
Flange Ring
Clevis Screw
Overflow
Clevis Strap
Spring Clip
Drain Body/Tailpiece
Pivot Rod
Pivot Ball
Retaining Nut
Coupling Nut
Shutoff Valve
Trap

HOW TO REPLACE A BATHROOM FAUCET & DRAIN

Bathroom faucets are sold with and without drain assemblies. Because the lift rod that operates the drain's pop-up plug is installed through the faucet-body cover, it's a good idea to replace both when changing out an old faucet. The procedures described here and on pages 104–107 assume that your sink is installed in a vanity cabinet.

To remove the old faucet and drain assembly, shut off the water, and drain the faucet lines. If you have to shut off the main valve and you have plumbing fixtures on the floor above this bath, open those fixtures as well. Otherwise, water from upstairs will drain onto you.

Removing Old Bathroom Faucets

How you remove the old faucet depends on whether it's a top-mounted or bottom-mounted faucet. Top-mounted faucets are the most common.

Top-Mounted Faucets. A top-mounted faucet is held in place from below by threaded shanks or fastening bolts, which fit through the basin's deck holes. If the faucet is a two-handle model, expect to find jamb nuts tightened onto the shanks from below. In this case, you'll connect the supply tubes to the ends of the shanks. If your faucet has copper tubes instead of brass shanks—usually the case with single-control faucets—the faucet will be held in place by threaded bolts.

When working within the cramped spaces behind a sink, you'll find loosening and tightening nuts a lot easier with a basin wrench. A basin wrench is really just a horizontal wrench on a vertical handle. Lay its spring-loaded jaws to one side to loosen the nut. Lay it to the other side to tighten. The extended handle allows you to work high up under a basin or sink deck without having to reach.

Loosen the two coupling nuts that secure the supply tubes to the faucet. Then, using a standard adjustable wrench, disconnect the lower ends of the tubes from their compression fittings. Set the supply tubes aside, and use the basin wrench to remove the fastening nuts from the faucet shanks or fastening bolts. Finally, lift the old faucet from its deck holes.

Bottom-Mounted Faucets. If the old faucet is a bottom-mounted model, with the body of the faucet installed below the sink deck, the initial approach is from above. Leave the supply tubes in place until after you've completed your work on top. Start by prying the index caps from the faucet handles. Remove the handle screws, and lift the handles from their stems. This will give you access to the flanges (or escutcheons) threaded onto the stem columns. Thread these flanges off; disconnect the supply tubes; and drop the faucet from its deck. Some bottom-mounted faucets consist of three isolated components, including two stems and a spout. These components are joined with tubing and are easy to disassemble from below.

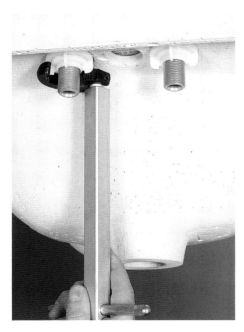

To remove a top-mounted faucet, use a basin wrench to unscrew the jamb nuts from the faucet shanks.

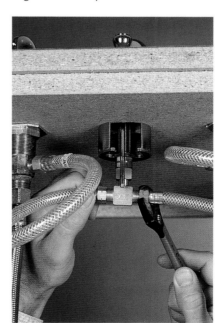

To remove a bottom-mounted faucet, undo the connecting tubing and remove the jamb nuts.

Installing Sinks & Related Equipment

REMOVING THE SINK DRAIN

project

Sink drain lines are installed in two basic ways: through the floor, which is the standard approach in older houses, and through the wall, which is the approach used in most newer ones to make installing bathroom vanities and kitchen cabinets easier. Drain assemblies consist of three basic components. The tailpiece hooks to the bottom of the sink opening. A long elbow attaches to the house waste system. A simple trap joins the first two. Once installed, the trap is filled with water that prevents sewer gas from entering the house.

TOOLS & MATERIALS

▌ Groove-joint pliers
▌ Bucket or shallow pan
▌ Putty knife

1 To remove a sink drain, first slide a bucket or pan under the trap to catch any spilled wastewater. Then loosen the two nuts that hold the P-trap in place. Use groove-joint pliers or a pipe wrench. Cover the jaws of each with a rag to prevent scratching the chrome surfaces.

WATER CONNECTIONS FOR FAUCETS

ONCE YOU'VE REMOVED THE FAUCET and drain from a bathroom sink, you'll likely install a new faucet and drain. (See "Installing a New Bathroom Faucet and Drain," on page 106.) Faucets have a number of possible water connections (below), including threaded shanks, which serve the dual purpose of connecting the faucet to the sink base and connecting to a water coupling; copper tubes with coupling nuts already attached; and short or long copper tubes, which connect to water piping by soldered couplings or compression nuts. (See "Connecting the Water," page 124.)

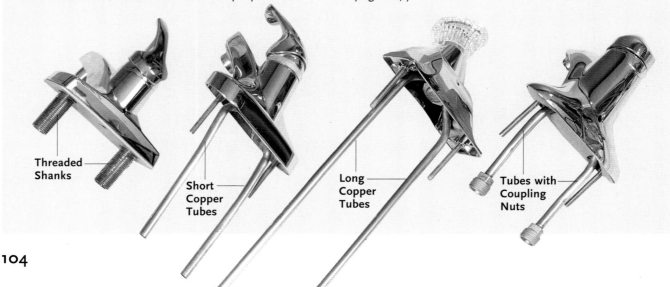

Threaded Shanks

Short Copper Tubes

Long Copper Tubes

Tubes with Coupling Nuts

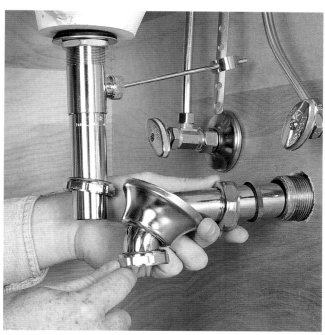

2 Once the trap is removed, slide the escutcheon away from the wall; unthread the friction nut that holds the drain elbow to the main waste line; and pull it and its the washer away from the drainpipe. Set aside the drain elbow. If this pipe is corroded, now is a good time to replace it.

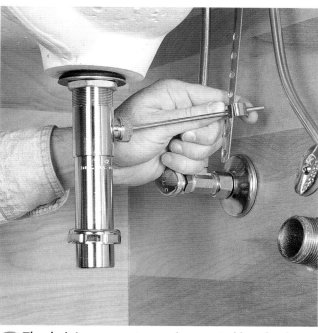

3 The drain's pop-up stopper is operated by a knob on top of the sink. This knob is connected to a lift rod under the sink, which is joined to the pop-up linkage (in the side of the drain body/tailpiece) with a tension clip. To remove the clip, just squeeze the ends together and slide it off the pop-up rod.

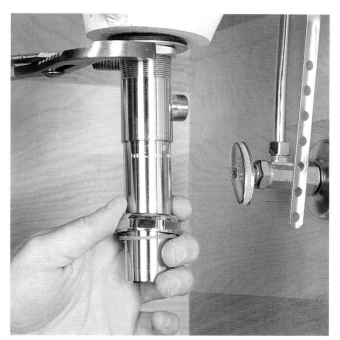

4 After you disconnect the pop-up linkage from the side of the drain body/tailpiece, loosen the jamb nut that holds the drain gasket in place using groove-joint pliers. If the drain body/tailpiece spins, backhold it using a second pair of pliers.

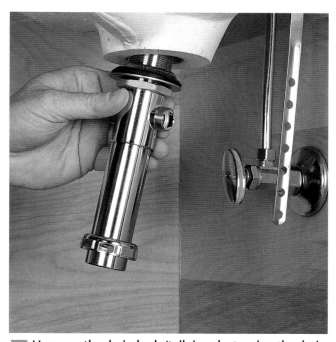

5 Unscrew the drain body/tailpiece by turning the drain flange ring from above. (Having some help makes this job much easier.) Pull the drain body/tailpiece away, and lift the flange ring out of the sink drain hole from above. Clean up any putty residue from the where the drain flange ring was seated.

INSTALLING A NEW BATHROOM FAUCET & DRAIN

Installing a new faucet is not particularly difficult because most come with a baseplate gasket to seal the joint between the faucet and the sink. Even if your faucet doesn't come with a gasket, just fill the base of the faucet with caulk or plumber's putty before installing it. Then attach the supply lines to the bottom of the faucet. The drain lines are nearly as easy. Install a drain body/tailpiece in the sink drain hole and a long elbow in the wall's drain opening. Then connect these two pipes with a P-trap. Tighten all the nuts, and check for leaks.

TOOLS & MATERIALS

▌ New faucet ▌ New drain assembly
▌ Plumber's putty ▌ Adjustable wrench
▌ Pipe joint compound ▌ Groove-joint pliers

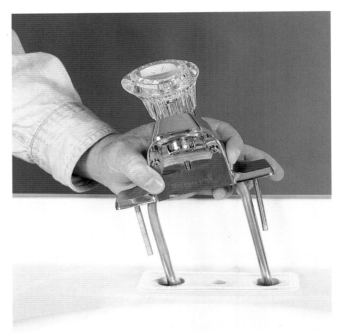

1 Begin the installation by cleaning the top of the sink where the faucet will rest. Then put the baseplate gasket in place. Lower the new faucet into the sink deck openings, and center it on the gasket. Temporarily hold the faucet in place with a couple of pieces of masking tape.

4 Insert the drain body/tailpiece into the drain hole from below. Have a helper hold it in place while you thread the drain flange ring onto the drain body/tailpiece from above. The flange should have enough plumber's putty in place to fill the depression around the drain hole.

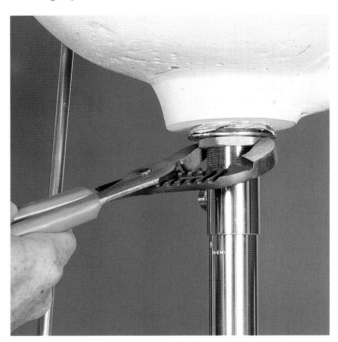

5 Once you have threaded the drain flange ring in place, carefully pull the drain body/tailpiece down from below. Make sure that the putty spreads evenly around the drain hole. Then tighten the jamb nut with groove-joint pliers, and trim away the putty that squeezes out from around the flange in the sink.

Installing Sinks & Related Equipment

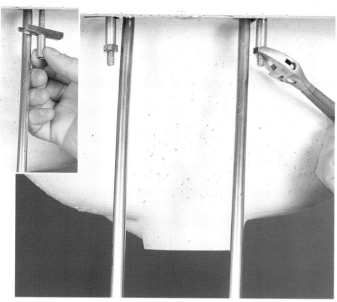

2 The faucet is held in place by wide rectangular washers (inset), steel bushings to act as spacers, and hex nuts threaded onto the faucet's fastening bolts. Use an adjustable wrench, and only tighten the nuts until they are snug. Don't overtighten them.

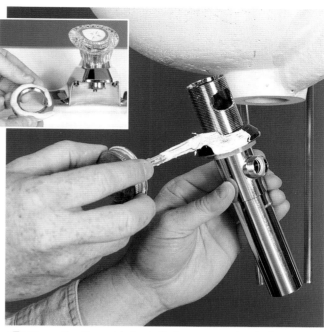

3 Once the faucet is installed, prepare the drain parts for installation. First, create a rope of plumber's putty, and spread it around the underside of the drain flange ring (inset). Then install the cone-shaped gasket on the top of the drain body/tailpiece, and coat it with pipe joint compound.

6 To install the drain plug assembly, first feed the lift rod down through the hole in the back of the faucet body. Then connect the bottom of this rod to a clevis piece that extends the lift rod and provides several different holes to receive the pop-up assembly lever.

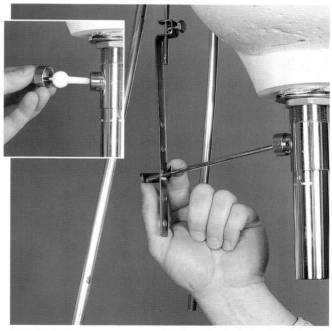

7 Slide the pop-up assembly into its hole in the side of the drain body/tailpiece (inset), and tighten it in place with its nut. Then slide the pop-up lever into one of the lower holes in the clevis, and hold the two together with a tension clip. Make sure the lift rod is in the down position (pop-up stopper open).

REPLACING A WALL-HUNG BATHROOM SINK WITH A VANITY & MOLDED TOP

Vanity cabinets come in a variety of standard widths and depths. They range from 18 to 60 inches wide and 16 to 24 inches deep. Unless your bathroom is unusually small or oddly shaped, you're sure to find one that will fit. You'll find everything from well-built hardwoods to unassembled particleboard.

Sinks and Vanity Tops. When it comes to sinks and vanity tops, you'll find choices in two basic categories—a molded top with an integral sink basin or a plywood top that is finished in tile or plastic laminate. In the latter case, you fit the basin into a hole cut into the top. (See "Installing a Sink in a Plywood Top," pages 112–113.)

Wall Repair. You may see some minor wall damage, ranging from torn drywall paper to screw holes to a gaping hole in the wall. To repair drywall, start by knocking down

any high spots, including those around screw holes. Use a hammer or the end of the knife handle to batter high spots into slight depressions. Wipe on a thin coat of drywall joint compound with a 4-inch taping knife. This first application is just a base coat, so don't worry about the finish. Just put it on, and walk away. Better to apply three or four skim coats, which dry quickly, than one or two thick coats. After the first coat dries, knock down any high spots with a sanding block, and give it at least one more skim coat, feathering the edges around the perimeter. When the final coat has dried, sand it lightly. Finally, paint the area with primer and a top coat.

At this point, you'll also need to remove the base trim. Pry the base moldings loose using a pry bar. Place a small square of plywood under the bar to protect the wall.

FIXING PLASTER

IF A WALL-HUNG SINK IS MOUNTED on a plaster wall, the final coat of plaster may have been applied after the sink was installed. Expect to break a little plaster in the sink-removal process. You can fix minor plaster damage with standard drywall joint compound, but deeper damage, including missing chunks of plaster, require some perlite plaster or quick-setting drywall compound. Avoid filling deep holes with standard joint compound; it may take days to cure, shrinking as much as 30 percent.

Where plaster is loose or missing, carefully break

out the loose material without enlarging the hole any more than necessary. Use a surface-forming plane or a rasp to knock down the perimeter of the hole. Mix plaster for the job in a plastic bucket, using enough cold water to bring the mixture to the consistency of toothpaste. Wait a few minutes for the plaster to stiffen; then stir and apply it directly over the wood or metal lath. Trowel on enough perlite plaster to bring the surface to within 1/8 inch of finish, and let it dry completely. Follow with several skim coats of drywall joint compound; then prime and paint the new surface.

1 Water-spray the area before applying the rough plaster.

2 Follow the plaster with several coats of drywall mud.

3 Seal the area with primer and a coat or two of paint.

REMOVING THE WALL-HUNG BATHROOM SINK

project

In most new homes, kitchen sinks are installed in base cabinets, and bathroom sinks are installed in vanities. Even contemporary pedestal sinks are supported by a freestanding base. But many older bathrooms feature simple one-piece sinks that are hung directly on the wall. While removing these sinks is not usually difficult, and can be handled by one person, larger models are difficult to grasp and too heavy for some people. The best approach is to get some help lifting the sink from its support bracket and carrying it outside.

TOOLS & MATERIALS

- Screwdriver
- Groove-joint pliers
- Utility knife
- 4-to-6-inch taping knife
- Sandpaper
- Drywall joint compound

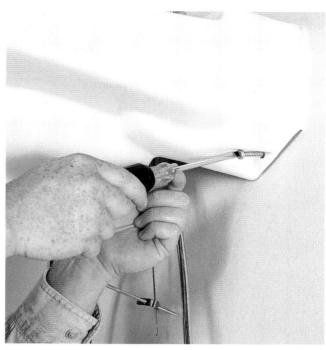

1 Begin this job by turning off the water at the supply valves or main valve. Then remove supply lines from the shutoff valves and the trap and wall pipe from the drain assembly. Then remove the fastening screws, and cut the caulk around the perimeter of the sink with a sharp utility knife.

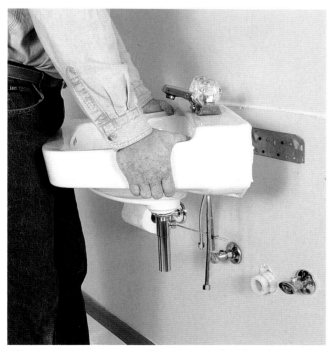

2 The sink is supported by a hanger bracket that is screwed to the wall framing through the drywall. To remove the sink, firmly grip it on both sides; lift it up; and set it aside. Once the sink is free, remove the bracket from the wall.

3 Removing the sink and bracket invariably damages the surface of the wall. In most cases the repair is simple: just scrape away any debris so the surface is smooth; then cover the entire area with two coats of joint compound and sand smooth.

INSTALLING THE CABINET AND TOP

Floors and walls are seldom absolutely level and square. A floor may have sagged over the years, or a wall may well have been installed with a corner at a slightly acute or obtuse angle. However, vanity cabinets and the tops that are made for them are typically perfectly square. That makes installing them a challenge. Expect to have to trim under a corner a corner of a cabinet to make it level or to trim a side of the vanity top so it will fit tightly against the corner of a wall. Taking the time to make these adjustments makes for a better-looking job.

TOOLS & MATERIALS

▌ Vanity cabinet ▌ Cultured-marble top
▌ Saw or utility knife ▌ Level ▌ Belt sander
▌ Variable-speed drill ▌ Shims ▌ Pencil

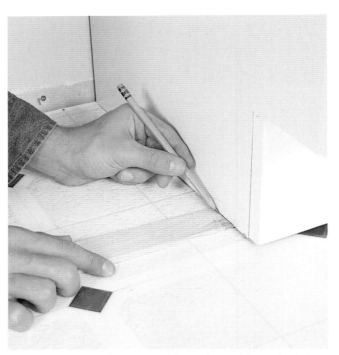

1 Once you put the vanity cabinet in position, check it for level in both directions: front-to-back and side-to-side. In low spots, slide a shim under the cabinet to raise it until it's level. Then cut the shim to length; cover it with yellow glue; and slide it in place.

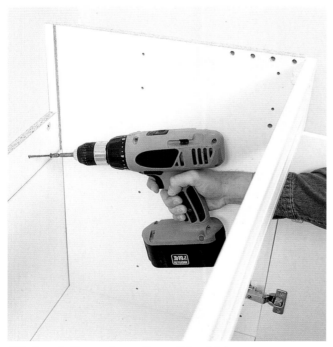

2 Attach the vanity to the wall by driving screws through the back of the cabinet into the wall studs behind the drywall. Two drywall screws will work. This will be easier if you pre-bore clearance holes. Fasten the top to the cabinet using construction adhesive.

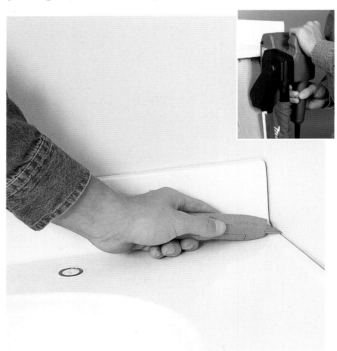

3 If an out-of-square corner makes for a bad fit, scribe the cultured-marble top for the amount of material to remove for a good fit, and sand it (inset). If the top won't sand well or you need to remove too much material for a resonable fit, cut a recess in the drywall with a utility knife.

Installing the Cabinet and Top

When walls and floors are plumb and level, vanity installations are easy. First, locate the wall studs behind where the cabinet will be. An electronic stud finder works best. Set the cabinet in place, and screw its back brace to the wall studs. Use 3-inch drywall screws, and predrill holes through the brace to clear the screw threads. You'll need to hit at least one stud. You may also want to trim out the toe-kick with cove or baseboard.

Unfortunately, though, most floors and walls are a little out of square, either because the house has settled or because it was built that way. To check, move the cabinet into position and look for gaps along the wall. If you see a gap near the top, shim the bottom of the cabinet. Slide pine or cedar-shingle shims under the cabinet until you've closed the gap. Then mark the shims where they meet the cabinet, and pull them back out. Cut the excess from each one, and apply a spot of construction adhesive to the top side. Finally, slide the shims back in place, and screw the cabinet's back brace to the wall studs using 3-inch drywall screws. When you're finished, trim around the bottom of the vanity using baseboard or vinyl cove.

Attach the Top. Always install the faucet and drain assembly in the sink basin portion before setting a molded top. In fact, you can even install the supply tubes if you'd like. With the faucet and drain attached, lift the molded top over the cabinet, and set it in place. If the vanity sits out in the open, center the top on the cabinet and push it back against the plumbing wall. When it looks right, hook up the drain and the water supplies, and turn on the water for a test. If you don't find any leaks, secure the counter to the cabinet. With a cultured marble top, it's best to glue it down with construction adhesive. Lift the front of the top about ¼ inch, and pump several daubs of adhesive between the top and cabinet along each side. With a plywood-based laminate or tile countertop, fasten the top using screws through a corner bracket (photo at right.)

Dealing with Out-of-Square Wall Corners. What if your vanity is supposed to go into a corner and the corner isn't square? There are actually three practical solutions here, depending on the severity of the problem. With a minor gap of ⅛ inch or so, tub-and-tile caulk will disguise the joint. More-serious gaps can be handled by shaving the top to fit the space or by notching the drywall to accept the top.

A belt sander works best in shaving one side of the top. Test-fit the top; mark the estimated stock removal; and

sand the edge of the counter to this mark. You may have to sand and test-fit several times, but with patience, you're sure to find an acceptable compromise.

If you don't have a belt sander or if the top is made of a material that doesn't sand well, you can notch the drywall. Set the top in place, and push it tight against the corner. Using the top as a guide, draw or score a line on the wall indicating the upper limit of the notch. Remove the top, and cut neatly along this line. Make a corresponding cut along the top of the cabinet, and dig out enough paper and gypsum to accept the edge of the top. Slide the top into the notch; fasten it to the vanity; and caulk the joint between the top and the wall.

Use deck screws to secure a plywood-based countertop to the cabinet. Measure for the right screw length.

INSTALLING A BATHROOM SINK IN A PLYWOOD-BASED TOP

Plywood or high-density-particleboard tops are finished with plastic laminate or tile. You can buy them pre-laminated or install your own laminate or tile. Unlike a molded top, plywood and particleboard tops can hold screws, so it makes sense to screw them down. The standard approach is to screw from the bottom up, through the cabinet's corner brackets. Choose your screws carefully, however. (Don't use a screw long enough to pierce the laminate glued to the top of the counter.) Set the top in place; then measure from the bottom of the bracket to the bottom of the counter. Add the depth of the countertop (not the edge band) to this measurement, and subtract ⅜ inch to determine an overall length. Plastic brackets can arch upward when you draw the screw tight. This lets the screw travel too far. Stop when the screws begin to bind.

Cut the Sink Opening. If you have a new plastic-laminate top and would like to install a china, steel, cast-iron, or plastic-resin sink, don't be intimidated by the prospect of cutting an opening. (To cut an opening in a plywood countertop that you plan to tile, cut the opening before installing the tile.) You'll need a saber saw and a drill. If the new sink comes with a paper template, tape the template to the top, and cut along its dotted line. If you don't have a template, turn the sink upside down on the top; center it; and trace around it with a pencil. Remove the sink, and draw a second line roughly ½ inch inside the first. You'll cut along this interior line.

Drill a ⅜-inch hole through the top, just inside the line. Install a medium-course blade in a saber saw, and lower the blade into the hole. Cut along the line until you're within 6 inches of completing the circle. At this point, have a helper support the cutout. Just a little support from below will keep the cutout from falling abruptly and perhaps breaking the laminate in the process. (If the top's backsplash interferes with your saber saw, finish the cut using a utility saw.)

When you set the sink, adhere it with latex tub-and-tile caulk. In addition to having great adhesive qualities, latex caulk comes in a variety of colors.

Start by squeezing a liberal bead of caulk all along the joint. Draw a finger along the bead, smoothing it and forcing it into the joint. Finally, wipe away the excess caulk using a damp sponge or rag.

project

Cutting a sink hole in a plywood-based countertop is not a difficult job. You can make the cut before or after you install the top. But most installers do the job after the top is screwed in place because it's more stable, which reduces the amount of saber saw vibration during the cut. Keep in mind that a saber saw blade cuts on the upstroke, so it tends to chip the top surface of the counter when you cut from above. In most cases, just using a medium-coarse blade ensures that any chipping will be minor. Just don't use a very coarse blade.

TOOLS & MATERIALS
- New sink ▪ Countertop
- Pencil ▪ Tub-and-tile caulk
- Saber saw ▪ Sponge

3 Remove any plywood chips from the top of the cut; then lower the sink into the hole, and center it from front-to-back and side-to-side. Connect the water supply and drain lines; then apply latex caulk around the rim of the sink.

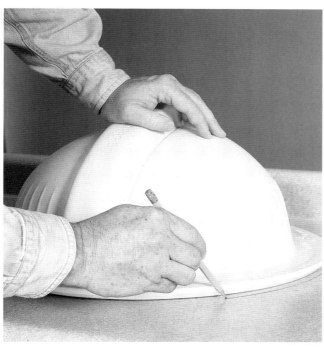

1 Many sinks come with a paper template for locating the sink hole on the countertop. But if your model didn't, just trace the perimeter of the sink in place. Invert the bowl on the counter; center it; and trace it with a pencil.

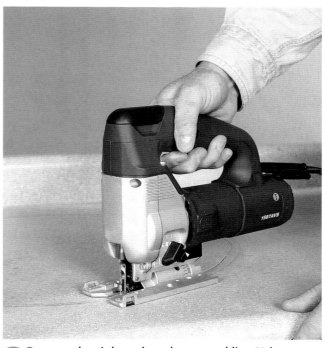

2 Remove the sink, and mark a second line ½ in. inside the first line. Then bore a blade clearance hole inside the second line, and cut this line with a saber saw. Support or have someone hold the waste section from below until the cutting is finished.

4 Once the caulk is applied, use your finger to force it under the rim of the sink. Most tub-and-tile caulks clean up with soap and water, so you don't have to be careful about spreading the caulk. Any mess will be easy to clean away.

5 Clean up the caulk with a sponge and slightly soapy water, rinsing the sponge frequently. Then clean the sink with household cleanser. When the countertop and sink are dry, buff them with a clean cloth to remove soap residue.

HOW TO REPLACE A KITCHEN SINK

Kitchen sinks come in several rim styles: self-riming (cast iron, porcelain, solid-surface resin, and stainless steel), metal-rim (enameled steel and cast iron), and rimless (stainless steel, cast iron, and solid-surface resin).

Self-Rimming Sinks. The most popular is the self-rimming variety. A self-rimming sink has a rolled lip, which rests directly on the counter. All that holds the cast-iron and solid-surface sinks steady is a bead of latex tub-and-tile caulk. Stainless-steel self-rimming sinks have a slightly different rim style. The sink rests on its rim, but special clips secure it from below. These clips slide into channels welded to the bottom of the sink rim. (See the photo below.) While this rim style is popular, a common complaint is that the rolled edge doesn't allow you to sweep food crumbs directly from the countertop into the sink.

Metal-Rim Sinks. The second rim style is more traditional. In this case, a separate metal support rim allows the sink to lay almost flush with the countertop. This rim, which is usually made of stainless steel, clamps over the unfinished edge of the sink. You place the entire assembly into a counter opening and use a series of metal clips to bind the rim to the counter. A metal-rim sink requires a more precisely cut opening. Most of these sinks are inexpensive enameled-steel models.

Rimless Sinks. And finally, there are rimless cast-iron, stainless-steel, and solid-surface resin sinks made to be in-stalled in tiled or solid-surface countertops. Cast-iron sinks are installed similarly to self-rimming sinks. The depth of the sink perimeter lip approximates the thickness of ceramic tiles. This makes for a more uniform appearance, especially when you match the tile and cast-iron sink colors. Under-mount stainless-steel sinks and solid-surface resin sinks are attached from below, with the finished joint all but invisible in solid-surface installations. This makes for an integral solid-surface look because sink and countertop are the same material. Note: to maintain the manufacturer's warranty, a professional must install solid-surface countertops.

Removing the Old Kitchen Sink

Tearing out the old sink is well over half the battle. How you proceed will depend on how the sink is mounted and whether you'll be leaving the countertop in place. If the old countertop is headed for the dumpster, don't bother separating the sink. Pitch it all. But if you plan to save the old countertop, you'll need to remove the sink and its components carefully. Start by turning off the water at the supply valve or main valve.

Stubborn Drain Fittings. When all goes well, it's easy to remove a kitchen sink drain. Just disconnect the tailpiece; grip the spud nut with large groove-joint pliers or a spud wrench; and unscrew the nut from the drain spud. (See the photo opposite, bottom left.) But if the drain is old, the pot-metal spud nut may be fused to the brass with corrosion. If a spud wrench or large pliers won't do the trick, spray the nut with penetrating oil. Give the lubricant 10 minutes to

Fastening clips are used in both stainless-steel self-rimming sinks (left in photo) and metal-rim sinks.

Seen from underneath the sink, this is how a rim clip for a rim-style sink will look. Use a special sink-clip wrench or a nut driver with an extension to remove it.

work, and try it again. If lubrication doesn't help, try driving the nut loose with a hammer and cold chisel. Set the chisel against one of the tabs on the nut, and drive it in a counterclockwise direction. This will usually break the nut free, allowing you to finish with pliers or a wrench. If that doesn't work, switch to a hacksaw. Position the saw at a slight angle across the nut, and cut until you break through it. (See the photo below right.)

Stubborn Sink Faucets. A similar approach works on old faucets whose fastening nuts won't break loose. Try a basin wrench and penetrating oil first; if they don't work,

saw through the fastening bolts or the inlet shanks. A hacksaw won't fit in these cramped spaces, so use a hacksaw blade by itself. If you're the impatient type, a reciprocating saw fixed with a long metal-cutting blade will slice through faucet components in no time. Reciprocating saws are common rental items.

The Sink Itself. If the old sink is a self-rimming cast-iron or porcelain model, undo the water and waste connections and slice through the caulk between the sink and countertop. A sharp utility knife works best. Then just lift out the sink. If the old sink is a self-rimming stainless-steel unit or a rim-style model, you'll first need to remove the rim clips. (See the photos opposite and left.) The easiest way to remove clips is with a special sink-clip wrench, called a Hootie wrench, though a long screwdriver or nut driver will also work. If the old sink is made of cast iron, take note of the sink rim's corner brackets. These brackets are all that keep an extremely heavy sink from falling into the cabinet space. Leave the brackets in place until you've removed the sink. In fact, it's good practice to prop up the sink with a short piece of lumber when undoing the clips from a cast-iron sink. Stainless-steel sinks are much lighter and don't present a problem.

To remove the rim clips, unscrew the hex-head bolt from each one. When lifting a cast-iron sink, you may have difficulty getting a starting grip. Try lifting by the faucet column until you can get a hand under the rim. If that feels too awkward, remove the drain fittings, and reach through the openings.

REMOVING STUBBORN DRAIN FITTINGS

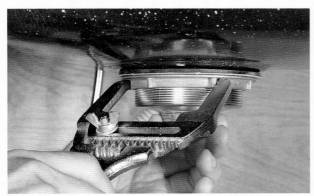

Use a spud wrench or large pliers to remove a drain's spud nut.

If the spud nut won't budge, remove it by cutting it in two using a hacksaw.

Installing Sinks & Related Equipment

ATTACHING DRAIN FITTINGS TO THE SINK

project

You must install a basket-strainer drain in each of the drain outlets in a typical two tub kitchen sink, unless you install a waste-disposal unit in one of the holes. Drain kits come with a removable basket strainer, drain body, large spud nut, paper washer, rubber washer, and coupling nut. The only tool you'll need for this job is a spud wrench (shown below) or a pair of large groove-joint pliers. The only extra supplies you'll need are some plumbers putty for sealing the drain flange and some light oil for lubricating the threads on the drain body.

TOOLS & MATERIALS

▌ Drain kit
▌ Plumber's putty
▌ Spud wrench

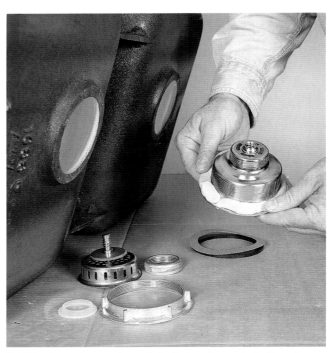

1 The best time to install drain hardware is before the sink is in place. Just prop up the fixture so you can work at a comfortable height. Begin by forming a ½-in. roll of plumber's putty and pressing it to the bottom of the drain flange.

2 Press the drain body into the opening from the top of the sink, making sure that the plumber's putty fills the depression around the hole. Then install the rubber gasket, the paper gasket, and the spud nut in that order.

3 While holding the threads of the drain body, tighten the spud nut using a spud wrench or large groove-joint pliers. When the nut is snug, trim the excess putty that squeezed out from around the flange, and tighten the spud nut again.

ATTACHING A FAUCET AND SPRAY ATTACHMENT

project

If the kitchen sink faucet that you are installing has a separate spray hose, as many do, install this hose at the same time you install the faucet and the drain hardware. Like bathroom sink faucets, kitchen sink faucets usually come with a rubber or plastic gasket that is installed between the bottom of the faucet and the top of the sink. This gasket keeps the sink from leaking around the base of the faucet. If your faucet doesn't have one of them, fill the base of the faucet with plumber's putty before sliding it into the faucet holes.

TOOLS & MATERIALS

▌ Faucet ▌ Spray attachment
▌ Adjustable wrench
▌ Pipe-sealing tape

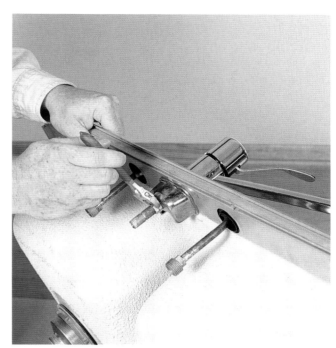

1 Insert the faucet with deck gasket included into the main three holes of the sink deck. Tighten the center nut under the spout using an adjustable wrench.

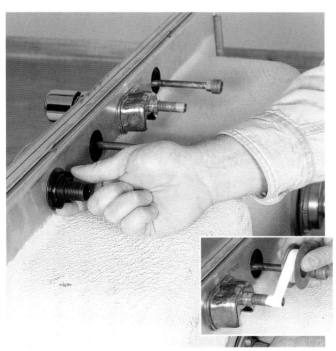

2 Fit the sprayer's deck fitting through the sink's fourth deck hole; slide the jamb nut onto the fitting's shank; and tighten it. Wrap two to three layers of pipe-thread sealing tape clockwise around the male threads of the faucet's center diverter nipple (inset).

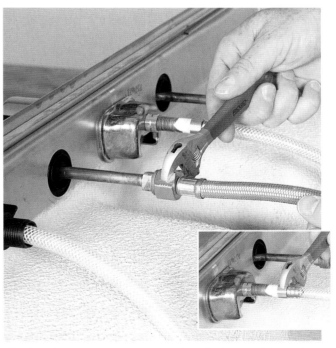

3 Feed the hose through the deck fitting, and thread its fitting onto the faucet's diverter fitting, tightening it with an adjustable wrench (inset). Attach the sprayer assembly to the deck (left in main photo). Then install the supply tubes to the faucet stubs.

ATTACHING A CENTER-COLUMN FAUCET

project

Like sinks, cabinets, and countertops, faucets are expensive. And they can be easily damaged, at least cosmetically, so it makes sense to take care when installing them. The best protective material is cardboard. It can shield flat surfaces from dropped tools and at the same time provide a smooth material for spreading out parts. Also keep in mind that faucet designs do differ. For instance, some have separate knobs while others have a single lever. They also differ slightly in the way they're installed, but most are approached the way we show here.

TOOLS & MATERIALS

- New faucet
- Screwdriver
- Groove-joint pliers
- Latex tub-and-tile caulk

1 Center-column faucets can be installed on sinks with one or three holes. If you have a single-hole sink then the installation is a snap. But if you have a three-hole sink, install a baseplate (that usually comes with the faucet) to cover up the unnecessary holes.

4 On a typical single-column faucet, two water supply risers, and a chrome-plated hose that connects to the head of the faucet all fit inside the faucet body. To hook the faucet hose to the supply lines, first install a supply adapter on the faucet nipple that has an O-ring seal.

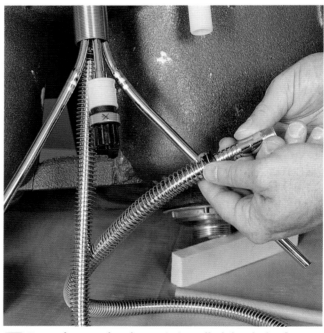

5 Once the supply adapter is installed, hook the faucet hose to it. To do this, first prepare the hose by pulling back the tension spring on the end of the hose. This will expose the male attachment threads that will attach directly to the bottom of the adapter.

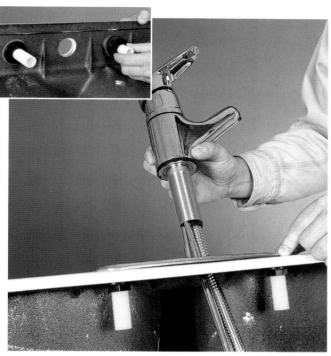

2 Fasten the plastic base-plate support/gasket from below with plastic jamb nuts (inset). Make them finger-tight. Snap the decorative base plate in place over the plastic support, and insert the faucet's column through the center hole in the sink deck.

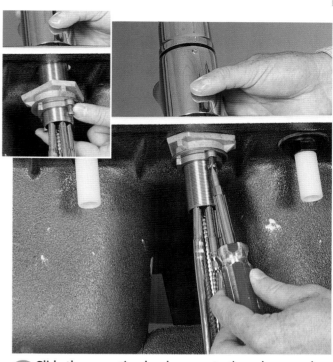

3 Slide the mounting hardware onto the column and tighten it (inset). The unit shown has a plastic spacer, steel washer, and brass nut. Use a screwdriver to drive the setscrews against the large washer. Stop when the screws feel snug.

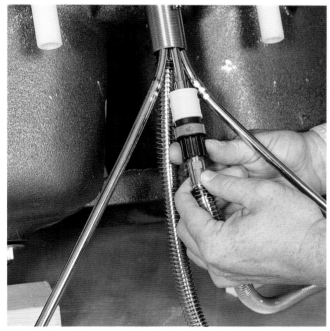

6 Hold the faucet hose so the tension spring is compressed. Then thread the end of the hose into the bottom of the supply adapter. Use only hand pressure to install the hose; don't use pliers. Stop when the hose feels snug. You'll connect the water supply (pages 124–125) after installing the sink (pages 120–123).

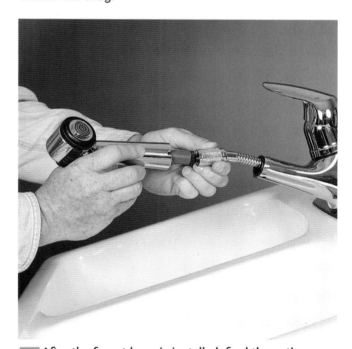

7 After the faucet hose is installed, feed the entire length of hose up through the outlet hole at the top of the faucet. Attach the spray head to the top end of the hose. Again, use only hand pressure to tighten the joint between the two. Check the hose for smooth operation.

INSTALLING A METAL-RIM SINK

Plumber's putty has long been used to seal sinks with metal rims, like the one shown here. However, plumber's putty can turn dark over time—sometimes it even turns black—and grow brittle, so it cracks and chips, leaving behind voids that can leak. These days, a better choice is one of the popular tub-and-tile caulks. The best kinds to use are PVA (polyvinyl acetate) caulk, acrylic latex caulk, or silicon caulk. Any of these will last a long time and remain flexible for years. Just make sure to buy a brand that has mildew-resistant additives.

TOOLS & MATERIALS

▌ Sink and rim ▌ Pencil ▌ Saber saw
▌ Rim clips ▌ Screwdriver
▌ Sink-clip wrench

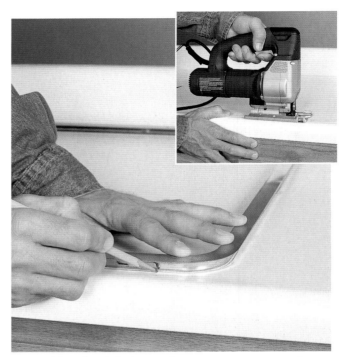

1 Position the sink rim on the countertop, centered over the sink base and parallel with the back wall, and trace around it. Drill ¼-in.-or-more holes in one or more corners of the marked cutout (inset), and follow the pencil mark using a saber saw to cut the hole for the sink.

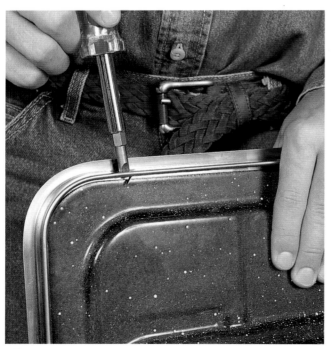

2 To support the sink rim until you can install the sink clips, you have to bend inward the precut tabs on the vertical band of the rim. Use a flat-blade screwdriver for this job. If a tab doesn't move, strike the top of the screwdriver with a hammer until the tab breaks free.

3 Lower the sink and rim into the countertop opening, and begin installing the hold-down clips around the perimeter of the sink. Use a sink-clip wrench (shown above) or a long nut driver or screwdriver to do this job. The clips bear against the underside of the counter.

Installing a Metal-Rim Sink

Sinks that require separate metal rims have the advantage of a neat, flush-fit appearance. The rim is T-shaped, with a horizontal support flange and a vertical extension. The bottom of the vertical extension is rolled over to form a lip. You need to cut a more precise opening in the countertop, and you'll have a dozen or so rim clips to install. As noted earlier, the easiest way to install clips is with a special clip wrench. Lacking such a wrench, a long, slotted screwdriver or nut driver will work.

Lay Out and Cut the Opening. When laying out the countertop opening, you won't need a paper template. Instead, the vertical extension of the rim serves as a template. Just position the rim on the countertop, with the flange on top, and trace around the outside of the vertical portion of the rim. Then cut carefully on this line using a saber saw.

Install the Sink Rim. Sink rims are made to work on cast-iron sinks and on enameled-steel sinks. (Stainless-steel sinks have built-in clip channels.) As you inspect the rim, you'll notice that it has two sets of knock-in tabs running around its vertical band. These tabs, when folded in, support the rim of the sink before the clips are installed. The row of tabs closest to the flange are for steel sinks, while those farther from the flange are for thicker cast-iron sinks.

When buying a rim for a cast-iron sink, choose one that has separate corner brackets. Unlike steel sinks, cast-iron sinks are too heavy to be supported safely by thin metal tabs. If you can't find corner brackets, make a homemade support. (See "Supporting a Heavy Sink," page 124.)

To install a sink rim, set it over the sink and use a screwdriver to punch in all the perimeter tabs.

Install the Sink Clips. The sink clips (page 114) and bolts will come unassembled in a plastic bag. Thread a bolt into each clip, and lay the clips within reach on the cabinet floor. Look closely, and you'll see that the clips have one hooked end and one flat end. Install the clips so that the hooked end fits into the rolled edge in the sink rim, and the flat end grips the bottom of the counter. Tighten each bolt until the clip drives the sink up against the rim flange and pulls the flange down against the counter. A typical installation requires about a dozen clips, spaced roughly 8 inches apart. If the corners of the sink do not draw down completely, use an additional clip at each corner.

Installing sink clips is usually fairly straightforward. In some cases, however, the standard ¾-inch spacers that

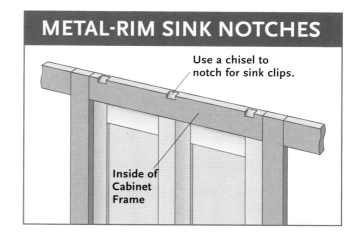

METAL-RIM SINK NOTCHES

Use a chisel to notch for sink clips.

Inside of Cabinet Frame

separate the counter from the cabinet rails or the rails themselves will block a portion of the clip space along the front of the sink. If you find yourself unable to fit clips along the front of the sink, chisel a few clip slots in the rail or spacer material. (See the illustration above.)

INSTALLING A SELF-RIMMING CAST-IRON SINK

CAST IRON IS HEAVY AND EXPENSIVE, so call a friend if you think you need help. With one of you on each side, lift the sink over the counter and into the sink opening. When it's set, carefully move it around until you have it centered and straight, relative to the backsplash. Don't be too concerned if the sink rocks in place. Some sinks become a little warped in the firing process. A warp of ¼ inch is acceptable, because when you've balanced and shimmed the sink corner to corner, you'll be able to fill the two ⅛-inch gaps with caulk. A warp greater than ¼ inch is too much.

CONNECTING THE TRAP

project

With two sink basins (each with a strainer drain), you'll need a plastic waste kit and P-trap to complete the waste hookup. Depending on the position of the permanent piping, you may also need a tailpiece extension with a female hub and compression fitting on one end. The new sink waste kit will consist of two flanged tailpieces, a baffled T-fitting, a 90-degree extension tube, and assorted nuts and washers. Two of these washers will be insert washers with a L-shaped profile. Insert one of these into the top of each flanged tailpiece.

TOOLS & MATERIALS

▪ Hacksaw ▪ Sink trap kit (1½-in.)
▪ Groove-joint pliers ▪ Marker or pencil
▪ PVC primer, cement ▪ Deburring tool

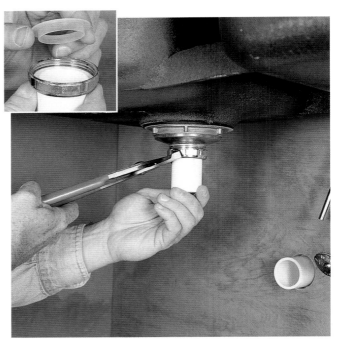

1 The joint between the tailpiece and the sink drain is sealed with a plastic insert washer and a chrome-plated nut. To install these parts, press the washer into the tailpiece (inset), and press the tailpiece against the drain. Tighten the nut that joins the two using groove-joint pliers.

4 Once the extension tube is cut and deburred, install a compression nut and washer on the cut end, and slide the tube into the T-fitting. Attach the other end to the second tailpiece. Then slide the nut and washer at the cut end against the T-fitting, and hand-tighten the nut.

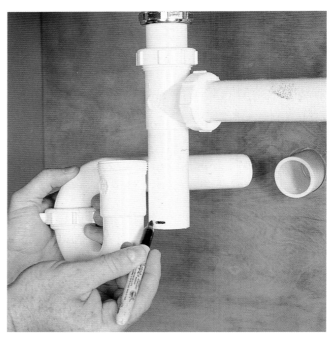

5 Like the horizontal extension tube that joins the two sink tailpieces, it's best to mark the length of the T-fitting while holding the mating parts in place. To do this, connect the sink trap to the wall elbow, and hold it next to the T-fitting, in line with the drainpipe. Mark the length to the bottom of the trap hub.

2 The baffled T-fitting, which connects the two sink basins, is attached directly to one of the drain tailpieces. To make this joint, slide a compression nut and a beveled washer on the bottom of a tailpiece. Then push the T-fitting up against the washer and hand-tighten the nut.

3 Install another tailpiece on the other sink drain (inset). Then temporarily attach a horizontal extension tube on this tailpiece. Hold this tube up against the side the T-fitting, and mark the length of the tube. Make the cut using a hacksaw, and remove any plastic burrs with a heavy rag.

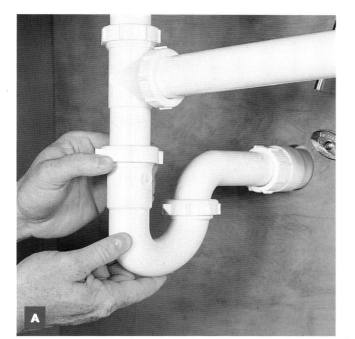

A

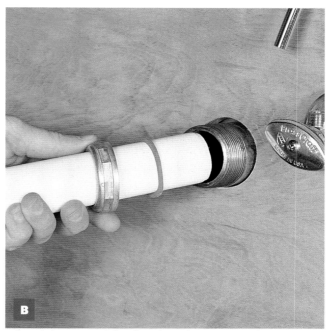

B

6 How you make the trap-to-drainpipe connection depends on the piping material at hand. A plastic ground-joint adapter (A) is best with plastic pipe, and a flat, rubber washer and metal nut (B) are best when the drain is made of galvanized steel. A banded coupling is another option, but it costs much more. (See page 48.)

Connecting the Water

Almost all water pipe-to-supply-tube connections these days are made with compression fittings, either adapters or valves. Only older homes have friction (cone) washer connections, and even these are easy to convert to compression fittings. Just clean the supply-pipe threads with a wire brush; coat them with pipe joint compound; and screw on new compression valves.

Connection Types. The kind of supply tube you use will depend, in part, on the type of connection on the faucet. Some faucets come with threaded shanks. Others have copper tubes fitted with threaded adapter nuts, and still others have copper tubes with no fittings. For those with threaded fittings, choose a ball-head supply tube or a prefitted stainless-steel-and-polymer tube. If your faucet has threaded adapters on copper tubes, be sure to back-hold the adapters when tightening the coupling nuts.

Of the faucets that come with simple copper inlet tubes, some have short tubes—about 6 inches long—and others have longer ones. The longer ones are usually long enough to reach water stub-outs in the wall. In that case, no supply tubes are needed. Just trim the inlet tubes to length, and join them to compression fittings or valves. If the copper inlet tubes are not long enough, the best approach is to buy braided-steel supply lines or supply tubes that have a ⅜-inch compression fitting at one end. Join the supply tube to the inlet tube with this fitting, and join the bottom end to the shutoff valve's compression fitting.

smart tip

SUPPORTING A HEAVY SINK

IF YOU'RE PLANNING TO INSTALL A METAL-RIM CAST-IRON SINK AND CAN'T FIND A RIM WITH CORNER BRACKETS, RIG YOUR OWN SUPPORT USING A 2X4 AND ¼-INCH ROPE. AFTER INSTALLING THE RIM ON THE SINK, LAY A 36-INCH 2X4 ACROSS THE SINK.

RUN THE ROPE DOWN THROUGH ONE DRAIN OPENING AND UP THROUGH THE NEXT (OR AROUND A SECOND, SMALLER, 2X4 SET ACROSS THE DRAIN WITH A SINGLE-BASIN SINK). TIE OFF THE ROPE TIGHTLY AROUND THE 36-INCH 2X4, AND SET THE SINK IN THE OPENING. AFTER SECURING THE CLIPS AND CONNECTING THE WATER-SUPPLY PLUMBING, UNDO YOUR HOMEMADE SUPPORT.

Installing a Cast-Iron Sink

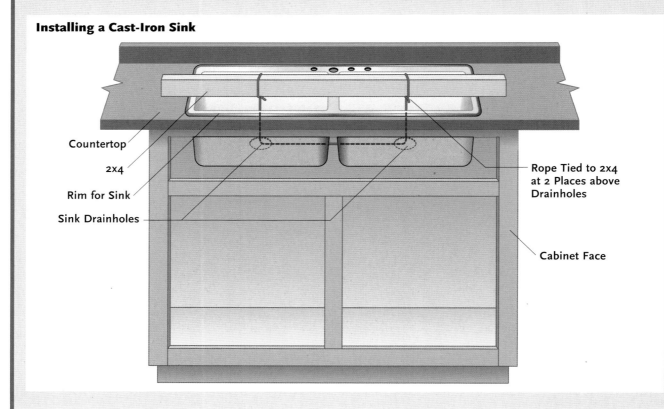

Countertop

2x4

Rim for Sink

Sink Drainholes

Rope Tied to 2x4 at 2 Places above Drainholes

Cabinet Face

CONNECTING THE WATER

project

Most plumbing fixtures, such as kitchen and bathroom sinks, have clearly visible shutoff valves underneath, so if there is a problem, you can quickly shut off the water supply. These days, this job usually is done with inexpensive chrome-plated shutoff valves. On the wall end they are tied to the house water supply lines. On the other end they have a valve handle. In between there is an opening for a flexible supply tube that carries water up to the fixture. These joints are almost always sealed with compression fittings like those shown here.

TOOLS & MATERIALS
▎ Water supply tube
▎ Tubing cutter ▎ Pipe joint compound
▎ Adjustable wrench

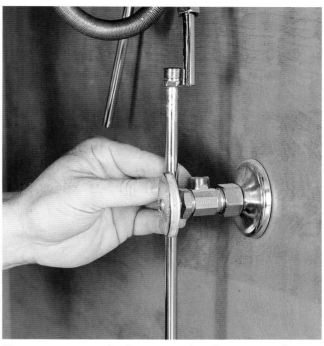

1 In most cases, the faucet supply tubes are too short to reach the shutoff valves. These tubes can be extended with new braided-steel supply lines or supply tubes, which have a ⅜-in. compression fitting on one end and require a bit more work. Hold the tube in place, and mark the length.

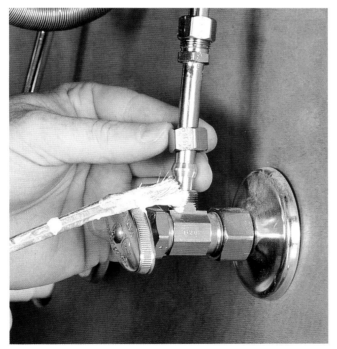

2 Cut the supply tube extension using a tubing cutter. Then slide this new tube into place between the sink tube and the valve. Coat the threads and ferrule on each joint with a thin layer of pipe joint compound.

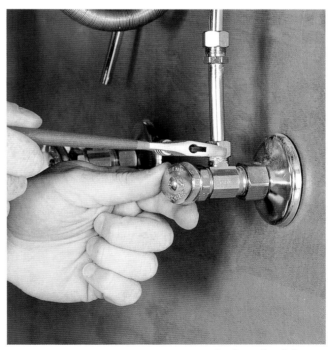

3 Tighten the compression nuts on both ends of the tube using an adjustable wrench. If the nut at the other end starts to turn, back hold it with a second wrench as you tighten the first nut. Make the joint snug, but don't overtighten.

HOW TO INSTALL A LAUNDRY SINK

Plastic and fiberglass laundry sinks come in several forms. Some are freestanding, some are wall-hung, and some are counter-mounted. Counter-mounted sinks are made to drop into cabinets, like self-rimming kitchen sinks. Fiberglass is sturdier than plastic.

Laundry sinks require lower piping connections than bathroom and kitchen sinks. While a kitchen sink drain is roughed-in 18 inches above the floor, a laundry sink's connection shouldn't be higher than 13 inches, measured from the center of the drainpipe. You can install the water-pipe stub-outs 15 to 16 inches off the floor.

Other differences involve the drain fittings and mounting methods. Drop-in models require that you install a basket-strainer drain, like those used on kitchen sinks, while wall-hung and freestanding models usually come with a drain fitting molded right into the bottom of the sink. These drains come with a rubber stopper, like those used in older bathtubs.

Installing a Freestanding Laundry Sink

The deck holes on laundry sinks have 4-inch center spreads, so kitchen faucets, which have 8-inch spreads, won't work here. Many people use lavatory faucets on laundry sinks because they're affordable, but special 4-inch-spread utility faucets are a better choice. They're often made of heavy brass and have spouts fitted with hose threads.

HOW TO REPLACE AN S-TRAP

AS DISCUSSED in "Trap Variations," pages 288–289, a sink that is drained through the floor via an S-trap is no longer legal, because the trap can't be vented. If your fixtures now drain through S-traps, you won't be required to change them, however, because they're covered by the grandfather clause.

Still, if you're replacing a sink that has an S-trap, you may as well do all you can to improve the way that new fixture performs and replace the trap. It consists of installing an automatic vent device inside the sink cabinet. (For more on automatic vents, see "Automatic Vent Device," page 282.)

Begin by attaching a banded coupling to the drain at floor level. (Or if possible, thread a 1½-inch PVC female adapter onto the drain's threads. These are the same threads used to connect the S-trap.) From the top of the coupling (or adapter), use two 45-degree PVC elbows to offset a riser to the back wall of the cabinet, about 4 inches left or right of center. Bring the riser up to trap level, about 18 inches off the floor, and install a sanitary T-fitting. Using a PVC ground-joint trap adapter, pipe the trap into the riser. Out of the top of the T, extend the riser up 6 inches, ending with another 1½-inch female adapter. Thread an automatic vent device into this adapter.

Automatic vent devices don't last indefinitely, so check the vent every two years. Its spring-loaded diaphragm should be held firmly against its seat. If the diaphragm is down even slightly, or if the rubber has deteriorated, install a whole new unit.

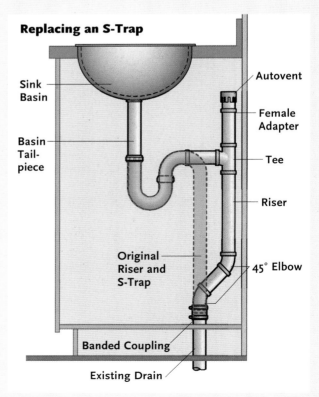

Replacing an S-Trap

Sink Basin

Basin Tailpiece

Original Riser and S-Trap

Autovent

Female Adapter

Tee

Riser

45° Elbow

Banded Coupling

Existing Drain

INSTALLING A FREESTANDING LAUNDRY SINK

project

A freestanding laundry sink is a great addition to any laundry room. They're lightweight, inexpensive, and easy to install. Their only real shortcoming is a tendency to be knocked out of alignment. As a result, the water supply and drain lines can loosen and start to leak. This problem is simple to fix; just screw the legs to the floor. Holes in the feet are provided for this job; all you need are a few drywall screws and a screwdriver. If your sink is located on a concrete floor, use plastic or lead anchors to do the job. Drill holes in the floor using a masonry bit.

TOOLS & MATERIALS

▮ Laundry sink kit
▮ Screwdriver ▮ Hacksaw
▮ Adjustable wrench

1 First, decide a good location for the sink. Generally, placing it close to the washer is the best idea. This makes a short trip for pretreated laundry that is wet. Begin by turning the sink upside down and attaching the legs.

2 While the sink is turned upside down, install the plastic drain tailpiece. Place the nylon washer on the drain body; then put the flange end of the tailpiece on the washer, and slide the nut against the body threads. Tighten the nut.

3 Install the faucet in the sink; then turn the whole assembly right side up. Position the unit where you want it; then attach the legs to the floor. Attach the water supply lines to the faucet and the drain trap to the tailpiece and wall drainpipe.

WASTE-DISPOSAL UNITS

Despite the claims of some manufacturers, the list of things that a disposal unit can safely handle is fairly short. Soft food items like boiled potatoes and oatmeal, or crispy vegetables such as lettuce, carrot or potato peels, and the like are easily ground into a pulp that can be flushed away with enough water. Hard or stringy food items, on the other hand, are troublesome. Celery, egg shells, coffee grounds, and even apple seeds are common sources of trouble. And of course, you should keep trash such as paper, plastic, twist-ties, and so on from making its way into a waste-disposal unit.

When a waste-disposal unit won't work, the problem is most likely either a jammed drum or a burned out motor. It's easy to clear a jammed drum, but unless the motor is still under warranty, it may not be worth repairing. The labor cost to diagnose and fix a used unit is often almost as much as the cost of a new unit.

Restarting a Jammed Waste-Disposal Unit. Manufacturers expect their units to stop once in a while, so they build in two useful features. One is a wrench slot in the unit's motor shaft; the other is an electric restart button.

Both are located on the underside of the motor housing.

If you can't see or feel the obstruction from above, find the wrench that came with your waste-disposal unit or a large Allen wrench, and move to the bottom of the unit. Unplug the disposal; insert the wrench into the shaft at the bottom-center of the unit; and crank the motor back and forth. This will almost always clear the obstruction. You'll know you've made progress when the motor spins freely, without continuous resistance. Plug in the unit, and press the reset button to allow it to run again. Once the unit starts up again, test it.

Routine Maintenance. Use cold water when grinding food scraps. To sharpen impeller blades, fill the waste-disposal unit with ice cubes, and turn it on. Do this every couple of months. To keep a unit from developing a bad odor, use it often and with lots of running water. If your unit already smells, pour lemon juice into the drum, and let it stand for a few minutes; then flush it. Run the unit with plenty of water thereafter. To clean the inner workings, quarter a potato, toss it in, and run the unit with cold water. When the drum is empty, run the unit with lots of hot water. Avoid pouring sodas into the unit. Carbonated drinks contain carbonic acid, which is corrosive.

smart tip

SWITCHING OPTIONS

IF CUTTING A NEW SWITCH BOX INTO A KITCHEN WALL SOUNDS LIKE MORE OF A PROJECT THAN YOU'D CARE TO TACKLE, THINK AGAIN. THE CONVENIENCE OF A READILY ACCESSIBLE SWITCH MAY BE WORTH THE EXTRA WORK. AS AN ALTERNATIVE, YOU COULD INSTALL A SURFACE-MOUNTED SWITCH INSIDE THE CABINET, BUT REACHING INTO THE CABINET TO TURN THE UNIT ON AND OFF WON'T BE HANDY. ANOTHER WAY TO AVOID CUTTING A SWITCH INTO A WALL IS TO BUY A BATCH-FEED WASTE-DISPOSAL UNIT. TO ACTIVATE THIS TYPE OF UNIT, YOU PRESS THE STOPPER INTO THE DRAIN AND GIVE IT A TWIST. THESE UNITS CAN BE ORDERED WITH THREE-PRONG PLUGS, SO ALL YOU'D NEED TO PROVIDE IS A 120-VOLT, 15-AMP GROUNDED RECEPTACLE INSIDE THE CABINET. THIS ARRANGEMENT STILL REQUIRES A NEW CIRCUIT.

INSTALLATION METHODS

FOLLOW THE LOCAL CODE when installing a waste-disposal unit. The photos below show two possible ways to hook up the drainage piping. For one-bowl sinks and where it is required by code for two-bowl sinks, install the waste L-fitting that comes with the unit (photo on the left), and attach it to a P-trap that is connected to the main waste line of the house. For two-bowl sinks, buy a disposal-unit waste kit that includes a straight extension pipe that you connect to a tailpiece extension located above the trap on the other bowl (right).

project

RESTARTING A JAMMED WASTE-DISPOSAL UNIT

Waste-disposal units come in two designs: continuous feed and batch feed. The continuous units allow you to push the waste directly into the top opening so it can be ground up and flushed away into the house drain system. They are switch operated, usually from a wall next to the sink. Batch feed models grind up and flush away the waste in the same way but don't use externally mounted switches. Instead, they are activated when the cover is pushed into the disposal unit's opening. Waste is ground up and washed away one batch at a time.

TOOLS & MATERIALS

▌ Hex wrench

smart tip

WHEN TO REPLACE A DISPOSAL UNIT

IF, WHEN USING A WRENCH TO FREE UP A JAMMED WASTE-DISPOSAL UNIT, YOU CAN FEEL OR HEAR A BEARING GRIND OR SEE LATERAL MOVEMENT IN THE SHAFT, IT IS PROBABLY TIME TO REPLACE THE UNIT. ALSO, IF AFTER PRESSING THE RESET BUTTON YOUR UNIT MAKES A LOW HUMMING NOISE AND THEN TRIPS AGAIN, YOU SHOULD REPLACE IT. AS A LAST-DITCH EFFORT, YOU MIGHT TURN OFF THE POWER SUPPLY, DISCONNECT THE DISPOSAL UNIT, AND DROP THE UNIT OUT OF THE SINK. ONCE FREE, REMOVE THE LARGE RUBBER GASKET AT ITS TOP. THIS ALLOWS YOU TO SEE DIRECTLY INTO THE DRUM TO CHECK FOR PROBLEMS. YOU MIGHT FIND A PIECE OF STRING OR SOME OTHER OBJECT BINDING ONE OF THE IMPELLERS. IN MOST CASES, HOWEVER, THE SYMPTOMS JUST MENTIONED SIGNAL A DEAD OR DYING UNIT.

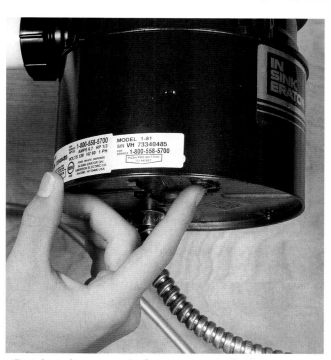

1 To free up a jammed disposal unit's motor, first shut off the power to the unit at the electrical service panel. Then insert the tool that came with the unit (or a standard Allen wrench) into the end of the motor shaft, and turn it back and forth.

2 When the motor shaft starts to spin freely, remove the tool and turn on the power to the waste-disposal unit at the service panel. Then press the reset button to restart the stalled motor. Repeat the entire process several times, if needed.

Installing Sinks & Related Equipment

REMOVING A WASTE-DISPOSAL UNIT

project

Unlike drain fittings, waste-disposal units don't usually become stuck to the bottom of sinks. The reason has to do with their mounting mechanisms, which range from simple hose-clamp fasteners to threaded plastic collars to triple layer bolt-on assemblies. The triple-layer mechanism described here is the most common—and the most complicated. To remove this kind of installation, start by turning off the power to the circuit that supplies the disposal unit at the service panel. Then follow the instructions shown in these photos.

TOOLS & MATERIALS

- Screwdriver
- Groove-joint pliers

1 Typical installations call for hooking the drain hose from the dishwasher to a side port on the unit. This makes sure that any food washed from the dishes will be ground up before it enters the waste piping. Remove this hose by loosening the hose clamp on the port.

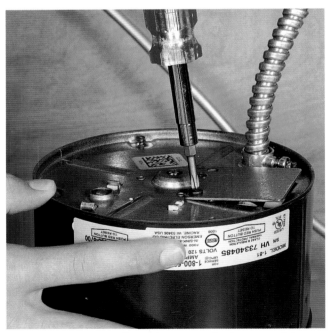

4 Once the unit is free, place it on the bottom of the cabinet, and turn it over. Remove the cover to the electrical box with a screwdriver to expose the wires underneath. Make sure the power to the unit has been turned off at the electrical service panel.

5 Underneath the box cover are five wires. Two wires (black and white) that are attached to the disposal unit and two more (black and white) that come from the circuit cable. All four are attached with wire connectors. The fifth wire is a green ground wire from the circuit cable.

130

2 After removing the dishwasher hose, remove the pipe that connects the disposal unit to the drain system. Some drain pipes are bolted to the side of the unit. Others feature compression fittings, as does this one. If you can't loosen this nut by hand, use groove-joint pliers.

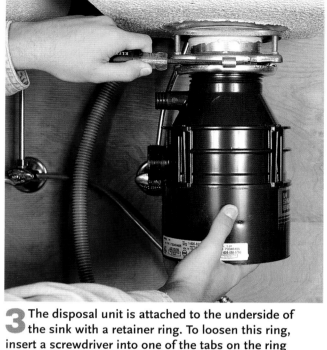

3 The disposal unit is attached to the underside of the sink with a retainer ring. To loosen this ring, insert a screwdriver into one of the tabs on the ring and turn counterclockwise. Support the disposal unit with your second hand so it doesn't fall when you have loosened the ring.

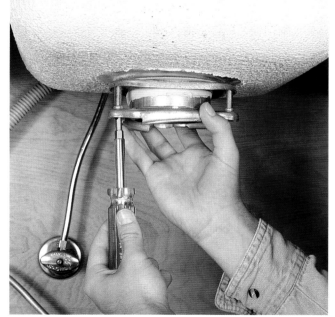

6 To remove the disposal unit's drain attachment assembly, start by loosening the three bolts that hold the retaining ring in place. Turn these bolts in a counterclockwise direction, using a flat-blade screwdriver, until all the components are loose and easy to turn.

7 The drain assembly components are held together with a snap ring. To remove this, push the retaining ring up against the bottom of the sink; then pry out the snap ring with the tip of a flat-blade screwdriver. Once this ring is removed, the rest of the parts just fall away.

INSTALLING A WASTE-DISPOSAL UNIT

project

If you are doing a simple one-for-one swap, then installing a new disposal unit will be easy. Just hook it up it in the reverse order of the way the old one was removed. This job is easy because most of the time you can use the existing electrical and drain hookups for the new disposal unit, which are two of the hardest parts of the job. But if you are installing a unit in a sink that didn't have one before, you have a bigger job on your hands. Follow the instructions, shown here, for doing the whole thing.

TOOLS & MATERIALS

- Waste-disposal unit
- Plumber's putty
- Switch, box, pencil
- Screwdriver
- Hacksaw
- Utility saw
- Wire stripper
- Electrical cable
- Wire connectors

1 Start the job by installing the drain flange, which comes with the disposal unit, in the sink drain hole. Take a golf ball size lump of plumber's putty in your hands, and roll it into a ½-in.-diameter rope. Press this around the bottom of the flange so its thickness is fairly uniform.

4 The disposal unit comes with a mounting ring installed at the top. This ring engages with the mounting flange in the mounting assembly that's installed on the bottom of the sink. To join these parts, lift up the disposal unit, and rotate the mounting ring until it fits over the flange.

5 Install the drain line to the side of the disposal unit. If the unit is hooked to a single bowl sink, then this pipe is attached to a trap and the trap is attached to waste piping in the wall. If there are two sinks, attach the disposer to the tailpiece extension that's connected to the other sink.

Installing Sinks & Related Equipment

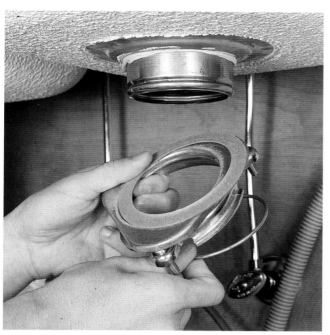

2 Press the drain spud into the drain hole so the putty squeezes out slightly around the flange. Then go under the sink and install the mounting assembly. This consists of a gasket, sealing flange, mounting flange, and split ring. Lift these parts over the spud.

3 As you hold the mounting assembly parts to the bottom of the sink with one hand, install the split ring with the other hand. This ring fits in a depression around the bottom of the drain spud and holds the other parts in place. Tighten these by turning the flange bolts using a screwdriver.

6 If you have a dishwasher next to the sink cabinet and want to hook its waste line to the disposal unit, there is a port supplied for this purpose. To use it, break out the plug that seals it; then push the hose onto the nipple, and attach it using a hose clamp.

7 To install an electrical switch for controlling the disposer, choose a convenient location; hold a cut-in retrofit box against the wall and trace around it. Pick a spot that is easy to reach while standing at the sink, and stay at least 1 in. away from any studs behind the drywall.

Continued on next page.

Continued from previous page.

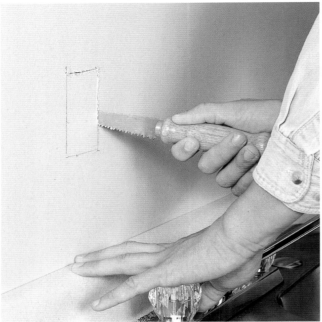

8 Use a drywall saw or a keyhole saw to cut an opening in the wallboard for the cut-in box. Work carefully to achieve a tight fit. Start on the waste side of the lines and test fit the box a couple of times, trimming the opening as necessary to make it fit.

9 Fish two electrical cables into the hole in the wall, one from the service panel, the other from the disposer unit. Feed these cables into the cut-in boxs and press the box into the wall. Engage the attachment tabs on the top and bottom of the box by tightening their screws.

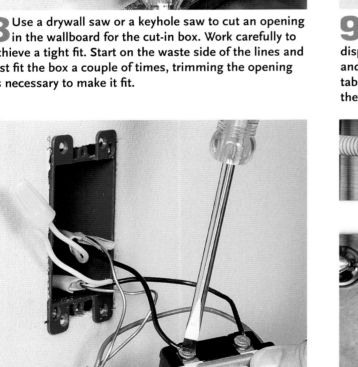

10 Strip the sheathing and paper from both cables; then strip about ⅝ in. of insulation off the end of each wire. Join the white wires with a wire connector and the ground wires and a green pigtail with another connector. Then, attach the black wires (and green pigtail) to the switch.

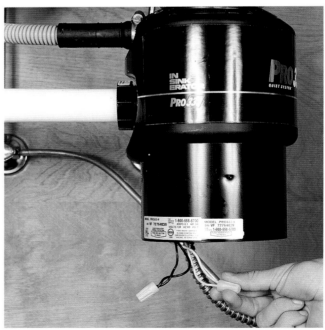

11 Use armored cable in exposed areas under the sink, and join it to the disposal unit with a cable connector. Attach the wires from this cable to the ones inside the unit's electrical box with wire connectors. Attach the green ground wire to the green grounding screw inside the box.

WIRING DISPOSAL UNITS

LOCAL BUILDING CODES VARY, but you should bring a new circuit into the kitchen for a new waste-disposal unit.

Shut off the power at the main electrical service panel, and install a new 15-amp circuit breaker or have an electrician do it.

Install the Switch. Install a new switch box above the counter as shown in step 7 on page 133. After cutting the box opening, bore a 1-inch hole below the countertop into the wall of the cabinet in which you'll install the unit.

Using electrical fish tape, feed the new cable into the stud space from the attic or basement, and pull at least 12 inches of cable into the kitchen through the new box opening. Again using fish tape, feed a second length of cable down through the switch opening, and pull it into the sink cabinet through the hole you bored. Leave at least 30 inches of cable showing in the cabinet and about 12 inches at the switch-box opening. Push the two cables into the retrofit box, and press the box into the wall. Tighten the bracket screws on the face of the box.

To wire the switch, strip ⅝ inch of insulation from each wire. Join the white wires using a wire connector, and connect both grounds with a 6-inch pigtail to the green screw on the switch. Then connect the black wires to the screws on the side of the switch. Mount the switch in the box.

Slide about 24 inches of flex conduit over the cable in the cabinet, and push the conduit through the wall opening several inches. Install a conduit box connector on the waste-disposal-unit end of the conduit, and strip the sheathing from the cable extending beyond the connector.

To wire the waste-disposal unit, remove its cover plate, and install a cable connector in the box. Trim the cable wires to length, and strip ⅝ inch of insulation from each wire. Feed the wires through the box connector, and with yellow twist connectors, join the unit's lead wires to the new circuit wires, white to white and black to black. Attach the circuit's bare wire to the grounding screw in the unit, and reattach the cover plate.

BUILT-IN DISHWASHERS

The best thing you can do for a dishwasher is use it. If you don't for weeks on end, the water held in the pump may evaporate, allowing the seals to dry out and leak the next time you use the machine.

Maintenance. Check for a slimy dirt buildup on the lower section of the door seal at least once every few months. It's hard to see this accumulation from above, so use a pocket mirror. If you see signs of a buildup, clean the seal with detergent. While you're at it, lift the float from the bottom of the unit to check for dirt. A dirty float can increase the water level enough to cause a leak. And finally, check the spray arms for bits of plastic and other debris. If you see any debris in these holes, pick it out with tweezers.

Installation Requirements. A dishwasher requires hot water and electricity in order to function. Most often, the unit is located as close as possible to the kitchen sink for access to water. Location doesn't have much of an effect on electrical needs.

To wire the dishwasher, your minimum electrical requirements will be a dedicated 15-amp circuit run in 14 / 2G NM-B cable (14-gauge, two-wire-with-ground nonmetallic cable). If the dishwasher has a preheater, which boosts the temperature of the water, you may need a 20-amp circuit with heavier 12 / 2G cable. Check the manufacturer's specifications carefully. Some local codes allow a direct connection, in which the cable is brought into the opening through the back wall or the floor and is connected directly to the dishwasher's electrical box through a standard box connector. In this case, no conduit is needed. Other codes require a disconnect switch inside the sink cabinet. Any cable that you install in a cabinet needs to be encased in flexible conduit.

Air gap fitting code updates may affect your installation or repairs (see page 292). Before replacing or repairing any section, consult your local building authority for applicable enforcement.

smart tip

HOSE ALTERNATIVE

MANY DISHWASHERS COME WITH A DISCHARGE HOSE ALREADY ATTACHED. IF YOU BUY ONE THAT DOES NOT, OR IF YOUR INSTALLATION HAS THE DISHWASHER FARTHER AWAY FROM THE SINK THAN IS NORMAL, YOU CAN SUBSTITUTE AUTOMOTIVE HEATER HOSE.

Installing Sinks & Related Equipment

REMOVING AN OLD DISHWASHER

project

If you're replacing a worn-out dishwasher, you'll need to remove the old unit before you can install a new machine. Start the job by turning off the power to the appliance at the electrical service panel. Then protect the floor in front of the old unit so it won't be damaged as you work. If you have prefinished wood flooring like the type shown here, generally you can pull out the old machine without any damage. But if you have vinyl sheet flooring, which is much softer than wood, place some cardboard over the floor and slide the unit onto it.

TOOLS & MATERIALS

▮ Screwdriver ▮ Nut driver
▮ Adjustable wrench
▮ Bowl or pan

1 The first step in taking out an old dishwasher is to remove the access panel below the main door. Look for a couple of screws on both ends of the panel that, when removed, will free it from machine's framework.

2 Once the access panel is removed, loosen the nut that holds the water supply tubing to the machine using an adjustable wrench. Place a bowl under the tube to catch any water that may drain out. Also remove the electrical cable.

3 Once the water and power connections are removed, the only things holding the machine are two screws driven into the underside of the countertop. These are usually located at the upper corners of the unit. Remove the screws.

BUILT-IN DISHWASHERS

WHAT IF YOUR KITCHEN WASN'T DESIGNED with a built-in dishwasher in mind? With custom cabinets, it's going to be tough to create an opening. But if your kitchen has stock cabinets, with standard-sized units screwed together to make a continuous run, it's possible to remove one of the base units to make room for a dishwasher.

A conventional dishwasher requires a cabinet opening 24 inches wide, 24 inches deep, and at least 34½ inches tall, as measured to the underside of the countertop's edge band. (Most companies also make compact 18-inch models, which fit 18-inch-wide cabinet spaces.) The target cabinet should be a door unit. A drawer unit would work, but most kitchens have too few drawers for convenient storage already.

In conventional installations the dishwasher sits next to the sink cabinet, but this is not the only possibility. If you need to, you can skip a cabinet's width,

but don't place the dishwasher more than 6 feet from the drain outlet. Not only will it be a hassle moving dishes from the sink to the dishwasher, but the unit's purge pump may not be up to the distance. If you do skip a cabinet's width, drill the side walls of the intervening cabinet, and pipe the water and discharge tubes straight through, tight against the back wall.

To free a cabinet, you'll have to remove three sets of screws, plus the toekick trim. One set of screws joins the cabinet stiles (the vertical hardwood framework). If you don't find them in the target cabinet, look at the stiles in the adjoining cabinets. The next set of screws will be in the corner brackets at the inside top of the cabinet. These secure the countertop to the cabinet. The final few will be in the back brace. These secure the cabinet to the wall studs. Toekick trim can be made of wood, hardboard, or vinyl. Work it loose using a pry bar.

INSTALLING A NEW DISHWASHER

Most appliances come in a few standard sizes so they will fit in typical kitchen layouts. If you are replacing an old dishwasher, you are practically ensured that a new one will fit in the space. Just make sure to measure the existing machine, and take these measurements with you when you shop for a new unit. All dishwashers come with adjustable feet. So if you have a tight fit, you can lower the feet to slightly reduce the dishwasher's overall height. Make sure to adjust all four feet the same amount.

TOOLS & MATERIALS

▌Dishwasher, fittings ▌Nut- and screw-driver
▌Wire strippers ▌Needle-nose pliers
▌Wire connectors ▌Adjustable wrench
▌Pipe sealing tape

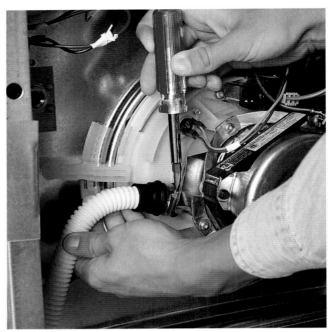

1 Before you slide the unit into its opening, attach the discharge hose that comes with the unit to the dishwasher pump. Use the hose clamp or the grip ring that's supplied for this purpose. This hose carries the waste water to the house drain system, usually through the side of the disposal unit.

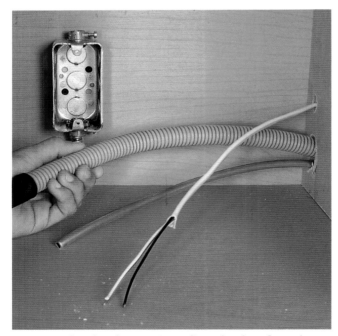

4 Some local codes require a direct hookup from the dishwasher to the service panel. Others require a switch in the adjacent sink cabinet so the power can be turned off easily. For the latter, install a switch box and drill holes for the drain hose, water line and electrical cable.

5 Remove the existing hot-water supply valve from under the sink, and install a new dual-stop valve. This fitting has a compression opening on the top for the sink supply tube and one on the side for the dishwasher supply tube. Install the fitting so the side port faces the dishwasher.

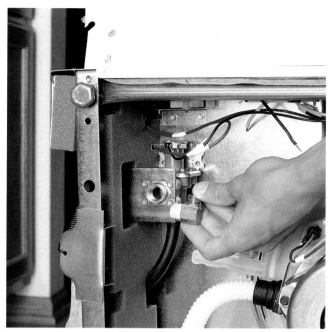

2 The solenoid valve is the entry point for the water supply line. It's designed to receive a standard dishwasher elbow; a fitting with pipe threads on one end and compression threads on the other. Wrap pipe-thread sealing tape around the pipe threads and install the elbow.

3 The electrical cable that brings power to the appliance must be installed in the electrical box that's provided. Remove the knockout plate on the side of the box, and install a cable connector. Wait to install the cable until after the unit is pushed into place.

6 Slide the dishwasher into its cabinet opening slowly to avoid scratching the floor and to keep from damaging the electrical cable, drain hose, and water supply tube. Have a helper pull these into the sink cabinet as you push the machine into place.

7 Once you have pushed the dishwasher completely into the cabinet, turn down the leveling legs at the front of the machine using an adjustable wrench. Make sure that the unit is level from side to side when you are done.

Continued on next page.

Continued from previous page.

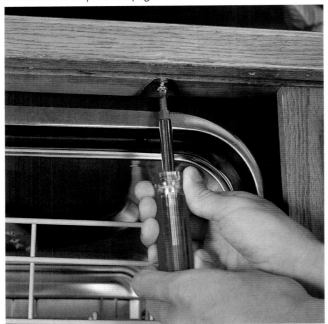

8 Dishwashers usually have two fastening brackets located at the upper corners of the machine. These are accessible only when the door is open. They are designed to be screwed to the underside of the countertop. Drilling screw pilot holes will make driving the screws much easier.

9 Water is delivered to the solenoid valve with ⅜-in.-diameter flexible copper tubing. Hook this tubing in place with a compression ferrule and nut. Tighten the nut securely with an adjustable wrench, but don't overtighten. This can ruin the ferrule.

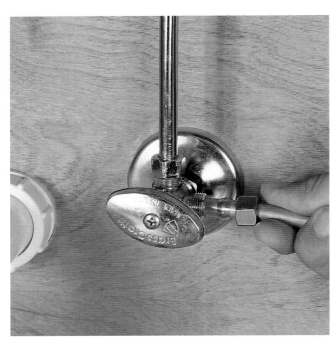

12 Install the sink supply tube to the top of the hot-water valve and the dishwasher supply tube to the side port. Carefully tighten the compression nuts so they are snug but not so tight that the ferrules are deformed. Once the water is tuned on, check for leaks and tighten if necessary.

13 Plumbing codes require overflow protection for the drain hose. This keeps the overflow from a backed-up drain from entering the dishwasher. The best way to do this is to install an air gap. You can also loop the hose to the top of the cabinet, and secure it with a piece of pipe strapping.

10 Insert the electrical cable into the electrical box, and tighten the cable connector screws. Then strip the sheathing and paper from the cable and about ⅝ in. of insulation from each wire. Join the cable wires to the dishwasher wires using wire connectors.

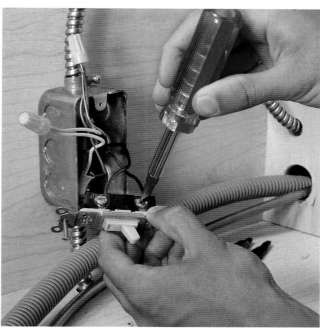

11 Slide the cables into the switch box, and tighten the box connectors. Strip the sheathing from the cables and insulation from the ends of the wires. Join the white wires with a wire connector, attach the ground wire to the box, and screw the black wires to the side of the switch.

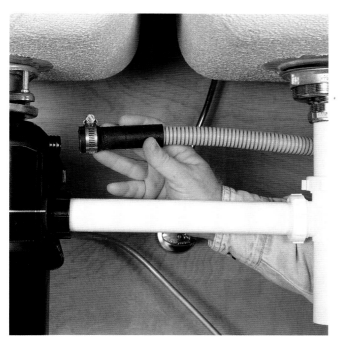

14 Disposers have a built-in side port for hooking up the discharge hose from a dishwasher. To use this port, first break out its plug with a screwdriver, then push the end of the hose over the ribbed nipple.
Attach the hose by tightening the hose clamp with a screwdriver or nut driver.

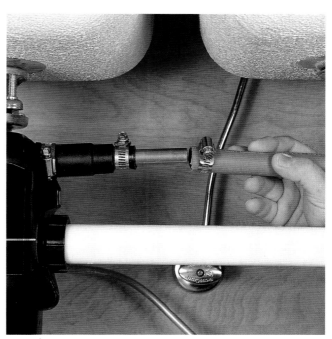

15 If the discharge hose is made of standard heater hose, you need a rubber adapter and a short length of copper pipe to make the connection. Check requirements in your area.

HOT-WATER DISPENSERS

Instant-hot-water dispensers are custom made for today's busy lives. At 190° F, the water that these pint-sized appliances serve up is most likely at least 40 to 50 degrees hotter than that delivered by your water heater. (Water heaters are dangerous when used above 140° F.) Water at a temperature of 190° F is just right for blanching vegetables, making instant soups, and brewing real coffee, one cup at a time.

Basic Considerations. Hot-water dispensers are easy to install and don't require a dedicated water line. If you have copper pipes, you can steal water from the cold-water riser under the sink through a self-piercing saddle valve. This is the same kind of valve used to supply icemakers. If your cold-water riser ends at floor level or the back wall, keeping you from installing a saddle tap, you might consider splicing into the supply tube using a T-fitting and a ⅜-to-¼-inch reducing coupling.

Most units come with three-prong plugs, so a grounded receptacle inside the cabinet will do. Although a shared small-appliance circuit can power the dispenser, a dedicated circuit is a good idea. Expect the unit to consume between 4 and 5 amps of power. That's roughly one-third the capacity of a 15-amp waste-disposal-unit circuit or one-fifth of a 20-amp kitchen circuit.

Begin the installation of an integral-tank-and-faucet instant hot-water dispenser by installing the self-piercing saddle valve on the cold-water riser pipe. Back the tapping pin out as far as it will go, and with the rubber tapping seal in place, bolt the two halves of the assembly together over the riser. Draw the two bolts down alternately, a little at a time, until they feel snug.

Next, install the tank. Feed its threaded shank up through the fourth sink-deck hole, and screw the mounting nut onto it from above. This will leave the unit hanging loosely from the deck. To secure it firmly, turn the jamb nut clockwise until it contacts the deck, and then tighten it with a basin wrench. With the unit in place, insert the chromed spout into the port atop the shank.

To make the water connection, carefully bend the unit's ¼-inch copper supply tube down to meet the saddle valve, and trim it to length. Then join the tube to the valve's compression fitting.

Hold a thermometer under the running water. It should read 190° F. If it doesn't, adjust the temperature by turning the adjustment screw counterclockwise.

project

Instant-hot-water dispensers are designed to be installed in the fourth deck hole of a standard stainless-steel kitchen sink. If you have a deck hole that is plugged with a cover, just remove the cover. If you have a stainless-steel sink without a fourth hole, you can cut one with a knockout punch. Unfortunately, you can't drill a hole in cast-iron or enameled-steel sinks. In these conditions your only choice is to drill a hole in the countertop. In most cases, this hole should be 1-½ inches in diameter and located so the spout extends over the sink.

TOOLS & MATERIALS

- Dispenser, fittings ▌ Screwdrivers
- Open-end wrenches ▌ Groove-joint pliers
- Tubing cutter

3 Once the heater unit is installed, push the bottom of the water spout into the top of the heater and hold it with a set screw installed from the side. Check the movement of the spout. It should turn freely from side-to-side.

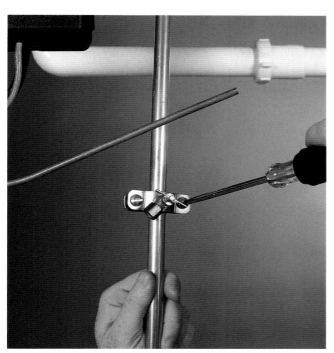

1 Dispensers come with a simple saddle valve for making the water connection to the sink cold-water supply tube. To install this valve, just hold both sides against the pipe and alternately tighten the screws until they are snug.

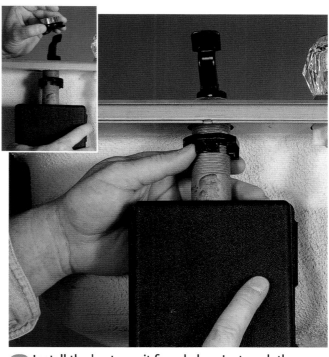

2 Install the heater unit from below. Just push the handle up through the sink hole, and have a helper tighten the mounting nut from above (inset). Then finish the job by tightening the jamb nut from below.

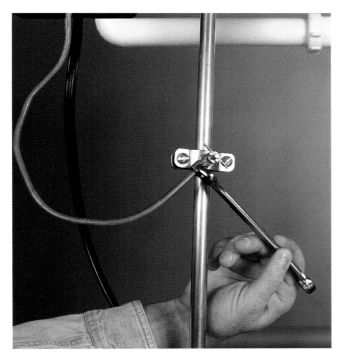

4 Attach the flexible copper tubing from the heater to the water supply line, using the compression fitting on the saddle valve. Slide the nut and ferrule onto the tubing; then tighten the nut until it's secure.

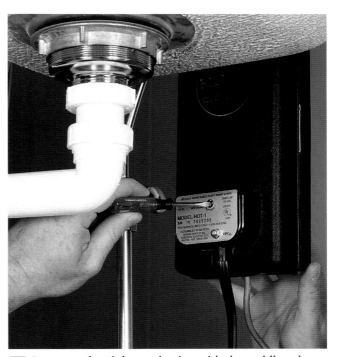

5 Puncture the sink supply pipe with the saddle valve, and check for leaks. Then plug in the heater unit, and test the operation. Adjust the water temperature with a flat blade screwdriver until it's approximately 190°.

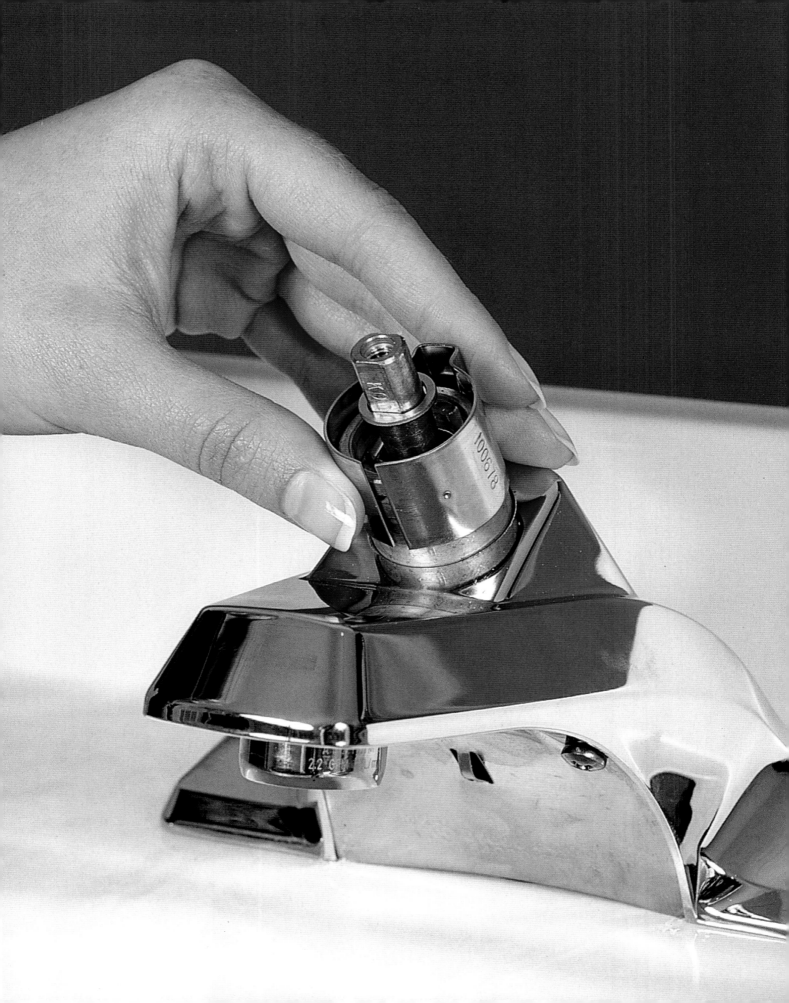

5
faucet repairs

FAUCET REPAIR is an ideal do-it-yourself project. It's light work, and the parts are affordable—often less than $5 for a complete repair. With compression faucets, the material costs can run closer to 5¢ if you just have to replace a washer. (Hire a professional plumber to repair a faucet, and you will find that labor accounts for most of the bill.) But the longer you wait to fix a faucet, the more expensive a repair will become. A steady drip will eventually destroy important parts in the unit, requiring complete replacement (at a much higher cost than a repair) and wasting hundreds of gallons of water in the process.

FAUCET OVERVIEW

Older single-inlet faucets turned a supply of water on and off, yet they provided only cold or hot water. To get warm water, you had to mix hot and cold water in a basin. Dual-inlet faucets, now universally standard, mix hot and cold water before sending it through the spout. You'll find dual-inlet faucets with four kinds of water-control mechanisms.
- Compression valve
- Cartridge
- Ball valve
- Ceramic disk valve (also usually in a cartridge)

Cartridge Faucet. With a cartridge faucet, repair usually consists of merely replacing the cartridge.

Compression Faucet. Compression faucets are fitted with a replaceable stem washer for low-cost repair.

Faucet Repairs

Leakproof Water Control. To make a positive seal against continuous water pressure, one of two things is required: 1) some form of resilient gasket material to fill the gaps between sealing surfaces or 2) companion sealing surfaces that are rock hard and precisely machined for a tight, leakproof fit. Compression, ball-type, and many cartridge faucets use metal or nylon moving parts with neoprene-rubber washers, seals, or O-rings. Other cartridge designs use ceramic disks, which have an extremely hard, smooth surface. Many top-of-the-line faucets use ceramic-disk technology for water control.

There are hundreds of faucet makes and models in use today, each with its proprietary twist. In fact, when working on some older models you might wonder why anyone would clutter such a simple device with so many extra gaskets, spacers, and sleeves. Don't be overwhelmed. Just replace the extra parts in reverse order of removal, and concentrate on the basics. Discussion of the four basic faucet designs follows.

Washer-Equipped Faucets

The most familiar faucet may be the washer-equipped compression-type faucet, sold under many brand names. Compression faucets control water by means of threaded, washer-fitted stems that move up and down over brass seats. Every compression faucet consists of a handle, a packing nut and/or bonnet nut, a threaded stem with washer, a washer screw, and a brass seat. The washer fits into the seat to shut off the water.

In two-handle dual-inlet faucets, each inlet port has its own stem. (The compression design doesn't allow for single-handle dual-inlet faucets.) Single-stem compression faucets, including hose bibcocks, sillcocks, boiler drains, and old-fashioned single-inlet faucets are not mixing valves.

Washerless Faucets

When cartridge faucets first appeared in the late 1950s, they were touted as "washerless" at a time when virtually everyone was familiar with leaky washer faucets. Since

REPAIRING COMPRESSION FAUCETS

Compression faucets have three basic problem areas, each with its own symptom.

- If the faucet drips from the end of the spout, you'll have to replace the seat washers and frequently the seats themselves. (See "Replacing a Seat Washer, pages 148–149, and "Replacing Valve Seats," page 150.)
- If the faucet leaks around its handles when the water is turned on but not when the water is shut off, the stem packing is worn, and you should replace it. (See Repairing Worn Stem Packing," page 152.)
- If a kitchen or bar faucet leaks around the base of its spout, replace the spout collar seals. (See "Repairing a Two-Handle Faucet Spout," page 169.) This problem can occur with any type of faucet.

Compression Faucet Repair. If you repair a faucet soon after it develops a drip, you can make a lasting repair with a 5¢ washer or two. But if you allow a faucet to drip for several months, the leak can do real damage. Pressurized water has incredible force. When allowed to seep past a defective washer, it actually cuts a channel across the rim of the brass seat. Stick a finger into the port of a relentlessly leaky faucet, and you'll feel the pits and voids in its

damaged seat. If the seat feels smooth, leave it. If not, replace it. If the seat is not removable, grind it smooth with a dressing tool. (See "Grinding Valve Seats," page 151.) A pitted seat can chew up a new washer in a matter of weeks.

"H" is for Left

The hot-side faucet handle, usually marked "H" somewhere on the handle, should always be on the left as you face the faucet. If it's not, the faucet may have been installed backward or the water lines under the sink may be reversed. To see whether the faucet was installed backward, check the stems. They may look alike, but they don't rotate the same. With compression faucets, you turn the water on by rotating the hot-water handle counterclockwise and the cold-water handle clockwise. If your faucet doesn't work that way, someone reversed the stems. A repair is your chance to correct them.

There probably isn't a technical reason why faucets that control the hot water should be on the left. Many people say that because most people are right-handed, they will reach for the right faucet first—the one that controls the cold water—avoiding a burn. Add that to the fact that plumbers like things neat, orderly, and consistent, and you come up with the convention of putting the hot-water faucet on the left.

then, other washerless designs have been developed, namely, ball-type and ceramic-disk faucets.

Virtually all washerless faucets operate according to the same design principle: inlet ports are moved into or out of alignment with companion ports in the faucet body. When the openings are aligned, water flows from pipes to spout. When rotated out of alignment, the flow is sliced off. The degree and angle of rotation dictates the volume of the flow and the temperature of the mix. These faucets tend to work trouble-free longer than compression faucets for one very good reason: you can't overtighten a cartridge faucet. While those who grew up with leaky compression faucets instinctively give faucet handles an extra twist, which hastens the destruction of the washer and results in a dripping faucet, washerless faucets can only be tipped or lifted (single-handle models) or rotated one-quarter turn (dual-handle models).

Cartridge. A cartridge faucet may have two handles, like compression faucets, or a single handle. If your faucet is an Aqualine, Moen, Price Pfister, or Valley brand single-handle model, it is probably a cartridge type. If it is a dual-handle model, you'll have to take it apart to tell what kind it is. For repairs, see pages 156 to 163.

Ball-Type. Ball-valve faucets are always single-handle units. The ball contains inlet ports that align with faucet-body ports to allow water flow. Movement of the ports alters the flow rate and hot-cold mixture. Delta and Peerless are the major ball-type faucet brands. For repairs, see pages 154 to 155.

Ceramic Disk. Another single-handle or two-handle design, ceramic-disk faucets contain a cylinder that houses two mating ceramic disks, one with inlet and outlet ports and one without. The disks slide into and out of alignment with each other to control water flow and temperature. If you have an American Standard or Reliant unit, it is likely a ceramic-disk faucet. For repairs, see pages 164 to 166.

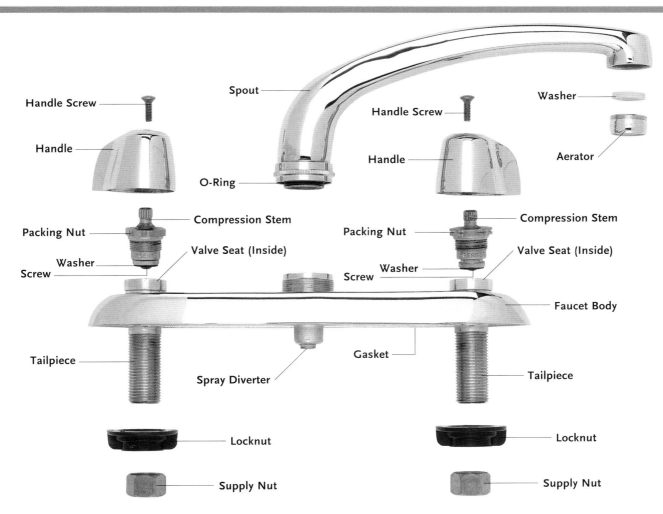

REPLACING A SEAT WASHER

project

Compression faucets need repair as soon as you notice them dripping. The cause of these leaks is almost always a worn valve stem washer. These get brittle as they age, and small parts break off. If you replace them immediately, the job is easy. But if you don't and you try to stop the leak by over-tightening the handle, you run the risk of damaging the valve seat inside the faucet and spoiling the whole unit. Some older faucets have replaceable valve seats, but new models don't. Seat repairs are much harder than washer repairs.

TOOLS & MATERIALS

▌ Utility knife ▌ Screwdriver
▌ Handle puller ▌ Adjustable wrench
▌ Repair kit

1 After turning off the water at the shutoff valve, pry off the index cap on top of the handle using a sharp utility knife. This will expose the screw that holds the handle to the valve stem. In most, cases only one valve will be leaking, but it makes sense to replace the washers on both faucets.

4 When the handle is off, the bonnet nut that holds the valve stem in place is accessible. Loosen this with an adjustable wrench. If water starts running out from around the stem, tighten the shutoff valve under the sink. This should stop the water flow.

5 Lift the valve stem from the faucet, and place it in the bottom of its handle to make it easier to work on. Inspect the washer. If it is cut, broken, or deformed the washer must be replaced. Remove it by backing out the brass screw that holds it in place (inset).

2 Use a screwdriver to remove the screw from the top of the valve stem. In most, cases this screw has a Phillips head. Just be careful not to strip its head. If you do, the only way to remove the screw is to drill a hole into the body of the screw, and remove it with a screw extractor.

3 Once the screw is removed, pry off the handle by sliding a flat blade screwdriver under the bottom and lifting up. If this doesn't work, use an inexpensive handle puller to do the job. Insert the stem of the puller into the screw hole, and tighten the threaded shaft.

6 Remove the old washer, and buy a replacement that is the same size and shape. Hardware stores sell inexpensive assortment packages that have a couple washers of all the common sizes. These assortments are especially helpful for people with a variety of older faucets in their home.

7 Push the brass screw through the hole in the washer; then drive the screw into the hole in the stem. Tighten the screw so the washer is secure but not distorted. Then coat the washer with heat-proof grease; remove the stem from the handle; and reinstall the stem in the faucet.

COMPRESSION FAUCET SUPPLIES

BESIDES BASIC TOOLS like wrenches and screw-drivers, you may need some special materials to make a successful repair of a compression faucet. Common repair materials include washers, heat-proof plumber's grease, stem packing, bonnet nuts, and packing washers.

Packing Washers

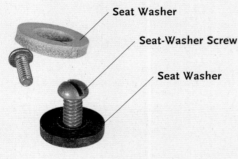

Seat Washer

Seat-Washer Screw

Seat Washer

Stem Packing

Heatproof Plumber's Grease

GREEN SOLUTION
REPLACING VALVE SEATS

WHEN THE VALVE SEAT in a compression faucet appears pitted or feels rough, replace it if possible. New washers installed over damaged seats won't last long, and each subsequent leak will worsen the seat condition. Replacing a seat is not difficult, but finding a replacement can be time consuming.

First, determine whether the faucet has replaceable seats. Most kitchen and bathroom faucets do, but some tub faucets do not. Shutoff valves and hose bibcocks almost never have replaceable seats. To determine whether a seat is replaceable, use a flashlight to illuminate the inside of the faucet and look for a wrenching surface in the throat of the seat. If the inner surface of the inlet is hex shaped or has four deep grooves, the seat is replaceable. If it's smooth, the seat was machined into the faucet body and is not replaceable. (To refurbish a permanent seat, see "Grinding Valve Seats," opposite.)

Use a seat wrench to remove a damaged replaceable valve seat by turning counter-clockwise.

Slide the new seat onto the wrench, and install it in the faucet, turning clockwise.

GRINDING VALVE SEATS

YOU SHOULD REFURBISH faucet or valve seats when the seats are not replaceable.

You'll need a seat-dressing tool. Inexpensive seat dressers include a threaded stem with a T-handle and several grinding blades. The blades are flat disks with cutting or burnishing surfaces stamped into them. A typical kit has one coarse cutting blade, a medium-coarse blade, and a burnisher.

Turn off the water at the shutoff valves, and remove the stem from the faucet. Attach the coarsest blade to the seat dresser, and insert the tool into the faucet port (top photo, below). With the blade resting squarely on the rim of the faucet seat, press down firmly and rotate the tool clockwise (bottom photo). Each twist of the blade will remove a little brass from the seat surface. The goal is to reduce the height of the seat uniformly to a depth just below the deepest void in the rim. This is usually only 1/32 to 1/16 inch, so it doesn't take much. When you reach this point, switch to a medium-coarse blade to remove the roughness. Finally, attach the burnishing blade, and smooth the seat surface.

Start with a coarse blade in the seat dresser, and feed it into the faucet port.

Hold the seat dresser straight, press down on it, and twist it clockwise.

smart tip

REPAIRING DAMAGED FAUCET STEMS

AS YOU MIGHT GUESS, NOT ALL COMPRESSION-FAUCET REPAIRS GO SMOOTHLY. IN SOME CASES, THE RETAINING RIM THAT KEEPS THE SEAT WASHER CENTERED ON THE END OF THE STEM WILL HAVE CRUMBLED. IN OTHERS, THE STEM OR THREADS MAY HAVE BEEN DAMAGED PREVIOUSLY. IF YOU FIND A DAMAGED, BENT, OR BADLY WORN STEM, YOU MAY BE ABLE TO REPLACE IT. PLUMBING SPECIALTY COMPANIES MAKE REPLACEMENT STEMS FOR MANY OLDER FAUCETS. TAKE YOUR DAMAGED FAUCET STEM TO A WELL-STOCKED PLUMBING OUTLET OR TO A PLUMBER WHO SPECIALIZES IN SERVICE WORK. CHANCES ARE, YOU'LL BE ABLE TO BUY A NEW STEM. IF NOT, IT'S TIME FOR A NEW FAUCET.

PLUMBING 101: Stuck Handle

ANY HANDLE CAN STICK, but those made of inexpensive pot metal are the most difficult to release. Steady contact between the pot metal and the brass in the faucet stem causes electrolytic corrosion, fusing the dissimilar metals. When faced with really stuck handles, you'll have two choices. You can cut the handle with a hacksaw, slicing along one side to release its grip. After overhauling the faucet, install universal handles. You'll find them at home centers and retail plumbing outlets.

Another option is to buy an inexpensive handle puller, available at hardware stores. (See page 149.) A handle puller looks and works a bit like an automotive wheel puller. It consists of a threaded stem, a T-handle, two side clamps, and a sliding collar. Insert the stem into the handle's screw hole until it bottoms out in the faucet stem. Press the side clamps under the handle, and slide the collar down to lock the clamps in place. Then twist the stem in a clockwise direction until you feel the handle break free.

REPAIRING WORN STEM PACKING

WHEN YOU ROTATE A FAUCET STEM into the "On" position, water rushes past the seat and into the spout. It also rises against the top of the stem, where it is held in check by an O-ring (opposite page) or, in older faucets and valves, a soft filler material called packing string. The packing is compressed between the stem and packing or bonnet nut, and the packing seal is completed with a cone-shaped leather or graphite washer.

When O-rings are used, the stem will usually not have a separate packing washer; rather, the O-ring fits between the stem and a brass sleeve, or stem nut, and is held in place by the bonnet nut. (See "Fixing O-ring Stem Leaks," opposite.) In some faucets, the stem nut replaces the bonnet nut entirely. On faucets and valves where stem nuts have male threads, a thin nylon or composition washer seals between the nut and the faucet.

Simple Fixes. When dealing with valves and faucets that have separate packing nuts, try tightening the nut about one-half turn. If that does not work, follow the directions shown here.

1 To repair a leak coming from around the packing washer, first shut off the water supply to the faucet. Then remove the handle screw; pull off the handle; and remove the bonnet nut with an adjustable wrench.

2 Once the bonnet nut is removed, pry up the old graphite packing washer using a flat blade screwdriver or a sharp utility knife. Clean any debris away from the valve stem; then slide a new washer into place.

3 Firmly push the replacement washer onto the stem. Then wrap packing string around the stem just above the washer. A couple of layers should do the job. Finish by reinstalling the bonnet nut and faucet handle.

GREEN SOLUTION

FIXING O-RING STEM LEAKS

OVER THE YEARS, most faucet and valve companies have gone to O-ring packing seals. The O-ring is held against the stem by a bonnet or stem nut. This nut may have male threads turned into a recessed body port. A large, flat washer made of nylon or a composition material seals the joint between the nut and faucet body. These assemblies are less susceptible to wear than graphite or leather packing washers, but both the O-ring and flat washer can fail. When they do, water appears under the handle, just as it does with older packing assemblies.

Neither O-rings nor flat washers are hard to replace, but you'll need a close replacement match. Take the stem to a well-stocked plumbing outlet.

To deal with O-ring stem leaks, shut off the water and drain the faucet or valve. Then remove the handle, and loosen the stem nut. Lift the stem from the port, and pull the nut from the stem. (See the photo below, left.) Roll or cut the old ring from the stem, and slide an exact replacement in place. If the old O-ring was seated in a groove in the stem, make sure the new ring seats as well. Lubricate the O-ring lightly with plumber's grease. (See the photo below, right.) Then press the stem nut back over the stem. As always with O-rings, try to find one that is made or at least recommended by the faucet manufacturer.

Although you won't usually need to, it's a good idea

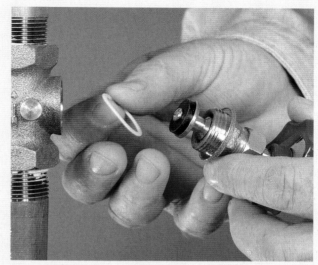

Modern shutoff valves have flat nylon washers to seal the stem against the fitting body.

to replace the flat washer (above) as a preventive measure, especially if one is included in the O-ring kit. If you find that the flat washer is leaking and don't have a replacement, you may be able to fortify the old washer with a thin layer of nonstick plumber's pipe-thread sealing tape. It works best if you stretch the tape around the washer, lapping it in the direction of the nut's rotation. Drop the washer into the recessed rim of the port; tighten the nut; replace the handle; and turn the water back on.

Replacing O-Ring Packing Seals

Expose the O-ring by lifting the bonnet nut from the faucet stem.

Roll a new O-ring onto the stem, and lubricate it using plumber's grease.

FIXING BALL-TYPE FAUCETS

Delta Faucet Company is one of the pioneers in alternative faucet design, with its proprietary ball-and-cam mechanism, and actually offers two name brands. The Delta trademark is sold through professional plumbers, while the Peerless line is sold at the retail level. You'll notice slight cosmetic differences between the two lines, but both use the original ball-and-cam mechanism.

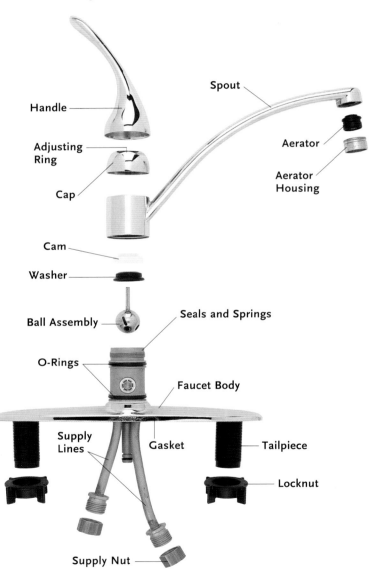

Handle
Spout
Adjusting Ring
Aerator
Cap
Aerator Housing
Cam
Washer
Ball Assembly
Seals and Springs
O-Rings
Faucet Body
Supply Lines
Gasket
Tailpiece
Locknut
Supply Nut

REPAIRING BALL-TYPE FAUCETS

project

What traditional ball-type faucets have going for them is affordable repair parts. You don't usually discard the entire mechanism. Instead, you can replace only those parts that are worn, which in many cases are the springs and rubber seals. Several repair kits are available. Some include only the inlet springs and seals; some include seals, springs, and cam cover; and some include all mechanical components, including a stainless-steel or plastic control ball and a special wrench needed to remove the old ball.

TOOLS & MATERIALS
- Screwdriver
- Groove-joint pliers ▌Allen wrench
- Ball-type faucet repair kit

3 After removing the cam nut, lift out the plastic cam to expose the ball assembly below. Although the cam may look like it has little wear, as long as you have the faucet apart it makes good sense to replace it.

1 To reach the handle screw, shut off the water supply to the faucet and tip back the handle. Clean any soap residue from the screw head; insert an Allen wrench or inexpensive faucet tool into the head; and turn counterclockwise.

2 Once the handle is removed, unthread the cam nut to gain access to the ball assembly. Delta faucets have slotted nuts (inset) that require a simple tool that comes with the faucet. Other faucets have flats on the sides for pliers.

4 Lift the ball from the faucet body, and set it aside. Some kits come with replacement balls and some do not, so choose accordingly. Use an Allen wrench (inset) or thin screwdriver to lift the rubber seals and springs from the inlet and outlet openings. Replace them all.

REASSEMBLY

ASSUMING YOU'LL BE REPLACING EVERY-THING except the ball, press each rubber seal onto its spring. Slide the seal and spring onto an Allen wrench or screwdriver, with the seal facing up. With an index finger holding the assembly in place, insert the spring and seal into the inlet. Install the remaining seal and spring in the same way.

With the new seals installed, press the ball into the body. The ball will have a peglike key on one side that matches a slot in the body, so there's no chance you'll get it wrong. Press the new cam cover over the ball, and align its key with the keyway on the faucet body. Push it down until the key engages, and then thread the cap over it. Tighten the cap until it feels snug, but don't overdo it by overtightening. Replace the handle; turn on the water; and test your work. If the faucet drips or water appears around the handle, remove the handle and tighten the cap a little more.

Faucet Repairs

REPAIRING CARTRIDGE FAUCETS

Repair of a cartridge-type faucet usually consists of merely replacing a long, self-contained cartridge.

Cartridge faucets offer an important benefit: if the water piping was installed backward, you can still have the

hot water on the left. All you do to reverse the hot and cold sides is rotate the stem 180 degrees. This is a handy feature in back-to-back bathrooms, where a shared set of risers always leaves one bath with reversed piping. A reversible faucet saves pipe and aggravation. But you must also make sure you replace the cartridge in the same orientation as the original.

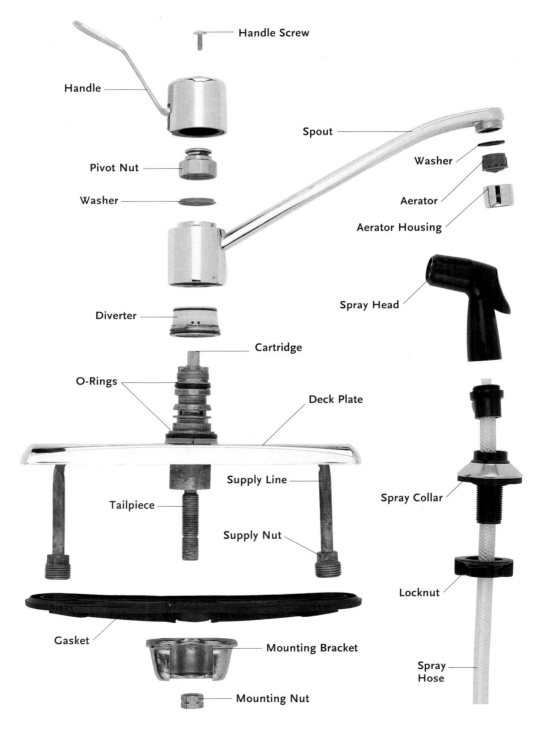

WATCH THE SLEEVE

MANY SINGLE-HANDLE cartridge sink and tub-shower faucets require an extra step before you can remove the retaining clip. With the handle off the faucet, you may see a stainless-steel sleeve installed over the cartridge and column. This sleeve is decorative, but it also keeps the clip from backing out. Pull the sleeve from the column; remove the clip; and replace the cartridge. As is the case with many single-handle cartridge faucets, reversed hot and cold sides can be corrected by rotating the stem 180 degrees.

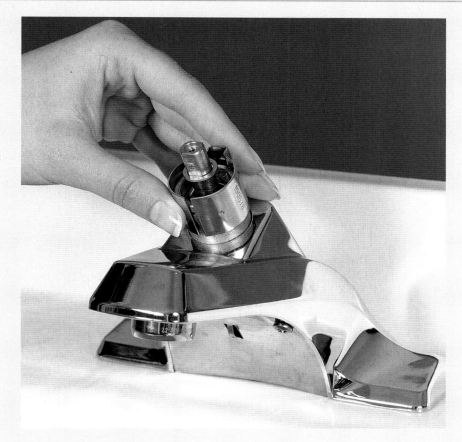

With single-handle cartridge faucets, you might find stem sleeves that you must remove before you can reach the retainer.

PLUMBING 101: Turn Off the Water

THE FIRST STEP in servicing any faucet is to turn off the water, usually at the shutoff valves under the sink, and open the faucet to relieve any remaining water pressure in the supply lines. If you don't find valves under the sink, shut off the water at the meter (municipal water) or pressure tank (private well). And if you have plumbing fixtures on a floor or two above the sink on which you're working, open all upstairs faucets and drain the system.

If you are unlucky enough to discover that the shutoff valve at the sink doesn't work, you have to replace it as well. For this job, turn off the water at the water meter and then open a faucet in the basement or bottom floor to drain the system.

smart tip

REMOVING A FAUCET SPRAYER

IF THE FAUCET SPRAY ATTACHMENT DOESN'T WORK PROPERLY AND YOU'D RATHER BE RID OF IT ENTIRELY, YOU CAN REMOVE IT AND IN THE PROCESS FREE UP THE SINK DECK HOLE FOR A SOAP DISPENSER OR HOT-WATER DISPENSER. YOU'LL HAVE TO CLOSE OFF THE FAUCET NIPPLE WITH A THREADED CAP, HOWEVER. (SOME NIPPLES ARE THREADED INSIDE AS WELL, SO YOU CAN BUY A THREADED PLUG TO CLOSE IT OFF INSTEAD.) TO ELIMINATE THE HOSE SPRAY, REMOVE IT, APPLY PIPE JOINT COMPOUND TO THE FAUCET NIPPLE THREADS, AND TIGHTEN THE PLUG OR CAP ONTO THE NIPPLE.

REPAIRING A KITCHEN CARTRIDGE FAUCET

project

Begin by turning off the water to the faucet at the shutoff valves under the sink. If the faucet handle cover has an index cap, pry it off using a utility knife. If it doesn't, like the one shown here, just lift off the decorative cap to gain access to the handle screw. Removing this screw frees up the handle and provides access to the faucet cartridge. In most cases, you can just lift out the cartridge. But sometimes mineral deposit prevents this. To loosen, pour a little vinegar over the top of the cartridge, and let it seep down around the sides.

TOOLS & MATERIALS

▌ Screwdriver
▌ Needle-nose pliers
▌ Adjustable wrench ▌ New cartridge

1 To get access to the handle screw, lift the decorative cap from the column. If this cap is held in place by the faucet handle screw, then there will be an index cap above this screw. Remove it with a utility knife, being careful not to damage the chrome plating.

4 Once the handle is off the faucet, use an adjustable wrench to remove the pivot nut, turning it in a counterclockwise direction. This provides access to the cartridge. You don't have to remove the faucet spout to make the typical faucet repair that is shown here.

5 After the pivot nut is removed, the top of the cartridge is accessible. It is held in place by a retainer clip that can be removed with needle-nose pliers. Be careful to avoid dropping this clip down the sink drain; the new cartridge you purchase may not come with a replacement clip.

2 Remove the handle screw using a Phillips-head screwdriver. This screw is threaded into the top of the faucet cartridge, so when the faucet is lifted or turned, the cartridge is lifted or turned. Be careful to avoid stripping out the screw head with the screwdriver when removing the screw.

3 Once the screw is removed, lift off the handle and set it aside. The top of the faucet cartridge and the pivot nut that holds it in place are now visible. If there is any soap residue on the cartridge or nut, wash it away and dry the surfaces before trying to remove the nut.

6 With the retaining clip removed, you should be able to grip the top of the cartridge and pull it out. If it's stuck, pull it out using needle-nose pliers. If this doesn't work, mineral deposits are holding the cartridge. To dissolve them, pour some vinegar around the cartridge.

PLUMBING 101: Eye Protection

WHEN WORKING UNDER A SINK or lavatory, wear goggles. These areas can get very dirty. And even if you are not cutting pipe or soldering a joint, old putty or caulk, rust, and debris may fall and lodge in the eyes.

smart tip

CHECK THE CARTRIDGE

AFTER REPLACEMENT OF THE CARTRIDGE, TURN ON THE WATER AND CONFIRM THAT THE HOT AND COLD WATER IS PROPERLY ORIENTED (HOT-LEFT, COLD-RIGHT). IF NOT, YOU CAN REMEDY THE SITUATION EASILY. TURN OFF THE WATER; REMOVE THE HANDLE; AND ROTATE THE CARTRIDGE 180 DEGREES. IMPROPER ORIENTATION MAY RESULT IN BURNS.

Faucet Repairs

REPAIRING A SINGLE-HANDLE CARTRIDGE FAUCET

project

Single-handle cartridge faucets are very common in today's kitchen because they make turning on the water much easier when your hands are full. Every cook can appreciate this convenience. This design may not be quite as common in bathrooms, but it's getting more popular all the time. Typically the bathroom sink faucet will require less maintenance than a kitchen unit because it is used so much less. But when a bath faucet does start to leak, the repair steps are basically the same as those used for a kitchen faucet.

TOOLS & MATERIALS

- Utility knife ▌Allen wrench
- Screwdriver ▌Slip-joint pliers
- New cartridge

1 Most single handle bath faucets feature an Allen screw that holds the handle in place. To find this screw, tip up the handle, and locate the decorative index plug. Pry out this plastic plug with the tip of a utility knife blade. Then remove the Allen screw with an Allen wrench.

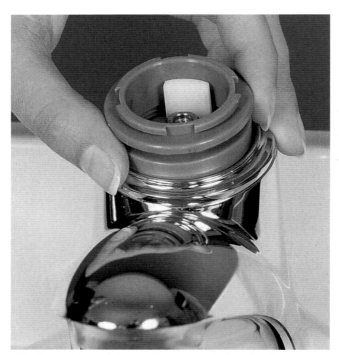

3 Once the metal cover is removed, the plastic ring that holds the cartridge in the faucet will be accessible. Remove the ring by turning it counterclockwise. Hand pressure is usually enough. But if the ring is stuck, use groove-joint pliers to do the job.

4 The white plastic cartridge nut is held in place by a simple U-shaped clip. Remove this clip using standard needle-nose pliers. Then remove the nut, and set it aside. In most cases you'll need to reuse these parts after you install a replacement cartridge.

2 The faucet cartridge is held in place with a plastic ring that's threaded into the faucet body. This ring is held captive by a metal cover that's screwed to the top of the cartridge. Remove this screw with a Phillips head screwdriver, and lift off the metal cover. Be careful to avoid stripping out this screw head. If the Phillips recess is spoiled, you'll have to remove this screw with a screw extractor. Also, don't drop the screw down the drain opening.

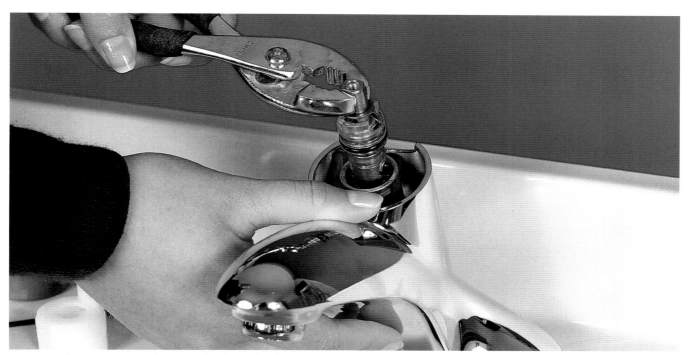

5 Use pliers to remove the cartridge from the faucet. In most cases, it will be easy to lift out. But in houses where the water is hard, the cartridge can be covered with mineral deposits, which will keep it from coming out. Pouring vinegar around the cartridge and letting it stand for a few minutes should dissolve these minerals. Once the cartridge is free, take it to a hardware store and buy an identical replacement part. Take this home and install it.

Repairing Two-Handle Cartridge Faucets

Many manufacturers offer two-handle cartridge faucets, most often as low-cost alternatives to their single-control models. They are modestly priced and perfect for the mechanically timid because anyone can overhaul them.

To fix a leaking faucet, shut off the water and pry the index cap from the handle. Remove the handle, and loosen the cartridge nut using an adjustable wrench, smooth-jaw pliers, or groove-joint pliers with the jaws wrapped in duct tape. Lift the original cartridge from the faucet body; throw it away; and stick a new one in its place. Restore the nut and handle. Be sure to work on only one side at a time to keep from accidentally reversing the cartridges. When you're finished, turn on the water to test your work.

Most dual-handle cartridge faucets have disposable cartridges similar to those just described. But some have spring-loaded rubber seals in the inlet ports like those you find in ball-type single-handle faucets. If upon lifting the cartridge you notice these accessible seals, keep the cartridge and replace only the seals and springs.

If you find that your faucet cartridges are hard to remove, hard-water calcification over the years may have stuck them in place. Some faucet manufacturers offer cartridge extraction tools, but it also helps to pour warm vinegar into the faucet port and around the cartridge. Give the vinegar a few minutes to work, and try pulling the cartridge again.

FAUCET STYLES

THE TYPE OF CONTROL MECHANISM in a faucet does not place limits on its appearance. In fact, you can't tell how a faucet operates by looking at it. From below clockwise, a two-handle cartridge faucet; a ball-type faucet, which you will only find in single-handle models; a single-handle cartridge; and a two-handle ceramic disk faucet.

REPAIRING A TWO-HANDLE CARTRIDGE FAUCET

project

Two-handle faucets have been standard equipment in both kitchens and baths for a very long time. They are examples of accepted technology, and they are relatively easy to manufacture, both in this country and offshore. Because people are comfortable with double-handle models, it's not hard to understand why some of these units are built with cartridge valves instead of the traditional compression valves. These newer valves are easy to replace and last longer between repairs, both without changing the traditional look of the faucet.

TOOLS & MATERIALS

▌ Utility knife
▌ Allen wrench ▌ Screwdriver
▌ Slip-joint pliers ▌ New cartridge

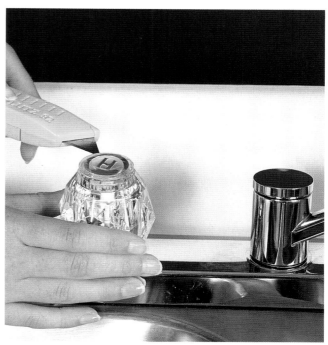

1 To replace a cartridge on a two-handled faucet, start by turning off the water to both the hot and cold faucets at the shutoff valves below the sink. Then remove the index caps on top of the both handles using a utility knife.

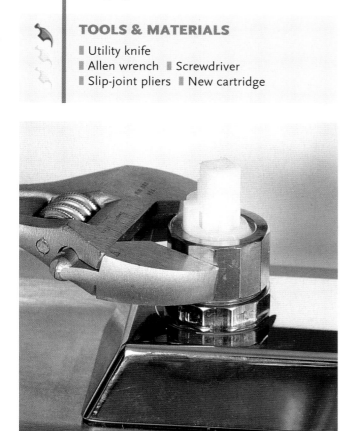

2 Once the index caps are removed, back out the screws that hold the faucet handles in place, and remove the handles. Then remove the chrome retaining nuts that hold the cartridges, using an adjustable wrench.

3 Remove both cartridges from the faucet, and take them to a hardware store or home center to buy exact replacement units. Coat the O-rings on the new cartridges with heatproof grease, and reinstall them in the faucet.

REPAIRING CERAMIC-DISK FAUCETS

Ceramic-disk faucets are particularly vulnerable to sediment accumulations. For this reason, don't assume that a dripping faucet needs a complete overhaul. When a ceramic-disk faucet develops a steady drip, remove the aerator, and move the handle through all positions several times. If sediment was the culprit, this should clear it. In general, a ceramic-disk faucet is not a good choice if you experience sediment problems with your water, especially if they are so severe that you require a filter.

smart tip

DON'T SHATTER THE CERAMIC

CERAMIC DISKS ARE EXTREMELY DURABLE, BUT THEY HAVE A WEAKNESS. WHEN YOU DRAIN A PIPING SYSTEM FOR REPAIRS AND THEN TURN THE WATER BACK ON, THE AIR IN THE SYSTEM ESCAPES THROUGH THE FAUCETS IN BURSTS AND SURGES. THESE PRESSURE SHOCKS CAN SHATTER A CERAMIC DISK, PREVENTING THE FAUCET FROM SHUTTING OFF COMPLETELY. TO AVOID RUINING YOUR CERAMIC-DISK FAUCET, TURN THE WATER BACK ON SLOWLY AFTER A PLUMBING REPAIR. ALLOW THE AIR TO BE PUSHED FROM THE SYSTEM GRADUALLY BEFORE TURNING THE SHUTOFF VALVE TO ITS FULL-OPEN POSITION.

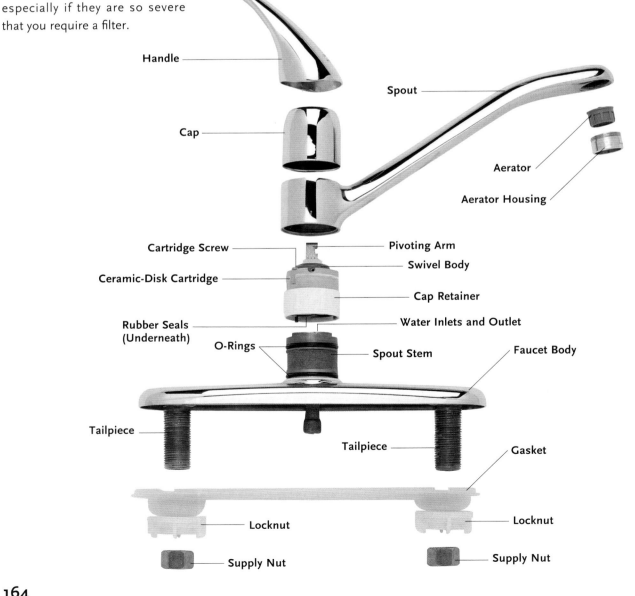

Handle
Spout
Cap
Aerator
Aerator Housing
Cartridge Screw
Pivoting Arm
Swivel Body
Ceramic-Disk Cartridge
Cap Retainer
Rubber Seals (Underneath)
Water Inlets and Outlet
O-Rings
Spout Stem
Faucet Body
Tailpiece
Tailpiece
Gasket
Locknut
Locknut
Supply Nut
Supply Nut

GREEN SOLUTION

FIXING A LEAK IN A CERAMIC-DISK FAUCET

project

Ceramic-disk cartridges are very durable and can last a long time unless they are attacked by sediment. If you have eliminated sediment as the cause of a leaky ceramic faucet, then your only repair option is to take apart the unit and replace the cartridge. These cartridges cost more, and are less available, than standard faucet cartridges, but they aren't any harder to install. If you have a replacement cartridge on hand the job should take less than an hour. To make sure you get the exact replacement part, take the old cartridge with you when you shop.

TOOLS & MATERIALS

▌ Allen wrench ▌ Flat-blade screwdriver
▌ Groove-joint pliers ▌ New cartridge
▌ Tweezers

1 To repair a ceramic disk faucet, first remove the handle. To do this, lift up the handle; find the screw; and remove it using an Allen wrench. Lift off the handle, and set it aside. Don't lose the Allen screw; you'll need it later.

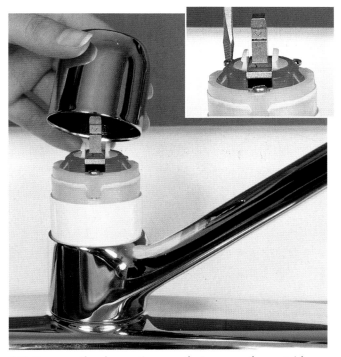

2 Remove the decorative cap that covers the cartridge, and set it aside. Then remove the screws that hold the cartridge in place using a flat-blade screwdriver (inset). Save the screws in case the replacement unit doesn't have any.

3 Lift off the cartridge, and inspect the bottom of it for sediment. If there's sediment at the inlet ports, you can try to clean it away, reassemble the faucet, and have a successful repair. But more often, the cartridge must be replaced.

Faucet Repairs

REPAIRING A TWO-HANDLE CERAMIC-DISK FAUCET

project

Dual-control ceramic-disk faucets are more substantial, but the repair sequence is similar to that for single-handle models. Start by turning off the water at the shutoff valves under the sink. Then repair one side of the faucet at a time. This will keep you from mixing up the hot- and cold-water parts, which are different but fit in the same-size opening. If you have the replacement ceramic disks on hand the job should take less than an hour. Just make sure the parts you buy are exact duplicates of the ones in the faucet.

TOOLS & MATERIALS

▮ Utility knife ▮ Screwdriver
▮ Adjustable wrench ▮ Slip-joint pliers
▮ New cartridges

1 Working on one side of the faucet at a time, start by removing the handle screw followed by the handle. Then use a Phillips screwdriver to unthread the cartridge cap screw. Save the cap and screw for reassembly.

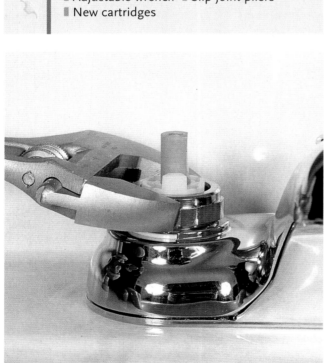

2 Once the cartridge cap is removed, use an adjustable wrench to loosen the retaining nut. Turn this nut in a counterclockwise direction, and when it's free, set it aside for reassembly. Avoid scratching the chrome on visible parts.

3 Remove the old cartridge with pliers. If it's stuck, pour some vinegar around the outside of the cartridge to dissolve any mineral deposits. Reassemble this side of the faucet; then repair the other side using the same approach.

Soak a mineral-hardened spray nozzle in warm vinegar to dissolve calcified minerals and enable it to work properly.

Faucet Repairs

FAUCET-RELATED REPAIRS

Besides the annoying drip-drip-drip of a leaky faucet, you may be faced with other faucet problems: leaking spray attachments, erratic or uneven water flow from an aerator, and leaks at the base of the faucet spout.

Dealing with Leaky Spray Attachments

The hose spray attachment is the weakest part of any faucet. It's not unusual for homeowners to replace one three or four times before the faucet wears out. In households with hard water, mineral deposits form in and around the spray nozzle. With enough calcification, the stop mechanism in the nozzle won't shut off completely, and the spray pattern becomes irregular.

With many troublesome attachments, the entire hose and nozzle assembly needs to be replaced, but when hard water is clearly the culprit, you might try cleaning the nozzle first. Heat about two cups of vinegar, and place the entire nozzle in a container with the vinegar. (See the photo above.) To hold the nozzle open, slip a rubber band over the release lever. After an hour or so, operate the nozzle under pressure, squeezing and releasing it repeatedly. You might also try rapping the nozzle on the counter. If you've cleared the stop mechanism but still have a partially clogged nozzle, unscrew the aerator from the nozzle and poke the mineral particles from the screen with a straightened paper clip.

HOW TO REPLACE A SPRAY ATTACHMENT

project

If cleaning the spray hose doesn't improve its performance, you need to replace it. These units are available at hardware stores and home centers and usually come in black plastic. But other colors are available, like white and chrome. It's best to buy a replacement unit made by the same company that produced the faucet. But universal kits will generally work. Start by shutting off the water at the valves under the sink. Then drain as much liquid from the hose as possible. Put a bowl under the sink just below the hose; then cut the hose in two so any water left inside will drain into the bowl. Pull one half of the old hose through the sink deck opening, and discard it. Then, disconnect the other piece of hose from the faucet nipple under the sink, and throw it away.

TOOLS & MATERIALS
▪ Screwdriver
▪ Pliers ▪ Adjustable wrench
▪ Spray replacement kit

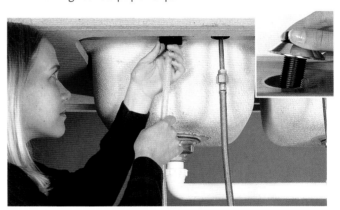

1 After removing the old spray attachment, slide the rubber gasket onto the new one's fitting shank (inset), and feed the shank through the sink deck hole. Slide the hose through the deck fitting from above, and install the jamb nut below. Tighten the nut until it's snug.

2 Finish by connecting the spray attachment's hose to the faucet's diverter nipple. Use a light coating of pipe joint compound on the nipple.

REPAIRING AN OLDER CERAMIC-DISK FAUCET

IF SEDIMENT IN THE FAUCET is not the culprit, you can often stop a leak in an older ceramic-disk unit by increasing tension on the disk. Pry off the index cap, and remove the screw and handle. You'll see a chrome cap concealing the ceramic water-control mechanism. Remove the two small screws that hold the cap in place, and lift the cap. This will reveal a large plastic adjustment nut with series of holes around its rim. The holes correspond to an identical set of holes in the faucet body. Using large pliers, rotate the nut clockwise until the next set of holes aligns. Replace the cap and handle, and test your work. If the faucet still leaks, advance the nut one more hole. Con-

tinue this routine until the leak stops or the nut feels too tight to move. If these quick-fix methods don't help, then it's time to replace the cartridge.

Shut off the water; remove any decorative cap and the handle from the faucet; and back the adjustment nut completely out of the faucet. Remove the screws that hold the old disk in place, and install an exact replacement. Thread the adjustment nut back into the faucet until its alignment is roughly the same as it was when you first opened the faucet. Turn the water back on, and test your work. If the faucet drips, continue tightening the adjustment nut, one hole at a time, until the leak stops.

CLEARING SEDIMENT AND MINERAL BUILDUP

Correcting Aerator Problems

If a sudden drop in pressure occurs at only one faucet or the water doesn't seem to flow properly, the aerator may be clogged.

As the name implies, an aerator mixes air into the flow from the spout and keeps the water from spiraling out at an angle or with too much force. Aerators also contain sediment screens. The screens hold a little water after you shut the faucet off, so with hard water they're prone to calcification, but the most common problem is sediment, usually from a line repair or a sandy well. When sediment is caught in an aerator, it shows as a pressure drop. If only one faucet in your home exhibits a pressure drop, suspect a clogged aerator.

Clearing Sediment and Mineral Buildup. Sediment can be easily cleaned from an aerator. Just grip the aerator with your fingers or with padded pliers, and unscrew it from the faucet. (See the photo below, left.) Carefully remove the various screens and disks, and lay them out in order of removal. Use a paper clip to poke through each hole in the plastic disk, and backflush the metal screens. Reassemble the components, and install the aerator. If this doesn't correct the problem, throw the old one out and buy a new one.

If a white crusty buildup from calcification is the problem, remove the aerator and soak it in warm vinegar for an hour or so. If the aerator holes and screens remain partially plugged, poke at them using a paper clip. (See the photo below, right.)

Mineral-encrusted aerators are easy to unscrew from the faucet spout for cleaning or replacement.

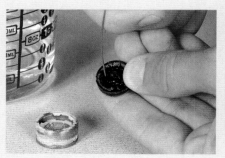

Soak the scaled-over aerator parts in vinegar, and clear the screens with a straightened paper clip.

REPAIRING A TWO-HANDLE FAUCET SPOUT

project

Leaks frequently appear around the base of a faucet spout, and should be fixed right away. If they are allowed to leak for a long time, the water can work its way under the base of the faucet and drop down into the cabinet below. It can ruin what's stored there and damage the cabinet in the process. This water can also seep into the countertop and cause enough damage that you'll have to replace a whole section of countertop. Fortunately, repairing a leaking spout it and easy job because most spouts are sealed with replaceable O-rings.

TOOLS & MATERIALS

▌ Strap wrench
▌ Utility knife
▌ Spout O-ring kit

1 To protect the chrome finish, use a strap wrench, or pliers with the jaws covered in duct tape, to remove the spout cap. It's usually threaded into the top of the spout and should be turned in a counterclockwise direction.

2 When you remove the spout cap, inspect the inside of the spout assembly. If it is covered with mineral deposits, flush the area with vinegar to dissolve them. Wait for a few minutes, then pull off the spout.

3 Once you have removed the spout and the old O-rings are accessible, cut them off with a sharp utility knife and install replacement rings that are the same size. Lubricate the new O-rings with heatproof grease; then reinstall the spout.

Faucet Repairs

REPAIRING SINGLE-HANDLE FAUCET SPOUTS

project

Repairing a single-handle faucet spout isn't much more difficult than fixing the spout on a double-handed faucet. (See page 169.) The obvious difference is that the spout and the cartridge valve occupy the same space. This means that once the handle and spout are removed the cartridge will be completely exposed. At this point, it's just as easy to replace the cartridge as it is to replace the O-rings. But if this valve is working properly, replacing the rings will save you some money, which is always welcome news.

TOOLS & MATERIALS

- Screwdriver
- Adjustable wrench
- Utility knife ▍ O-ring kit

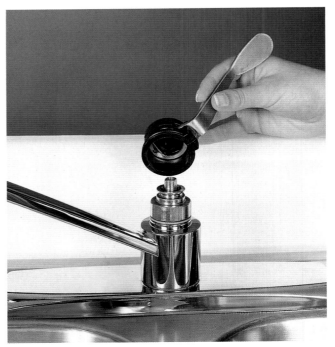

1 Start by turning off the water at the shutoff valves under the sink. Then remove the index cap on the top of the faucet handle with a utility knife. Unthread the screw that holds the handle to the cartridge, and lift off the handle.

2 The spout is held in place by a retaining nut located under the handle. Remove this nut with an adjustable wrench turned in a counterclockwise direction. Once this nut is gone, pull off the spout.

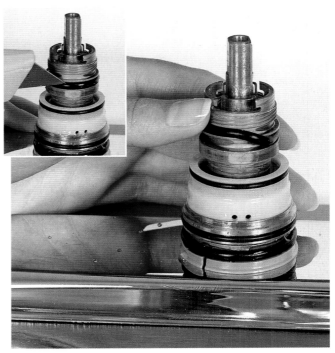

3 Remove the O-rings by cutting them off with a utility knife. Take the cut rings to a hardware or plumbing supply store and get replacements. Put these on the cartridge; cover them with heatproof grease; and reassemble the faucet.

REACHING RECESSED FAUCETS

WHILE ALL THAT STANDS between you and the inner workings of a typical sink faucet is the handle, bathtub and shower faucets have a deep-set mounting profile, making repairs a little more difficult. Full-skirt escutcheons or decorative chrome trim plates are indicators of deep-set faucets. You have to remove these escutcheons to reach the bonnet nuts, which hold the faucet stems in place. Escutcheons are easy to remove, once you determine which fastening method is being used. One is a nut located just behind the handle as shown here. The other has internal threads that are turned directly into the faucet stem. If you don't see a nut behind the handle, you have this type. To remove it, just grip it with your hand and turn counterclockwise.

TOOLS & MATERIALS

▌ Needle-nose pliers ▌ Adjustable wrench
▌ Deep-socket faucet wrench
▌ Screwdriver ▌ Faucet washers

1 Start by removing the handle index cover (inset) and screw; then remove the escutcheon nut from the stem or unscrew the escutcheon.

2 Press the deep socket wrench onto the recessed bonnet nut, and remove the nut and stem from the faucet body.

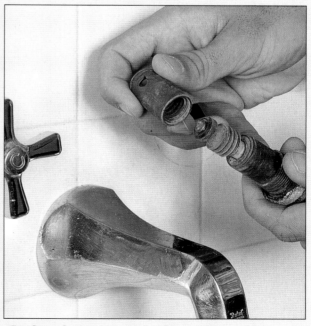

3 If you find a stem with a floating seat (an older design), unscrew the seat from the stem, and replace the washer.

FIXING SINGLE-HANDLE TUB AND SHOWER FAUCETS

project

Cartridge-type tub and shower faucets with single-handle controls have mechanisms similar to those used in their kitchen and bath counterparts. Their handles are attached in similar ways and their cartridges are usually held in place with the same U-shaped retaining clips as shown here. In fact, the biggest difference between the faucets is the size of the typical escutcheon plates behind the handles. Some models also have a separate sleeve between the handle and the cartridge as shown in photo 3 below.

TOOLS & MATERIALS

- Utility knife ▪ Screwdriver
- Needle-nose pliers
- Replacement cartridge

1 Start the repair by turning off the water to the faucet. Then, to gain access to the screw that holds the handle in place, remove the handle's decorative index cap. Pry it off with a utility knife or thin blade screwdriver.

3 Once you have removed the handle, you will usually find a decorative stainless-steel inner sleeve that covers the cartridge. It also shields the clip that holds the cartridge in place so it can't be removed. Pull out this sleeve.

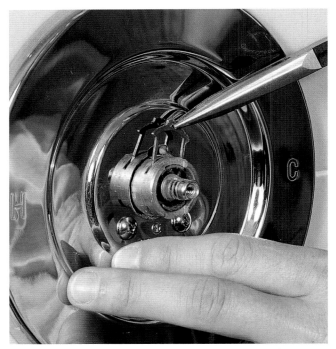

4 After removing the sleeve, use needle-nose pliers to pull the U-shaped retaining clip from the end of the cartridge. Be careful not to bend or lose this clip because they can be difficult to find at typical home center stores if you need a replacement.

2 Use a Phillips-head screwdriver to remove the handle, turning the screw in a counterclockwise direction. Remove the handle (and the plastic bushing if the faucet has one), and clean away and soap residue from inside.

5 Pull out the faucet cartridge; take it to a hardware or home center store; and buy an exact replacement. Matching the cartridge brand to the faucet brand is the best choice. Install the new cartridge unit, and reassemble the faucet.

Working with Scald-Control Faucets

Scald-control faucets have long been installed in hospitals and nursing homes, and in the past few years, many codes have been updated to require scald control for residential bath and shower faucets. It takes only a couple of seconds of exposure to 140° F water to produce a third-degree burn—and only one second at 150° F. Many homeowners have their water heaters set that high. (A 125° F setting is safer, and your heater will give you longer service at that setting.) Every manufacturer now makes affordable scald-control faucets for residential use. Most are single-control faucets.

Handle-Rotation Stop. Scald control is delivered in two ways. First is a temperature-limit adjustment, in the form of a handle-rotation stop. With a handle stop, you remove the handle and dial in a comfortable water temperature, with the water running, and then lock the setting and replace the handle. Thereafter, when you turn the handle to "Hot," it will rotate only to the stop position. You can reduce the temperature with cold water, but you can't exceed the hot limit. Because ground temperatures change with the seasons, affecting water temperatures, you may need to adjust these settings twice a year.

Pressure-Balance Spool. The second mechanism, a pressure-balance spool, is designed to accommodate a sudden drop in pressure on one side of the piping system. (See page 174.) Pressure drops are common. The most familiar scenario: you're taking a shower when someone in an adjacent bathroom flushes the toilet. The toilet diverts half the line pressure from the cold side of the faucet. The result is a sudden blast of hot water.

Balance spools come in several forms, but the most common is a perforated cylinder with a similarly perforated internal slide. The slightest drop in line pressure on one side of the faucet moves the slide over a bit, realigning the perforations and reducing intake from the high-pressure side. When pressure is restored to the weak side, the slide returns to its original position. These simple devices are so effective that water won't even flow through a faucet when one side is turned off. You'll find pressure-balance spools in two locations. Some are built into the faucet body—with front access—and some are built into the valve cartridge.

As an added benefit, many of these upgraded faucets are built with integral stops, one on each side of the control. Instead of shutting down the entire water system for repairs, you can just close the stops with a screwdriver.

CLEANING A PRESSURE-BALANCE SPOOL

PRESSURE-BALANCE SPOOLS are vulnerable to sediment problems. When a spool clogs with sediment, the faucet will deliver only a trickle of water, or only hot or cold water.

To deal with a gradual accumulation of sediment in a spool that's installed in the faucet body, remove the faucet handle and escutcheon. Close the integral stops by turning the exposed screwheads clockwise with a flat-blade screwdriver. (See the photo below, left.) You'll find the stops just behind the faucet's trim plate on each side of the faucet. Use a large flat-blade screwdriver to unscrew the spool from the faucet body, and remove it. (See the photo below, right.) Tap the spool several times on a hard surface, and then rinse it clean. The internal slide should easily slip back and forth when you tip the cylinder. If it doesn't, you'll need a new spool. Reinstall the spool; open the stops; and reassemble the faucet. Make sure both stops are fully open when you put the faucet back into service. It's easy to mistake a nearly closed stop for a clogged balance spool.

Of course, if the spool is built into the faucet cartridge, you may need a new cartridge. (See "Pressure-Balanced Cartridges," below.) Before replacing it, however, try to clear the sediment by rotating the handle through all positions with the water on.

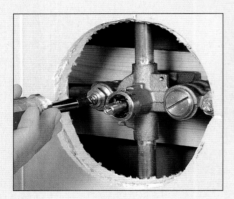

Some scald-control faucets have integral water stops. Use a screwdriver to shut off the water.

To clear a balancing spool of sediment, pull it out and tap it on the counter. Flush the spool with water, and replace it.

PRESSURE-BALANCE CARTRIDGES

SCALD-CONTROL MEASURES have changed the way you work on familiar cartridge faucets. Many cartridges and faucets look like standard units, but they now contain a balancing spool, and they are installed differently. You can't remove them from the faucet without a special tool, so don't try pulling them out, the way you would a standard cartridge. Either buy a metal twist tool, like the one shown here, or use the little tool that comes with each replacement cartridge. If you need to work on one of these faucets, proceed carefully until you learn how things work. Manufacturers normally provide toll-free numbers and Web sites with their instructions, so there's plenty of help available.

Cartridges with built-in balancing spools often require a special removal tool.

GREEN SOLUTION

FIXING YARD HYDRANTS

YARD HYDRANTS can be rebuilt like other compression faucets. But the seats are inaccessible, so be sure to make repairs before seat damage occurs.

The repair procedure for in-ground yard hydrants resembles that for freeze-proof sillcocks. (See pages 176–177.) The lever-type handle at the top of the hydrant connects to an extended stem that's 3 to 6 feet long. The bottom end of this stem is fitted with a large rubber stopper. To replace this stopper, turn off the water at the nearest shutoff valve, and lift the lever to the "On" position. Then, using two pipe wrenches, backhold the riser pipe with one wrench and grip the wrenching surface at the base of the head with the other. Rotate the head of the hydrant counterclockwise. When the head clears its threads, lift the stem completely out. You may need the assistance of a helper if the stem is longer than 3 or 4 feet. Remove the old stopper, and install an exact replacement. Return the stem to its riser pipe. As you tighten the head in place, be sure to backhold the riser to avoid breaking the hydrant away from its underground connection. Finally, lower the handle; turn the water on; and test your work.

TOOLS & MATERIALS

▌ Utility knife
▌ Screwdriver
▌ Needle-nose pliers

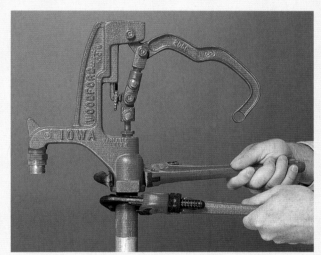

1 To gain access to the stem, shut off the water at the nearest shutoff valve; backhold the riser pipe; and rotate the head counterclockwise. To tighten the hydrant, hold the wrenches as shown above. To loosen, reverse the wrench directions.

2 Lift out the stem; unscrew the old stopper using pliers; and install a new one. Coat the stopper with plumber's grease, and reinstall the head.

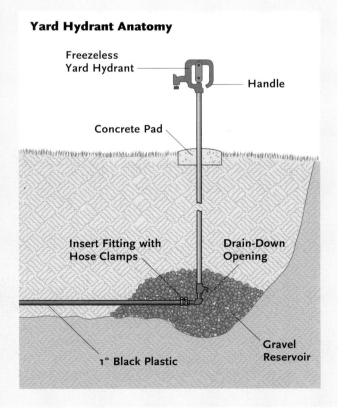

Yard Hydrant Anatomy

Freezeless Yard Hydrant

Handle

Concrete Pad

Insert Fitting with Hose Clamps

Drain-Down Opening

1" Black Plastic

Gravel Reservoir

Faucet Repairs

FIXING A FREEZE-PROOF SILLCOCK

Freeze-proof sillcocks prevent pipe damage in cold climates. (See page 33.) By turning off the water inside the wall where the temperature is above freezing, the pipe to the outside never has water in it and can't break when the temperature falls below freezing. The repair process is similar to that for other faucets, except that the packing nut (photo 2) sometimes has left-hand threads. This is usually marked on the faucet body. But if it isn't and you can't loosen the nut by turning it counterclockwise, try turning it clockwise.

TOOLS & MATERIALS

▌ Screwdriver ▌ Adjustable wrench
▌ Needle-nose pliers ▌ Sillcock repair kit
▌ Plumbers grease ▌ Packing string

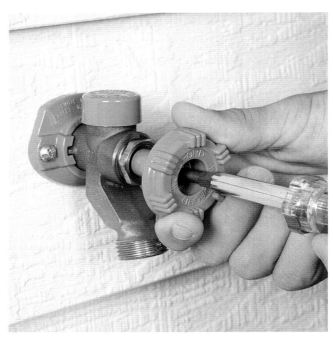

1 Start the job by turning off the water at the shutoff valve inside the house. Then unthread the screw that holds the handle to the valve stem, turning it counterclockwise. If the screw head becomes damaged, you can sometimes remove it with needle-nose locking pliers.

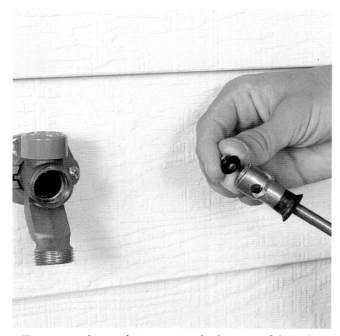

4 Remove the washer screw at the bottom of the valve stem, and pry out the old washer. Clean off the depression that holds the washer; then install a new one, tightening it with a new brass screw. Cover the washer and stem threads with heat proof grease, and reinstall the valve stem.

5 If the faucet was also leaking near the handle, spread two rounds of packing string over the packing that's still on the stem, and cover it with the packing nut. When you tighten this nut, it will force the string against the stem and prevent any water from leaking out.

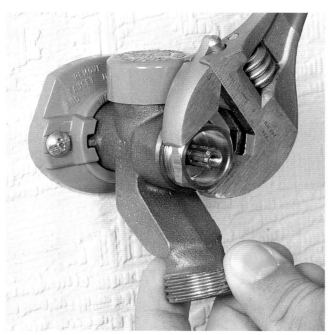

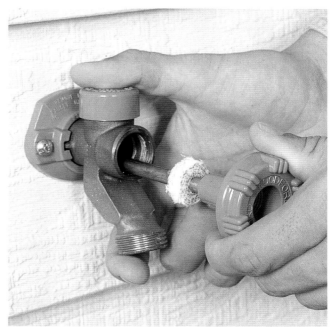

2 Before you remove the handle, use it to open the faucet valve completely. Then take it off; set it aside; and unscrew the packing nut that holds the valve stem in place. On most units, you turn this nut counterclockwise. But on some the threads are reversed, so turn the nut clockwise.

3 Temporarily replace the handle, and turn it to make sure the faucet is completely open. Then pull out on the handle to remove the entire valve stem. If the valve sticks, twist it from side-to-side a few times and then pull it out harder. Repeat as necessary until the stem is free.

6 To fix a leaking vacuum breaker, pry off the top cap (inset). If you can't remove the cap using just your finger, try to work it off with a screwdriver. Underneath the cap you will see a small rubber seal that needs to be replaced. Lift out the small plastic screen using needle-nose pliers.

7 Once the screen is removed, lift out the old vacuum breaker seal, and take it to a hardware or plumbing supply store to get an exact replacement. Install the new seal; then replace the plastic screen and the cap. Attach the handle to the valve stem, and turn on the water. If there are no leaks when you turn off the water, you're done.

6

clearing drainpipes

THERE ARE TWO BASIC KINDS of drainpipe clogs. The first is a localized obstruction, usually in a fixture trap, in a mechanical waste fixture or appliance, or at an abrupt change of direction in the piping. This kind of clog is the one so frequently depicted in chemical drain-cleaning commercials. It usually comprises a buildup of soap, hair, and cosmetics, or is caused by foreign objects such as toothpicks, cotton swabs, washcloths, or small toys. (See the illustration in "Types of Clogs," below.) This kind of clog is fairly easy to clear. You can often force it with a plunger or retrieve the object using an inexpensive drain auger or even a piece of wire.

The second type of drainpipe clog is an extended, sometimes pipe-length accumulation, which takes years to form.

Only a thorough reaming with a blade-head drain auger will clear this type of clog. Enzymatic drain cleaners can help, but nothing is as quick or thorough as a drain auger.

Horizontal and Vertical Clogs. If you were to cut through a pipe with an extended accumulation, you'd see that the clog formed in one of two ways, depending on whether the pipe was installed horizontally or vertically. (See the illustrations below.) In a horizontal line, the clog forms from the bottom up, until the water flows through only the top ¼ to ½ inch of pipe before it becomes completely stopped up. A clog in a vertical line accumulates from the outside in, leaving an ever-smaller drain hole in the center. Either way, a slow drain can abruptly stop flowing, with an object the size of a pea blocking the flow.

TYPES OF CLOGS

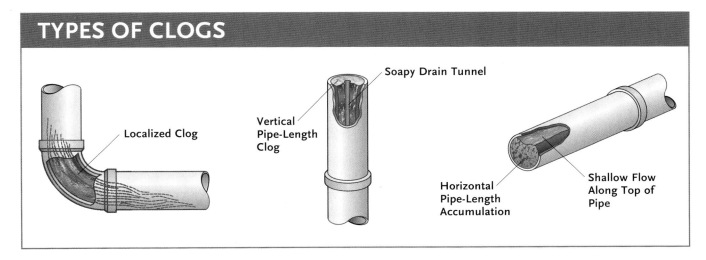

Localized Clog

Vertical Pipe-Length Clog

Soapy Drain Tunnel

Horizontal Pipe-Length Accumulation

Shallow Flow Along Top of Pipe

Clearing Drainpipes

smart tip

HOT-WATER DRAIN MAINTENANCE

EXTENDED DRAINPIPE ACCUMULATIONS USUALLY OCCUR IN KITCHEN AND BATH LINES, MOST OFTEN IN THE HOMES OF RESOURCE-CONSERVATIVE, CONSCIENTIOUS HOMEOWNERS. THE OPERATION OF AN AUTOMATIC DISHWASHER MIGHT EXPLAIN WHY: WHEN A DISHWASHER IS INSTALLED FOR THE FIRST TIME IN AN OLDER HOME, IT OFTEN QUICKLY CAUSES THE KITCHEN DRAIN TO BACK UP BECAUSE THE HOT WATER FROM ITS WASH CYCLE BREAKS UP EXISTING ACCUMULATIONS, WHICH THEN GATHER AT A CHOKE POINT AND BLOCK THE FLOW OF WATER BEHIND IT. CONVERSELY, A NEW DRAIN LINE THAT CARRIES THE HIGH-VOLUME HOT-WATER PURGE OF A DISHWASHER ALMOST NEVER CLOGS, NO MATTER HOW MANY YEARS IT'S IN SERVICE. THE LESSON HERE IS THAT REGULAR DOSES OF VERY HOT WATER KEEP ACCUMULATIONS OF GREASE AND COSMETICS FROM FORMING. SO IF YOU TAKE SHALLOW BATHS OR SHORT SHOWERS AND USE THE BARE MINIMUM OF HOT WATER IN THE KITCHEN, YOU INCREASE THE CHANCE THAT GREASE AND COSMETICS WILL CONGEAL AGAINST THE COOL WALLS OF DRAINPIPES. TO AVOID THESE PROBLEMS, RUN THE HOTTEST TAP WATER INTO EACH FIXTURE, WITH THE STOPPER CLOSED, EVERY TWO WEEKS OR SO. WHEN THE BASIN OR TUB IS NEARLY FULL, OPEN THE STOPPER, AND FOLLOW WITH 30 SECONDS OF HOT TAP WATER.

DRAIN CLEANERS

When you're faced with a stopped-up drain, you'll need to clear it, either by breaking it down and dissolving it chemically or by mechanically removing the obstruction.

Chemical Drain Cleaners

There are three general varieties of chemically based drain cleaners: alkaline, enzymatic, and acidic.

Alkaline Cleaners. Supermarket products, which you pour directly into a stopped sink or tub, are alkali-based products. These have copper sulfide, sodium hydroxide, or sodium hypochloride as their active ingredients. They work on the simplest of hair and grease clogs.

Enzymatic Cleaners. A less hazardous and more environmentally friendly version of drain cleaners uses enzymes to break down clogs. Enzymatic treatments often work, but they take a few days. They are especially good at preventive maintenance. Using an enzymatic drain treatment twice a year will help keep drains clear and will benefit your septic system if your house is not connected to a municipal sewer system.

Acidic Cleaners. Acids, the final group of chemical cleaners, are more troublesome. They can be effective, but they're enormously dangerous in the wrong hands. Sulfuric and hydrochloric acid will dissolve just about any blockage material commonly found in plumbing systems. But these are products of last resort, so read labels and proceed extremely carefully. Avoid using them if at all possible.

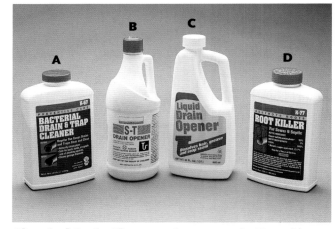

Chemical Drain Cleaners: A—enzymatic, **B**—acid-based, **C**—lye-based, **D**—copper root repellent

smart tip

REPELLING TREE ROOTS

COPPER SULFATE-BASED DRAIN CLEANERS ARE MOST EFFECTIVE AT KEEPING TREE ROOTS OUT OF SEWER LINES. COPPER REPELS ROOTS, SO IF YOU'RE LUCKY ENOUGH TO HAVE LARGE TREES IN YOUR YARD, A ONCE- OR TWICE-A-YEAR TREATMENT IS A GOOD IDEA. IT'S INTERESTING TO NOTE THAT OLD-TIME PLUMBERS SOMETIMES WRAPPED COPPER WIRE AROUND SEWER PIPE JOINTS.

GREEN SOLUTION

MECHANICAL DRAIN CLEANERS

MECHANICALLY CLEARING a clogged drain or cleaning a slow-running one is not complex work and doesn't require an armful of tools. Using a plunger, a screwdriver, and a piece of wire, you'll be able to handle the most common clogs. For those rare occasions where a plunger doesn't seem to work, you'll need additional tools, all of which can be rented. You need not own them unless you live where rental outlets are few and far between.

Plunger. Get the kind of plunger that has a foldout extension cup. When folded in, it works on sinks, showers, and tubs. When folded out, it fits a toilet outlet. Standard plungers have trouble making a seal in toilets, and without a good seal, plunging is not effective.

Drain Auger. A drain auger is a long coiled-wire cable with a springlike head on its working end. You'll see two kinds when you go shopping: one is just a coil of cable with a piece of offset tubing as a crank; the other has its cable spooled in a metal or plastic housing. Buy the one with the housing because its wider cranking arc provides better torque for stubborn clogs. The housing also minimizes the spraying of indelible black drain slime around the room as you retrieve the cable.

Closet Auger. A closet auger is also a drain snake, but its design limits its use to toilets (also known as water closets.) It consists of a 3-foot metal tube that bends at a right angle at the bottom. The tube houses a cable and crank-rod, which total about 6 feet. You insert the bend of the auger into the toilet outlet with the rod pulled back, then push the rod and cable forward while cranking in a clockwise direction. The cable is just long enough to reach through the trap to the drain opening in the floor. A closet auger is about the only effective way of dealing with stubborn toilet clogs.

Power Auger. If you try clearing a problem drain with a plunger or lightweight cable but don't seem to make much progress, you're probably dealing with an extended accumulation. Drain snakes, with their small springlike heads, don't make much of an impression on this type of clog. You may be able to improve the flow for a few weeks, but it won't last. Years of accumulation require a heavier cable with a more aggressive head. The head may be shaped like an arrowhead, or it may have caliper-like blades. While these features can be found on a few hand-crank models, most such snakes are power-driven by either a hand-held drill-like motor or a chassis-mounted motor. There are two basic sizes. The light-duty ones are designed for drains up to about 2 inches in diameter, which carry gray water. The larger heavy-duty ones are designed for sewers. These machines are too costly to purchase for occasional use. If you can't rent them, hire a professional plumber.

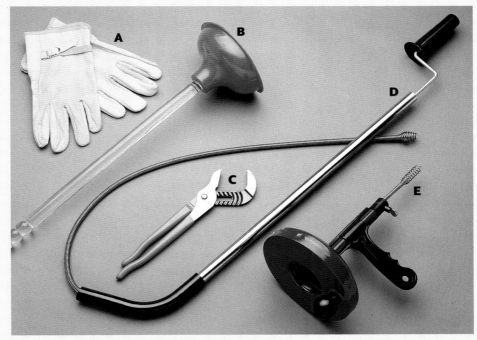

Mechanical Drain Cleaners: A—work gloves, **B**—plunger, **C**—groove-joint pliers, **D**—closet auger, **E**—drain auger

CLEARING TECHNIQUES

Drain-cleaning tools are not particularly difficult to use. Knowing how to apply a few special techniques, however, makes them more effective.

Getting the Most from a Plunger

A plunger works better when you add water. Many slow-draining fixtures already have water in them, of course (that's how you know they're slow), but if yours doesn't, add some. In addition, make sure that all overflow tubes are blocked. If you don't block these passages, they'll bleed off the pressure created by the plunger.

When plunging a tub, hold a wet rag firmly against the bottom of the overflow plate. With bath basins, stuff a rag into the overflow openings just below the rim. Hold it firmly in place. The same goes for dual-compartment kitchen sinks. When plunging one side, plug the other. But when a sink compartment with a waste-disposal unit backs up, plunge only that side. The problem here is usually a blocked T-fitting in the waste kit just under the sink.

Remember that suction is almost as effective as pressure, so try to maintain a good seal between the plunger and fixture surface through both the up and down motions. If the plunger doesn't seem to be creating enough pressure or suction because of the contour of the fixture, try coating the rim of the plunger with petroleum jelly. Follow every plunging by running hot water for at least a minute or two.

Effective Auger Techniques

Once you've gained access to the drainpipe, pull about 18 inches of cable from the spool and feed it carefully into the line.

Some drains reveal their clogs immediately. As soon as you pull the trap, you'll see it. Other lines appear clean but are clogged somewhere downstream. If you see no apparent obstruction, don't bother cranking the cable into the line. Just feed it in until you feel resistance. At this point, tighten the setscrew and push the cable forward while cranking in a steady, clockwise direction. When you run out of cable, loosen the setscrew; pull another 12 to 18 inches from the spool; tighten the setscrew; and crank forward again. Repeat this procedure until you break through the clog—it may take a few tries— or reach the next-largest drainpipe. (You'll know you have entered a larger pipe because you will feel the cable flop around in the larger pipe.)

When plunging a tub, plug the overflow fitting with a wet rag.

When plunging a bathroom sink, cover the overflow hole in the basin with a wet rag.

Meeting the Clog. You will know that you've snagged or forced open a clog when you feel the resistance—or lack of it—in the cable. This is an important consideration, especially with rented power equipment. With motor-driven cables, it's easy to break the cable off in the line when the head snags an obstruction. Always proceed slowly. When you feel sudden resistance followed by the sensation that you've broken through it, retrieve the line completely. This suggests that you've snagged an object, such as a rag. (In a sewer line, this behavior may indicate that you've hooked a bundle of threadlike tree roots.)

To retrieve a foreign object, pull the cable out of the line 1 or 2 feet at a time, feeding the cable into the spool as you go. As you pull back, continue cranking in a clockwise direction. Reversing direction will only release whatever you've snagged.

CLEARING PLASTIC DRAINPIPES

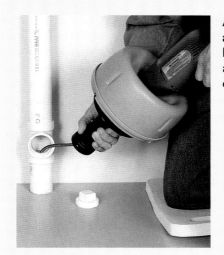

A motorized auger works best on really stubborn clogs.

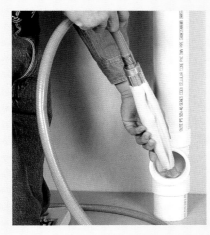

A flush bag forces the clog with water pressure.

Push and Pull. If you feel a steady, prolonged resistance after first hitting the clog, try a push-and-pull approach. Push the cable forward 3 feet, then back 2 feet, then forward 4 feet, then back 2 feet, and so on. Stop if you feel the cable flip over when pushing it through stiff resistance. If you keep cranking, the cable will tie itself in knots. Instead, reverse direction gently. When you feel the cable right itself, pull it out and check for damage. If you see no damage, crank the cable into the clog again, working forward and back. Many drain lines require several passes to break up enough of the clog to allow water to flow through.

Whenever you clear a line, follow up with lots of hot water. It's common for a newly opened line to immediately become clogged again because the loose debris settles into a new mass. To clear the line this time, just use a plunger and lots of hot water.

ACCESS TO CLEANOUT FITTINGS

If your home's drainage piping is made of cast iron, then removing a cleanout plug will require effort. The problem is that traditional brass plugs and cast-iron fittings are not compatible. The two metals have different electrical charges, so electrolytic corrosion occurs. (See "Electrolysis," page 26.) Iron, the more-negative side, breaks down, locking the plug in the fitting. The fact that brass is relatively soft and gets softer with age doesn't help: often the wrenching surface crumbles.

It's usually easier to chisel the plug from the fitting. Position a cold chisel against the exposed threads of the brass plug, and drive the plug in a counterclockwise direction. This will force the threads to break free, allowing you to turn the plug the rest of the way with a wrench. The method works on floor-drain cleanouts, too.

Replace the old plug with a plastic plug. If the cast-iron threads are too badly corroded to accept a new threaded plug, buy a rubber expansion plug instead. Insert it into the opening, and tighten the wing nut. The nut pulls a draw bolt, which expands the plug against the fitting.

If the drainage system is made of PVC (white) or ABS (black) plastic or copper and brass, opening a cleanout fitting is easy. Just remove the plug using a wrench.

With the plug removed, you can use a drain auger to clean out the drain. A motorized auger works best, but a hand-operated model will handle most small drain clogs. With larger pipes you can use a flush bag, sometimes called a blow bag, to clear a clog. Thread the bag onto a garden hose, and insert the bag into the drain. Push the hose into the pipe until you reach the clog, and then turn on the water. Water pressure should dislodge the clog and open the drain. Flush bags can only be used downstream of all branch lines.

When a brass plug won't budge, break it free using a hammer and chisel. Replace the plug with a plastic one.

ACCESS TO SINK DRAINS

Often the most perplexing part of cleaning a sink drain is gaining access to the line. Snaking directly through a sink opening almost never works because sink drains are designed to keep things out. For example, the flow in a kitchen sink is restricted to three or four openings, roughly ¼ inch in diameter, by the drain grate. Any drain snake with a large-enough head to get the job done won't fit through these openings. Even if it could, you'd have trouble working against the resistance of a fixture trap. If the clog is in the trap, you'll have better luck using a plunger or taking the trap apart.

The only time you will be able to fit a snake cable directly through a sink drain is when the sink is a bathroom basin with a removable pop-up plug. Simply removing the pop-up and lever would likely solve the problem without snaking because this is where most basin clogs occur.

Removing the Trap. Tubs and showers should be snaked from above, but not sinks. Instead, remove the trap. You can remove a sink trap easily, because all the connections are made with friction washers and slip nuts. If your fixture trap is made of plastic, you should be able to loosen the nuts by hand. If it's made of chrome-plated brass, use groove-joint pliers or a pipe wrench to free the slip nuts. Place a pail or pan under the trap to catch any spills; then loosen the nuts that secure the trap to the sink and drainpipe.

With the trap removed, push the auger cable directly into the waste pipe in the floor or wall. When reassembling a chrome trap, replace the rubber washers. Coat the washers with pipe joint compound, and tighten the nuts with pliers or a wrench. Tighten plastic nuts only hand tight, and avoid using pipe joint compound.

Dealing with Bath Sink Pop-Ups. Most bathroom-sink clogs occur around the pop-up drain mechanism, when hair gets caught on the lift lever and accumulates until most of the drain is choked off. To check the pop-up mechanism, remove the lift lever to free the pop-up plug. Just loosen the knurled nut securing the lever to the back of the drain tube under the sink basin, and pull the lever out partway. Don't pull it out all the way if you can help it, as this will let the hair clog fall into the trap. If it does fall, chances are it will be flushed out anyway, but it's always better to remove it. Next, remove the pop-up plug. If you see hair that could be clogging the drain, pull it out using needle-nose pliers or a piece of wire.

project

REMOVING THE TRAP

The best way to snake a sink drain is to remove the trap and insert the snake right into the waste line that is located in the wall or floor. Snaking from above is almost impossible because the trap will keep the snake from making its way to the clog. And if you are not sure where the clog is located, removing and cleaning the trap may be all you need to do to get your drain running freely again. With plastic drain systems, simply unscrew the nut by hand; you will need groove-joint pliers if the drain system is made of chrome-plated brass.

TOOLS & MATERIALS

- Groove-joint pliers
- Pail or bucket
- Drain auger

project

DEALING WITH BATH SINK POP-UPS

Clogs in bathroom sinks are usually caused by hair and soap scum that become trapped by the pop-up plug mechanism. In fact, the plug usually prevents the clog from ever making it to the trap. If the drain is simply running slowly, try pouring boiling water into the drain to clear the obstruction. If that does not work, remove the pop-up plug, and then use needle-nose pliers or even a straightened out wire clothes hanger to dig out the clog. Then replace the plug.

TOOLS & MATERIALS

- Groove-joint pliers

Clearing Drainpipes

1 To gain access to a sink drain, remove everything from under the sink, and place a bucket under the drainpipe. Unscrew the nuts on the trap and trap arm. Plastic is hand tight, while chrome requires a wrench to loosen the nut (inset).

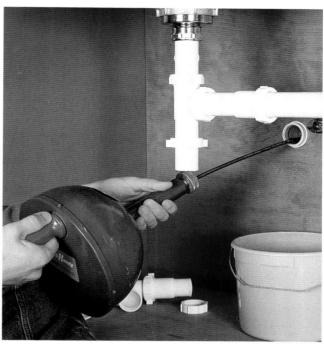

2 The drain line will either be in the floor of the cabinet or in the wall. Feed the auger cable into the drain until you feel resistance; then crank through the clog, forward, and back, always cranking clockwise. Once you break through, reassemble the trap.

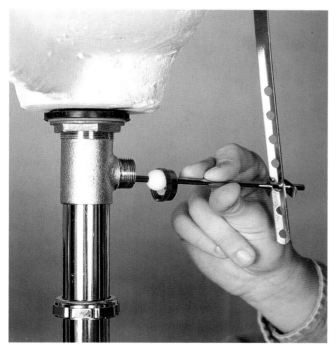

1 To check for a hair clog on the pop-up plug and lever, loosen the nut, and slide the lever out enough to free the plug. Try not to pull the lever all of the way out because the clog may drop into the trap.

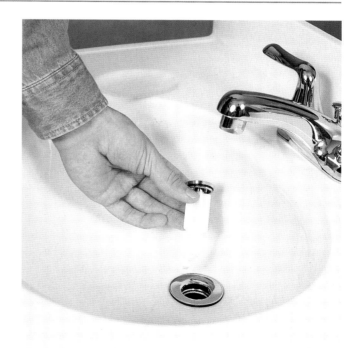

2 Lift the pop-up plug. You may see a clump of hair clinging to its bottom. Shine a flashlight into the drain, and retrieve a hair clog using a piece of wire that is bent into a small hook on its end.

CLEARING BATHTUB DRAINS

The best way to cable a tub drain is through the overflow tube. That's because the piping connection is not under the tub's drain fitting but 6 inches forward, under the overflow tube. If you try cabling the drain opening, you'll hit the drain T-fitting immediately.

If the tub has a screen over the drain opening, then the tripwaste mechanism consists of a cylindrical stopper inside the overflow tube. Hair accumulations often collect just below the screen. Remove the screen to clean this area. If the drain seems to flow well with the tripwaste removed but slows when it's back in place, you need to adjust the stopper.

To further complicate matters, most tubs that drain slowly do so not because the drain line or trap is clogged but because hair clings to the tripwaste linkage. In addition, an out-of-adjustment tripwaste will also cause a tub to drain slowly. With the plug linkage removed, you can clean the tripwaste and make any necessary adjustments. Don't assume you have a clogged line until you've checked out the tripwaste mechanism.

smart tip

CLEARING SEWER LINES

CLEARING THE UNDERGROUND SEWER LINE BETWEEN YOUR HOUSE AND STREET IS TRICKY WORK—NOT FOR MOST HOMEOWNERS. A POWER-OPERATED AUGER IS BIG AND POWERFUL, AND THE MOTOR IS ALWAYS STRONGER THAN THE CABLE. TO KEEP FROM BREAKING THE CABLE OFF INSIDE THE SEWER SERVICE, THE OPERATOR MUST LISTEN CAREFULLY TO THE MOTOR, FEELING AND LISTENING TO THE RESISTANCE OF THE CABLE. THE OPERATOR NEEDS TO BE ABLE TO SENSE THE DIFFERENCE BETWEEN TREE ROOTS, A BROKEN PIPE, PIPING OFFSETS, AND THE JUNCTURE OF PRIVATE SEWER SERVICE AND PUBLIC SEWER MAIN. ALL THINGS CONSIDERED, YOU SHOULD PROBABLY HIRE A PRO FOR THIS WORK.

CLEARING A STOPPER MECHANISM

project

If you have tried cleaning the area under the drain screen, but the tub still drains slowly, remove the overflow cover plate. This is usually held in place by two screws. Pull the plate, the lift wire, and stopper from the overflow tube. These parts are all connected to one another, so expect them to come out together. Insert the auger into the overflow opening. Crank down to break through the clog. If you reinstall the tripwaste mechanism and the tub still drains slowly, you need to adjust the stopper by shortening the lift wire.

TOOLS & MATERIALS
- Screwdriver
- Drain auger
- Needle-nose pliers

CLEARING A POP-UP MECHANISM

project

For bathtubs with a pop-up drain, simply pull the plug and the linkage attached to it from the drain. This is usually covered by a mat of soapy hair clinging to the rocker lever. Remove the hair. If the tub still drains slowly, remove the screws from the overflow plate, and pull the plate, lift wire, and coil from the overflow tube. Use an auger to clear the clog. If the drain flows well with the tripwaste removed but slows when it is back in place, you may need to adust the lift wire.

TOOLS & MATERIALS
- Screwdriver
- Slip-joint pliers

1 Remove the two overflow screws (inset), and pull out the tripwaste mechanism. Insert the auger cable through the overflow. Expect resistance when the cable enters the trap, about 18 in. down. Crank through the blockage.

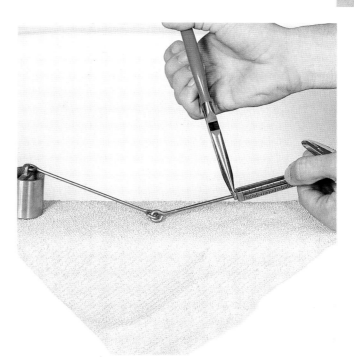

2 Adjust the tub drain's tripwaste linkage (usually making it shorter). If it is threaded, loosen the locknut, and thread the wire up about ¼ in. Secure it by tightening the locknut using needle-nose pliers or a wrench.

1 To remove the linkage of a bathtub pop-up drain that's not working properly, grip the plug, and pull the linkage out of the drain hole. The problem may be a mat of soapy hair clinging to the rocker lever.

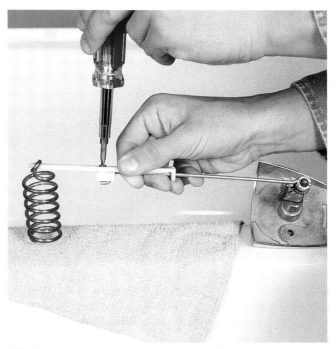

2 Some tripwastes have a plastic adjustment mechanism. Set it to the slot you want, and tighten the locknut with a screwdriver.

CLEARING OTHER TRAPS

Clearing Bathroom Drum Traps

Bathrooms plumbed in the early half of this century did not have dedicated fixture traps. The tub and sink drained directly into a drumlike canister called a drum trap in the bathroom floor. From there, a third pipe carried the water into the toilet drain piping.

If you have a house built prior to the 1950s, look for a cleanout plug in the bathroom floor to determine whether you have a drum trap. If you don't see a cleanout plug, there are other ways to tell whether you have a drum trap. If the bathroom sink backs up into the tub or you have a freestanding vintage tub with legs, a drum trap is likely. Another indicator is when the sink sits between the toilet and the tub. Modern methods favor the toilet in the center. To make sure, run a cable through the tub's overflow.

The only sure way to clear a drum trap is to open it up and clean it out. Use a wrench to remove the threaded cleanout plug. If this doesn't work, tap the plug counterclockwise with a sharp cold chisel and hammer. With the trap open, auger each of the three pipes; clean out any debris in the bottom of the trap; apply pipe joint compound to the threads; and replace the cleanout plug.

Clearing Shower Drains

Shower drains often clog because hair catches on the drain screen or a ragged pipe or fitting edge. These are simple, easy-to-clear clogs. Start by removing the screen that covers the drain opening. If the screen is held in place by two screws, remove them. If not, it will have friction tabs. Pry up the screen with a knife or flat-blade screwdriver, and shine a flashlight into the drainpipe. If you see a clog, pull

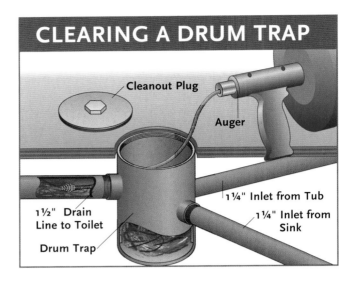

CLEARING A DRUM TRAP

Cleanout Plug

Auger

1½" Drain Line to Toilet

Drum Trap

1¼" Inlet from Tub

1¼" Inlet from Sink

it out and look for the sharp edge that caused it. If possible, file down the edges before replacing the screen. If you don't see the clog, insert an auger cable and clear the entire line, back to the stack.

Clearing Floor Drains

Look for a 1¼-inch cleanout plug, usually made of brass, threaded into the side of the drain bowl. Remove the plug, and insert the auger cable directly into the line. When you've finished, replace the brass plug with a plastic plug or rubber expansion plug. If you seldom use the floor drain, you may experience a sewer-gas problem. Gas will flow into your living space when the trap water evaporates. To solve the problem, add water once a week or install a float-ball kit. A kit consists of a ball and matching seat. Drop the ball into the drain; thread the seat in place; and add water. The ball allows water but not gas to flow through.

CLEARING SHOWER DRAINS

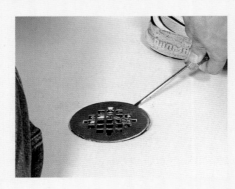

Pry up the drain screen with a knife or flat-blade screwdriver if it is not held in place by screws.

Insert an auger cable directly into the trap, and turn the handle clockwise as you feed the cable.

CLEARING FLOOR DRAINS

To gain access to a floor drain, remove the cleanout plug.

Insert the auger cable directly into the cleanout opening.

Install a float ball in a seldom-used drain to prevent gas leakage.

CLEARING TOILET DRAINS

Clearing Toilet Drains

If you see that your toilet is about to overflow, quickly remove the tank lid and lift the float. If the water is already up to the rim, reach into the tank, close the flapper, and lift the float. Turning off the shutoff valve is another alternative, but it usually takes longer.

Plunger. First, use a plunger. Fold the extension cup out, and press the plunger into the outlet. Pump the plunger vigorously until the water level drops. While remaining ready to lift the float if necessary, flush the toilet to see whether the obstruction has been forced over the trap. If the toilet flushes sluggishly, with large bubbles rising late in the flush cycle, the drain is still partially blocked.

Auger. If a plunger won't clear the blockage completely, then it's time to use a closet auger. Pull the handle and cable back, and insert the bend of the tube into the outlet. Slowly push the cable forward while cranking in a clockwise direction. When the crank handle bottoms out, retrieve the cable, and repeat the process. Run the cable through at least three times, forcing it left, right, and center.

If this method doesn't work, empty the bowl with a paper cup, and hold a pocket mirror in the outlet. Shine a flashlight into the mirror so that it illuminates the top of the trap. This should reveal what's causing the problem. Bend a length of wire into a hook, and pull the blockage back into the bowl.

To clear a toilet clog, first use a plunger that has a fold-out cup.

Use a closet auger if you can't clear a clogged toilet using a plunger.

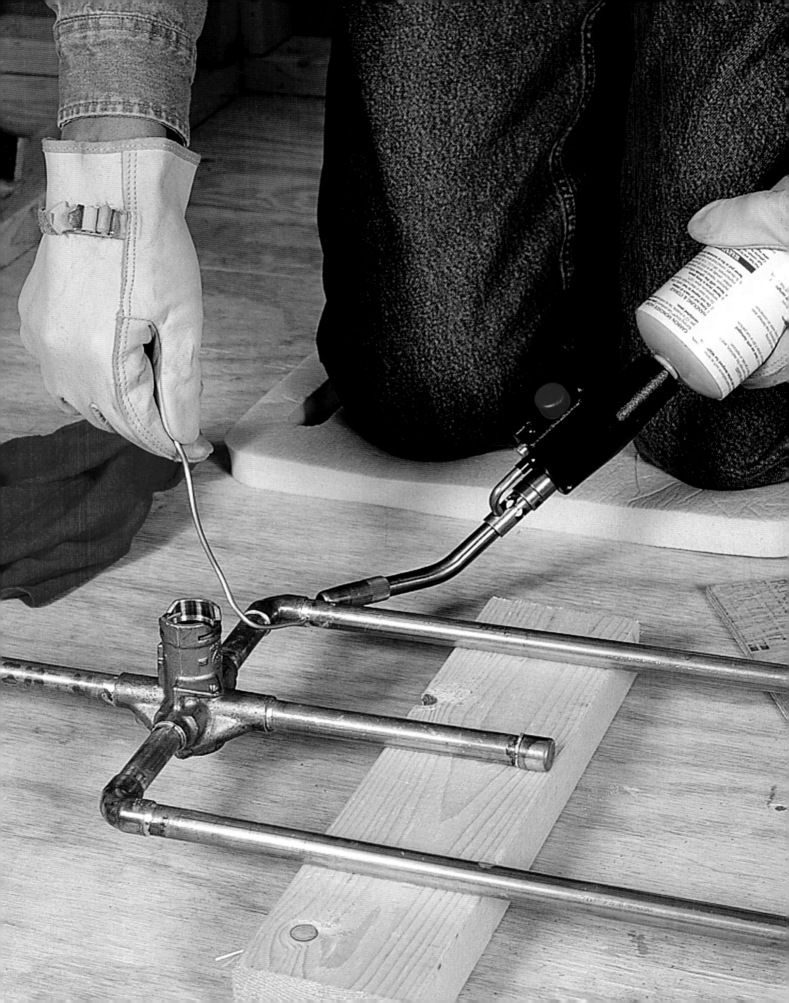

7

repairing & installing
tubs &
showers

WASTE & OVERFLOW REPAIR & REPLACEMENT

Bathtubs have two outlets: drain and overflow. Most also contain a tripwaste mechanism, an internal linkage that controls a drain stopper. Traditional tripwastes take two forms. One has a pop-up drain stopper, readily visible in the drain, while the other has an internal brass plunger that is not visible. (Plastic versions of both types of traditional drains are also available.) If your tub drain has a screen over it, expect to find an internal plunger. If it has nothing in it, you either have a very old drain or the handiwork of a frustrated homeowner. Many a homeowner has pitched the mechanism in favor of a rubber plug. Tripwastes require periodic tinkering, and if they're ignored, they grow less and less functional.

A third type of drain—a European design—has an external cable. External linkage is less likely to clog with hair. These sturdy plastic drains are also available in longer lengths than other types of conventional drains, making them ideal for jetted tubs, which are often taller than conventional tubs.

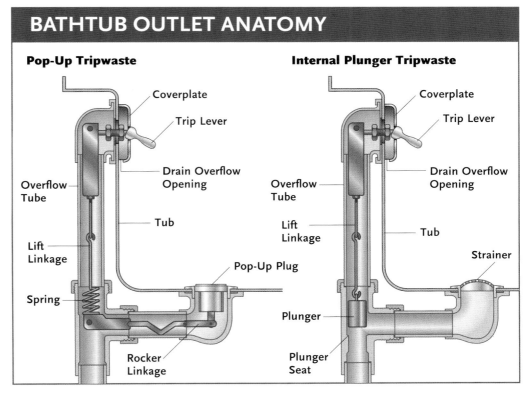

BATHTUB OUTLET ANATOMY

Pop-Up Tripwaste

- Coverplate
- Trip Lever
- Drain Overflow Opening
- Overflow Tube
- Tub
- Lift Linkage
- Spring
- Pop-Up Plug
- Rocker Linkage

Internal Plunger Tripwaste

- Coverplate
- Trip Lever
- Drain Overflow Opening
- Overflow Tube
- Lift Linkage
- Tub
- Strainer
- Plunger
- Plunger Seat

REPLACING A TUB DRAIN ASSEMBLY

There comes a time when every drain assembly needs to be replaced. The reason could be that the finish on the assembly has deteriorated, or the linkage is broken, or you want to make an aesthetic change from chrome to polished brass. For this type of job, you will need to get to the area beneath the tub because most of the work involves removing and replacing drain pipes and other fittings. This is a good time to replace old cast iron with plastic drain pipes. The first step is to remove the tripwaste and the existing drain.

TOOLS & MATERIALS

- Screwdriver ▪ Groove-joint pliers
- Hacksaw ▪ Pipe joint compound
- New drain kit ▪ Plumber's putty

1 Begin by unscrewing the two screws that secure the tripwaste cover plate. Pull the tripwaste linkage from the overflow tube—the plate, lift wire, and coil or stopper are all connected, so they should come out all at once. They may be covered with a mat of hair and soap scum.

4 At this stage, the P-trap will be the only thing holding the overflow tube in place. You may be able to loosen the trap's friction nut, and lift everything out at once. Or loosen the nuts that hold the waste and overflow components together, and remove the components piece by piece.

5 Begin installation of the new drain by placing plumber's putty around the underside of the new drain's flange.

Repairing & Installing Tubs & Showers

2 If there is a pop-up stopper in place, remove it. Then unscrew the drain fitting by inserting the handles of a set of pliers into the drain opening, and engaging the crosspiece at the bottom of the drain fitting. Grip the pliers with an adjustable wrench, and turn counterclockwise to unscrew the fitting.

3 If the fitting won't budge, move to the underside of the tub. Located between the drain shoe and the tub is a rubber gasket. Use a hacksaw blade to cut through the gasket and the drain spud. You can also use a reciprocating saw with a metal-cutting blade.

6 Have a helper, who is working below, place the rubber gasket on the drain shoe and hold the shoe up to the tub's drain outlet. Thread the drain fitting into the shoe. Insert pliers handles into the drain opening; grip with an adjustable wrench; and turn until excess putty squeezes out and the fitting feels snug.

7 Install the rubber gasket on the bathtub overflow tube, and into a waste T-fitting. (Fitting not shown. See step 8, next page.) Slide the tube upward into the wall cavity from below.

Continued on next page.

Continued from previous page.

8 Join the drain and overflow pipes in the drain T-fitting, and then tighten all the nuts using large groove-joint pliers.

9 Install the brass tailpiece that connects the drain system to the plastic drainage pipe. It will be a tight fit, so try pulling the plastic trap down far enough to get the tailpiece in position. Apply a light coat of pipe joint compound on the threads of the tailpiece, and turn the tailpiece into the bottom of the T-fitting.

10 Slip the compression nut and washer onto the tailpiece; insert the tailpiece into the trap riser; and tighten the nut to secure the tailpiece in the trap riser.

11 From inside the tub, slip the tripwaste linkage into the overflow tube until the cover plate sits against the tube's gasket, and secure the plate. If your tripwaste has a pop-up plug and lever, feed them into the drain hole with the trip lever in the open position. If your drain has a screen, install it now.

TUB & SHOWER FAUCETS

To replace a tub/shower faucet that's behind tile or a plastic wall surround, cut out a section of the back side of the plumbing wall, and make the repair from the rear. In this way, you can leave the tile or plastic surround undisturbed. Of course, your new faucet will need to match the wall openings left by the old faucet. A two-handle faucet, for example, would need a two-handle replacement. Most tub faucets have 8-inch spreads.

When the tub doesn't have a shower, consider removing the few rows of tile above the tub and making new tiles part of a general upgrade. With back-to-back tubs, which prohibit rear access, you'll need to remove at least several tiles surrounding the faucet. Cut through the wallboard exposed by the removed tile, and replace the faucet through this opening. When you're finished, repair the wall; cement the old tiles back in place; and grout the joints. While this method usually works, it's not easy, and there's always a chance that you'll break some tiles.

Finally, most manufacturers offer retrofit faucet kits for tubs and showers. The kit includes a standard faucet and an oversized trim plate. With care, you can cut out just enough tile or plastic surround to make the switch, and then cover the opening with the trim plate.

Oversized trim plates often come with retrofit faucet kits.

smart tip

TUB AND SHOWER DRAIN ACCESS

IF YOU WANT TO REPLACE AN INACCESSIBLE DRAIN (THAT IS, WHEN THE TUB IS INSTALLED ON A CONCRETE SLAB OR ABOVE A FINISHED CEILING BELOW), YOU'LL NEED TO DO THE WORK THROUGH THE BACK OF THE PLUMBING WALL. IF YOUR TUB HAS AN ACCESS PANEL ALREADY IN PLACE BEHIND THE TUB, YOU'RE IN LUCK. IF NOT, YOU'LL NEED TO CREATE ONE. TO GAIN ACCESS, CUT OUT THE DRYWALL BETWEEN THE TWO STUDS THAT STRADDLE THE CENTER OF THE TUB. WHEN YOU'VE FINISHED THE INSTALLATION, YOU CAN EITHER PATCH THE DRYWALL OR INSTALL A PERMANENT ACCESS COVER. YOU CAN MAKE THE COVER OUT OF PLYWOOD AND DOOR CASING OR PURCHASE A READY-MADE PLASTIC PANEL. WHEN THE OPENING IS IN PLAIN SIGHT, REPAIR THE WALL. WHEN IT'S IN A CLOSET OR CONCEALED BY FURNITURE, INSTALL A PERMANENTLY ACCESSIBLE PANEL.

HAIR CLOGS AND MINOR ADJUSTMENTS

WHEN A TRIPWASTE MALFUNCTIONS, it's usually because hair has accumulated around the stop mechanism or the internal lift wire has gone out of adjustment. (See "Clearing a Stopper Mechanism," and "Clearing a Pop-Up Mechanism," pages 186–187.)

If your tub is old enough to have a rubber stopper, or if you're just tired of dealing with tripwaste linkages, then I suggest an aftermarket drain insert. These toe-operated devices are made slightly larger than standard drain openings, for a friction fit. You just press the insert into the drain. Some new drains now come with toe-operated pop-ups.

A toe-operated drain plug just threads into its drain fitting. Press it down, and it seals the drains. Press it again, and it pops up to open the drain.

REPLACING A TUB-SHOWER FAUCET

project

Adding a new tub-shower faucet is a great way to upgrade your bathroom. You will be surprised at the number of finishes and styles available. Unfortunately, this type of project always incudes demolition work in order to reach the inner workings of the existing fixture. And, of course, any wall finishes you remove will need to be replaced once the installation is complete. If you are working on a ceramic-tile wall, such as the one shown here, try to save the tiles for reuse. Begin by shutting off the water to the tub.

TOOLS & MATERIALS

- Strap wrench ▪ Screwdriver
- Grout-removal tool ▪ Pipe wrenches
- New faucet ▪ Solder, flux, torch

1 If you'd like to save the old spout, use a strap wrench to unscrew it from the faucet and avoid damaging the finish. With the spout and escutcheons removed, use a grout saw (inset) to strip the grout from the tiles in the removal area.

4 When adding piping for the showerhead, you can fish Type L soft copper through the opening you made for the faucet, but you must still cut a second opening in the wall. Solder a drop-eared elbow to the top of the pipe. Secure the fitting to blocking you have nailed between the studs.

5 Complete the new faucet piping in copper. The type of faucet you purchase will determine whether you need to solder the joints or simply tighten unions. Install the spout pipe so that the spout will be 6 in. below the faucet body when installed.

2 Pry the caps from the handles, and remove the handle screws. Pull the handles off, and remove the escutcheons or trim plate. To remove the tile, start near the spout or faucet stems, and carefully pry each tile away from the wall. Set the tiles aside for reuse. It is almost impossible to match new and existing tiles.

3 Loosen the faucet unions with a pipe wrench, and lift out and discard the old faucet. If the shower riser is joined to the faucet with a union, remove it as well. For soldered joints, cut the pipes near the faucet connections, and pull the faucet out. Old cast-iron systems will be attached with unions as well.

6 Remove the nylon faucet cartridge from the faucet body, and sweat all copper fittings using lead-free solder. Protect the wall surface by using a layer of fireproof woven fabric made for this purpose.

GALVANIZED PIPES

For galvanized piping, thread ½-in. galvanized couplings to the old faucet supply tubes. Using pipe- thread sealing tape (not pipe joint compound), screw ½-in. copper male adapters or a dielectric union into the couplings.

PLUMBING 101:
Faucet and Piping Variations

When replacing a tub-shower faucet, there are a few different scenarios you might encounter. For example, old supply lines may be made of cast iron. In this case, you have to attach dielectric unions between the old iron and new copper pipe. If your piping system is already made of copper, just cut the existing copper risers, and solder new copper in place.

If the faucet you've purchased has threaded ports instead of sweat fittings, use copper male adapters to make the conversion to copper piping. Use pipe-thread sealing tape instead of pipe joint compound, but don't overdo it. Two or three rounds, stretched slightly over the threads and applied in a clockwise rotation, is ideal. Make sure the final wrap of tape covers the end threads of the fitting.

If you live where plastic is allowed for water piping, using CPVC is even easier.

BATHTUBS

As you shop for bathtubs, you'll see five basic types: traditional enameled cast iron and porcelain-coated steel tubs and the newer plastic-coated porcelain steel, fiberglass, and acrylic-plastic tubs.

Of the two traditional types, cast iron is the more durable. It retains heat better than steel and is much quieter when it's being filled. Expect cast-iron tubs to cost $300 to $400, compared with $100 to $175 for steel tubs. Steel tubs have always earned their keep in starter homes and, with care, hold up reasonably well.

The third type, a steel tub with a flexible-plastic coating separating the porcelain from the steel, is priced midway between traditional steel and cast iron and combines their best features: it is warmer, quieter, and more durable than traditional steel but without the weight of cast iron.

Fiberglass and acrylic-plastic tubs often come with matching wall surrounds. These tubs have better insulating properties than cast-iron but are generally less substantial. They're easier to install, however. Of these, acrylic is the most attractive, colorful, durable, and costly. Fiber-

smart tip

MATCHING FAUCET TYPES

WHEN YOU REPLACE A TUB/SHOWER FAUCET, THE EASIEST OPTION IS TO INSTALL A BRASS FAUCET THAT MATCHES YOUR OLD FAUCET'S SPREAD (ITS OPENINGS IN THE TILE), ESPECIALLY IF YOUR PIPING IS GALVANIZED STEEL. IN THIS CASE, YOU CAN USE THE NEW FAUCET'S UNIONS TO CONNECT THE SUPPLY PIPES. THIS APPROACH IS ESPECIALLY EASY IF THE FAUCET HAS A UNION CONNECTING THE SHOWER RISER PIPE. OF COURSE, YOU MAY NOT WANT THE SAME TYPE OF FAUCET. IF NOT, YOU'LL NEED TO BUY NEW TILES TO PATCH THE OLD OPENINGS, AND YOU MAY NEED TO MAKE MORE COMPLICATED PIPING CONNECTIONS. WHEN SWITCHING FAUCET TYPES, IT'S USUALLY BETTER TO CONVERT TO COPPER PIPE.

A replacement faucet with the same center spread as the original is the easiest to install.

glass is more affordable, and it can be repaired seamlessly and cheaply. The advantage of plastic, generally, is that it's easy to mold into attractive shapes and textures. Acrylic plastic is also available in a wide variety of colors, and unlike cast iron and steel, darker colors don't cost substantially more.

Fiberglass and plastic are also used to make one-piece tub-shower stalls. These units are generally too big for retrofits because they won't fit through doors or down hallways. Their chief advantages are durability, affordability, and ease of maintenance. When properly installed, they're virtually leak-proof. However, a frequent complaint with these tubs, especially fiberglass models, is that they're uncomfortable for bathing.

Tubs are available with right-hand and left-hand drains, as determined when you face the tub. A tub drain and faucet should be installed in the same wall, so order your tub accordingly. Standard tubs are 60 inches long, but special-order tubs are available in lengths ranging from 48 inches to 72 inches, graduated in 6-inch increments. Special-order lengths are naturally more expensive, as are deeper-than-normal soaking tubs.

Underwater lights and a stream of water that appears to flow from the ceiling provide custom touches to this tub.

BATHTUB INSTALLATION ANATOMY

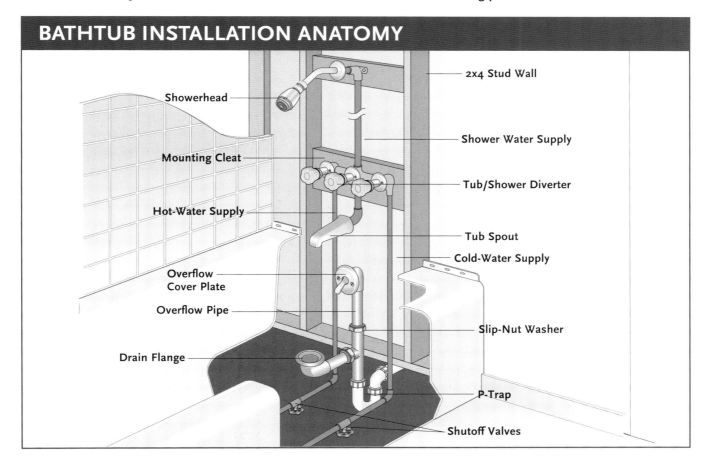

Showerhead · Mounting Cleat · Hot-Water Supply · Overflow Cover Plate · Overflow Pipe · Drain Flange · 2x4 Stud Wall · Shower Water Supply · Tub/Shower Diverter · Tub Spout · Cold-Water Supply · Slip-Nut Washer · P-Trap · Shutoff Valves

Whirlpool Tubs

There are two styles of whirlpool, or more properly, jetted, tubs sold today: apron-style tubs and drop-ins. Those with conventional aprons are typically smaller, and most are designed to fit a standard 32 × 60-inch tub opening. Drop-ins can be quite large, up to 48 × 84 inches. The term drop-in is a misnomer because these tubs are really free- standing. Drop-ins require a surrounding deck, however, so the impression is that of a tub suspended, like a sink. The tub and pump motor rest on a sturdy base, and the circulation piping wraps around the tub. The usual approach is to build a treated-lumber framework, cover it with backer board, and finish it with solid-surface resin, marble, granite, or ceramic tile. All support equipment, including the motor, pump, and piping, is concealed by the apron or decking. Apron models usually have a removable panel, while drop-ins require you to build an access panel into the deck. Make sure this service opening is large enough to allow you to replace a pump and motor.

Features. Most jetted tubs sold today are made of acrylic or a similar composite plastic. They range widely in price and are now sold just about everywhere. There are three types of motors in use today: brush-style motors, and single and multi-speed induction motors. Brush-

Jetted tubs not meant to be deck-mounted come with aprons, which often provide pump access.

style motors are used almost exclusively in builder's grade tubs. Brush style-motors are noisier and have a shorter expected service life. For a little more money, you can have a quiet, long-lasting induction motor, and for $300 to $400 more, a two- or three-speed motor. Pumps are sized between 1 and 3 horsepower. At the high end, manufacturers also vary jet speed with flow control devices. In any case, more speeds offer greater control and comfort. The high-volume jetting that feels so good when you first climb in can feel like a beating 10 minutes later, so variable speeds are a real benefit.

You'll hear a lot about jet technology, which is where most manufacturers try to differentiate themselves. If relaxation is what you're after, the most basic jets will do, but for massage therapy

Whirlpool Tub Anatomy

Air Intake

Faucet-Mounting Area

Jet

Water Intake

Water-Circulating Piping

Pump and Motor

GREEN SOLUTION

smart tip

RESURFACING BATHTUBS

IF YOU HAVE AN OLD BATHTUB, YOU CAN HIRE A COMPANY TO RESURFACE IT. THIS MIGHT BE WORTH THE COST WHEN YOU CONSIDER THE COMBINED COSTS OF REPLACEMENT, INCLUDING WALL REPAIR AND A NEW TUB, FAUCET, DRAIN, AND SHOWER SURROUND. THE RESURFACING PROCESS USUALLY CONSISTS OF REMOVING THE DRAIN FITTINGS, MASKING OFF THE TUB AREA, AND APPLYING AN ETCHING SOLUTION OR SOME OTHER CHEMICAL-BONDING AGENT. THE INSTALLER THEN SPRAYS ON SEVERAL COATS OF A PLASTIC TOP COAT, REMOVES THE MASKING TAPE, AND INSTALLS THE DRAIN FITTINGS. THE MATERIALS USED IN THESE PREPARATIONS HAVE BEEN IMPROVED, BUT THE KEY FACTORS IN DURABILITY STILL HAVE MORE TO DO WITH THE DILIGENCE OF THE INSTALLER AND THE CLEANING HABITS OF THE OWNER. A PROPERLY APPLIED AND MAINTAINED SYNTHETIC FINISH SHOULD LAST 10 YEARS OR MORE.

Refinishing a tub includes several applications of a plasticized finish.

Finishing touches include buffing out any imperfections in the final surface.

to specific areas of the body, specialty jets can make a difference. The better tubs also have built-in heaters. Injecting room-temperature air into churning water cools the water quickly, so booster heaters really improve comfort.

Hot-Water Concerns. Jetted tubs take a lot of hot water from the start, enough that your current water heater may not be up to the task. With the heater set properly, household faucets should deliver comfortably hot water at a ratio of two-thirds hot water to one-third cold water. At that rate, a 40-gallon gas water heater will deliver a maximum 60 gallons of hot water. For a larger tub, you'll either need a larger water heater or two standard heaters, plumbed in series. In this case, the first heater feeds the second, and the second feeds the house. Luckily, the water supply and drain hookups for these tubs are very similar to those of standard bathtubs.

Caution: *Large tubs of water are heavy. If you plan to install a large tub on an older wooden floor, check with an engineer. Building and safety officials may also be able to help. You may need to add floor joists to handle the extra weight.*

REMOVING A BUILT-IN BATHTUB

WHEN REPLACING A BUILT-IN TUB, expect the old bathtub to be locked in place by the plaster or wallboard above and in front of it. It may also be held in place by a layer of flooring. With a standard tub, you'll need to remove about 3 feet of drywall above the tub and 1 foot in front of it. In the case of a plaster wall, remove both plaster and lath. Use a reciprocating saw to cut away drywall or plaster.

With steel or cast-iron tubs, pry between the tub apron and the floor, using a flat bar. Pry from one end only, and when it pops loose, slide a shim under the apron and pry again. With a steel tub, you can then just lift that end and stand it up. In the case of a cast-iron tub weighing over 300 pounds, continue to shim until you can get a 2x4 under the apron. Pry with the 2x4, and block the tub until you have the apron about 2 feet off the floor on one end. Then, with the aid of a helper or two, lift it upright. If you'll be gutting the entire room, an easier method is to lift the tub with a cable ratchet. Loop one end over a ceiling joist and the other through the overflow hole.

In the case of a fiberglass or plastic tub-shower, cut it into pieces with a reciprocating saw and carry out the pieces. These units are usually too big to fit through doors.

INSTALLING A CAST-IRON BATHTUB

project

Whether working with new construction (in a new home or addition) or replacing an old tub, the process is similar. A new installation requires a few more steps, so this procedure assumes a new tub in a new bathroom. The time to install the tub is when the piping is roughed-in and before the drywall goes up.

First, you need to cut a drain opening in the floor. This should be an oversize rectangular hole, because the drain shoe will extend below the tub about 10 inches back from the wall. Center the hole in the tub space.

TOOLS & MATERIALS

- Reciprocating saw ▪ Groove-joint pliers
- Hammer ▪ Circular saw ▪ Screwdrivers
- New tub, faucet ▪ Solder, flux, torch

1 Cut the tub's drain opening (8 x 12 in.) with a reciprocating saw. The hole is so long because the drain shoe will extend below the tub about 10 in. back from the wall. This cutout needs to be centered within the tub space. To accomplish that, center the cutout 15 in. from the long wall.

4 Do as much of the plumbing work away from the wall as possible. Measure for the length of the supply lines and the shower riser. Assemble the shower-tree assembly on blocking laid on the floor, and when everything's straight, solder the copper tubing together, including a drop-eared elbow attached to the top of the riser. Set the tree in place so that the front of the faucet's rough-in plate matches the plane of what will be the finished wall. Then mark the locations of the supply tubes on the sole plate of the wall. Insert the supply tubes through ¾-in. holes in the soleplate.

2 Cast-iron tubs weigh over 300 lbs., but with a little effort you can maneuver one yourself. Uncreate the tub, and stand it up. "Walk" it into position by rocking the apron from corner to corner. Using this technique, move the tub into position.

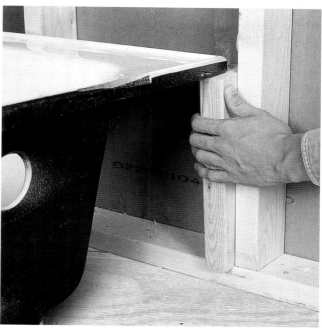

3 Use shims to level the tub if necessary. Cast-iron tubs are so heavy that there is no need to attach them to anything, but you can supply extra support by inserting 2x4s, vertically, under the long-wall-side corners of the tub. Attach the supports to the studs. These supports help spread the weight of the tub.

5 When positioned properly, the shower valve should be 28 in. from the floor. Cut 2x4 blocking to fit between the studs. Install backing boards behind the faucet, keeping the valve's rough-in plate even with the plane of the finish wall material. Secure the piping in place using straps and screws.

6 Attach a second brace to secure the top of the shower unit 72 in. to 76 in. above the floor. Position this one so that the drop-eared elbow is held behind the face of the studs. Screw the showerhead's elbow to the backing board, and install a temporary nipple in the fitting.

SETTING STEEL AND PLASTIC TUBS

UNLIKE A CAST-IRON UNIT, a steel tub should be light enough to muscle into position. There are two significant design differences between steel and cast-iron tubs. First, cast-iron tubs have support legs; steel tubs do not. Instead, a flat slab of polystyrene foam is glued to the bottom of the tub. On a level floor, this foam slab supports the tub. Secondly, a steel tub has a top lip, or flange, on three sides. On some models, the flange has a series of screw holes, allowing you to screw the tub to the wall. Don't drywall over this flange, however. Hold the wallboard about ⅛ inch above it.

Step into the tub, and walk around a bit. If you feel the tub rock, even slightly, investigate to determine why. If the floor is out of level—a common situation with concrete—then you can't rely on the polystyrene foam slab for full support. Place a level on the tub, along each wall, and make whatever adjustments are necessary. When you have it right, reach under the tub, mark each stud along the back wall where the tub meets the stud, and then pull out the tub. Cut a straight 60-inch 2x4, and nail it to the wall so that its top meets the marks on the wall studs. Push the tub back against the wall so that its deck rests on the 2x4. If the tub has flange holes, you can attach the flange to the back wall using 1½-inch-long drywall screws, though this is usually unnecessary.

With a steel tub, install the 2x4 support horizontally. Check it with a 4-foot level.

When setting a fiberglass or plastic (acrylic) tub-shower unit, use a spirit level to level and plumb the tub in all directions. Because the contours of the tub will yield a false reading, lay the level against the drywall flange on both the vertical and horizontal surfaces. When you get it right, use galvanized roofing nails or screws to secure the flange to the studs. Lay a piece of sheet metal or hardboard against the finished edge to avoid hitting it when nailing.

MAKING A SHOWER "TREE"

WHEN ASSEMBLING THE FAUCET PIPING for a tub or tub-shower, it's important to keep all the piping square and straight. Most plumbers assemble and solder the piping to the faucet on a flat surface, like a floor, and install the resulting "shower tree" in one piece.

Begin by laying three 16-to-24-inch pieces of 2x4 lumber on the floor. One piece should be near the showerhead, one just below the valve and one near the bottom of the supply tubes. For a tub, cut two lengths of type M rigid copper 36 inches long to be used for the supply tubes. Cut a second length for the shower riser, 44 to 48 inches long. Finally, cut two short stubs for the horizontal connections to the faucet inlets, and two 5-inch stubs for the spout piping. Clean and flux the pipe ends, and install a drop-eared elbow on the top of the shower riser. Insert the shower riser and pipe stubs into the remaining ports of the valve. Finish by joining the supply risers to the inlet stubs, using 90-degree elbows. Lay the assembly on the 2x4s to solder.

smart tip

INSTALLING A TUB ON AN EXTERIOR WALL

IF YOU PLAN TO INSTALL A TUB AGAINST AN EXTE-RIOR WALL, MAKE SURE THAT THE WALL IS WELL INSULATED AND THAT A PLASTIC VAPOR BARRIER COVERS IT. BE SURE TO OVERLAP THE SEAMS OF THE VAPOR BARRIER. AFTER YOU SET THE TUB, AND

BEFORE INSTALLING DRYWALL, STUFF MORE FIBER-GLASS-BATT INSULATION INTO THE CAVITY BETWEEN THE WALL AND THE TUB. THIS SIMPLE STEP CAN GREATLY REDUCE HEAT LOSS. IF THE TUB IS MADE OF PORCELAIN STEEL, THE INSULATION WILL ALSO MUF-FLE THE NOISE ASSOCIATED WITH STEEL TUBS WHEN THEY ARE FILLED. IT ALSO HELPS TO STUFF INSULA-TION BETWEEN THE TUB AND THE APRON. DON'T USE EXPANDABLE FOAM UNDER TUBS, HOWEVER.

CONNECTING TO A DRAIN BENEATH CONCRETE

WHEN A TUB WILL BE SET ON CONCRETE, a 2-inch drain line is usually brought to a boxed out area and taped off. The box, usually made of framing lumber, is then partially filled with gravel. If a tub will not be installed immediately, concrete is poured over the top of the gravel and finished. This keeps insects, radon, and moisture from entering the living space, but leaves a weak spot for easy access. If the tub will be installed as part of the primary construction, the box is usually left open.

Only after the tub and drain assemblies are installed is this pipe connected, using a PVC P-trap. If the waste pipe is made of cast iron and ends in a hub, press a rubber gasket into the hub. Then dull the end of a short length of 2-inch PVC pipe; lubricate both the pipe and gasket; and push the pipe into the hub. Immediately out of the hub, install a 2-inch plastic coupling with a 2 x 1½-inch reducing bushing. Pipe a 1½-inch P-trap into this bushing. All other aspects are the same as those of an upper-story connection.

If a cast-iron line is without a final hub, join the plastic to the iron with a banded coupling. If the drain line is also plastic, make the connection with solvent-cemented fittings. When you've completed the connection and tested it with water, mix a small batch of concrete and seal the opening around the trap.

Connecting a Tub Drain beneath Concrete

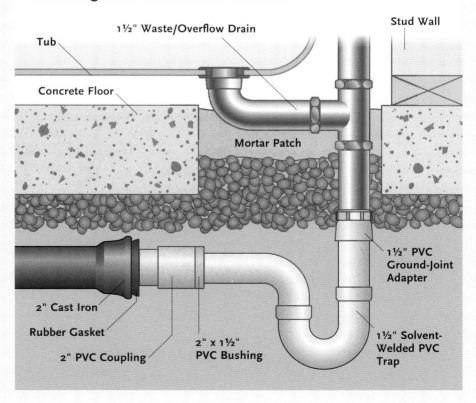

Tub

1½" Waste/Overflow Drain

Stud Wall

Concrete Floor

Mortar Patch

2" Cast Iron

Rubber Gasket

2" PVC Coupling

2" x 1½" PVC Bushing

1½" PVC Ground-Joint Adapter

1½" Solvent-Welded PVC Trap

SHOWERS

In a typical shower-stall installation, the in-wall faucet should be plumbed so that its spray hits a side wall and not the door or curtain. The showerhead should be roughed-in at around 76 inches off the floor, while the faucet should be positioned 48 inches off the floor. The faucet and shower head do not need to be on the same wall, though it's easier to plumb that way. The shower tree should be preassembled and soldered as described in "Installing a Cast-Iron Bathtub," pages 202–203.

Component showers can be made of several materials, but they usually start with a premolded plastic or fiberglass pan. These floor pans come in sizes ranging from 32 × 32 inches to 40 × 60 inches. You'll need to frame the opening to fit the pan dimensions. Side walls should extend beyond the front of the pan at least 2 inches so that drywall can be wrapped around the edge. As with a tub, you should install a shower pan before the drywall (and cement backer board if necessary). The drywall (or cement board), when installed, should be held just above the pan flange. You can buy pans alone or with surround walls.

Installing a Shower Stall

After framing the stall, you need to install the shower pan. Some shower pans come with the drain fitting packaged separately. The pan or stall is easier to ship that way. This means that you'll need to install the drain in the pan before you install the pan in the framed opening.

Set the Shower Pan. Before installing the shower pan in its opening, measure off the back wall and cut a 6-inch circular opening in the floor. A good way to make sure you spot the drain in the right location is to set the pan in place and mark the floor though the pan opening

Insert the rubber-gasketed drain spud through the opening in the pan. Then slide a rubber washer and paper washer over the spud from below. Thread the large plastic spud nut onto the drain, and tighten it. As with other spud-type drain fittings, tighten the nut until it feels snug. Don't overtighten it.

You'll need to plumb a 2-inch vented P-trap below the drain opening with a riser long enough to reach the pan drain. To determine the exact riser height, either set the pan in place, or measure down from the floor. It helps to lay a straightedge across the opening. There are two possible connections. Some drains require that the trap riser be brought through the spud, stopping just below the drain screen. (These drains may be plastic or metal.) The seal is made with a rubber gasket, which you tamp into the gap between the pipe and spud with a hammer and packing iron or flattened piece of ½-inch copper pipe. Don't lubricate this gasket. Just set it in place and tamp it down. Other drain fittings require that you cement the riser pipe right into the hub.

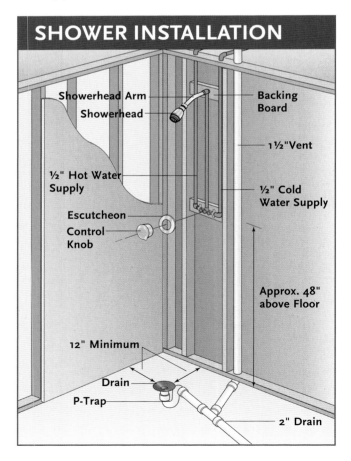

SHOWER INSTALLATION

Showerhead Arm
Showerhead
Backing Board
1½"Vent
½" Hot Water Supply
½" Cold Water Supply
Escutcheon
Control Knob
Approx. 48" above Floor
12" Minimum
Drain
P-Trap
2" Drain

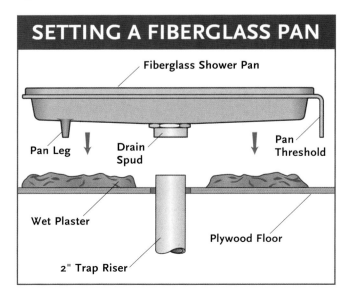

SETTING A FIBERGLASS PAN

Fiberglass Shower Pan
Pan Leg
Drain Spud
Pan Threshold
Wet Plaster
Plywood Floor
2" Trap Riser

INSTALLING A SHOWER STALL

Installing a shower stall such as the one shown here is a good way to add an extra shower to your home. It is important to follow the manufacturer's directions, but most kits consist of a shower pan, drain, and side walls. In some, the pan and surround walls are sold separately. You have to frame the area and, of course, add the faucet and fittings. You'll also need to connect the drain to a vented P-trap that will be located below the floor. Use cement backer board as a wall substrate.

TOOLS & MATERIALS

- Shower stall Measuring tape, pencil
- Circular saw Hammer Adhesive
- Groove-joint pliers Cement backer board
- Drill, hole saw Tub-and-tile caulk

1 With the framing for the sidewalls in place, set the pan in position, and mark the floor through the drain opening. Remove the pan, and cut a 6-in. floor opening using a saber saw or reciprocating saw.

2 Turn the pan on its side. The drain should have a rubber gasket around its flange. Insert the drain spud through the pan, and thread the spud nut over it. Insert from the top of the shower pan.

3 From what will be the bottom of the pan, slide a rubber washer and paper washer over the spud. Thread the large plastic spud nut onto the drain, and tighten with pliers or a spud wrench until it feels snug.

Continued on next page.

Continued from previous page.

4 Set the pan and its attached drain fittings to one side. This installation requires a 2-in. vented P-trap with a PVC drainpipe riser long enough to meet the drain fitting. Use a straightedge to check the riser's length.

5 This type of drain has the riser going through the spud and stopping just below the screen. With the drain installed over the riser, press the rubber gasket around the riser. Tamp it in using a hammer and packing tool. Another type of drain system requires that you cement the riser pipe in the hub.

8 Test-fit the panel. Peel the paper from the foam tape around the panel edges and wherever else it appears; then apply panel adhesive with a caulking gun. Apply a bead around the perimeter, around all plumbing holes, and with the body of the panel.

9 Tape strips of cardboard to the shower pan to maintain separation between the pan and the wall panels. Press the panels in place by placing the bottom against the wall first, while holding the top away. Still holding the top away, continue to press more of the panel against the wall, working upward.

6 Nail concrete backer board along the bottom 24 in. of the shower wall, and then continue with moisture-resistant wallboard. Use galvanized roofing nails on the backer board. At this point, you can line the shower with ceramic tile, or you can install a shower liner kit, as demonstrated here.

7 Place the panel that will cover the plumbing wall on sawhorses. Measure carefully to determine the locations of the plumbing holes. Take the measurements from the center of each plumbing fixture, such as a faucet cartridge or stem. Drill a faucet hole in the plumbing-wall panel using a drill and 3-in. hole saw.

10 After a few days, apply silicone caulk to the joint between the shower pan and panels. Caulk the vertical seams, too. Follow the manufacturer's directions on when to caulk. Apply just enough caulk to fill the gap. Lay out a bead, and then smooth it using your finger.

11 Caulk the shower faucet trim with a thin bead of silicone, and then wipe away all but an inconspicuous line.

ENERGY SAVER

8

maintaining & installing
water heaters

WHILE THE RANGE OF WATER-HEATER OPTIONS has grown in recent years, most of the units in current use are similar to those of generations past. The reason is that conventional electric and gas-fired water heaters are affordable and fairly inexpensive to operate, and they require little maintenance. High-tech, high-efficiency models use less fuel, but they are more expensive to buy and maintain.

ELECTRIC & GAS WATER HEATERS

Conventional tank heaters are basic-technology appliances, though these days they are more efficient than they used to be. Because of their no-frills mechanics, they are the most affordable and most popular heaters sold today.

Electric Water Heaters

An electric water heater consists of a welded steel inner tank covered by insulation and a metal outer cabinet. The inner surface of the steel tank is coated with a furnace-fired porcelain lining, often described as a glass lining. The bottom of the tank is slightly convex, which helps to control sediment, and a drain valve sits just above the bottom of the tank. The top of the tank has two water fittings and sometimes a separate anode fitting. The top (or upper side) of the heater also contains a fitting for a temperature-and-pressure (T&P) relief valve.

Two resistance-heat electrodes, called elements, heat the water in the tank. Each element is controlled by its own thermostat. The thermostats are joined electrically so that the elements can be energized in sequence: the bottom element comes on only when the top one shuts off. The elements are threaded or bolted into the unit, and the thermostats are surface-mounted next to the elements, covered by access panels and insulation. To help keep the tank from rusting, a magnesium anode rod is installed through the top of the heater. And finally, a dip tube usually hangs from the inlet fitting and delivers incoming water to the bottom of the tank.

Plastic Tanks. Electric water heaters with plastic tanks carry a lifetime warranty and cost about double the price of standard water heaters. Plastic makes an ideal tank because it can't corrode. Sedi-

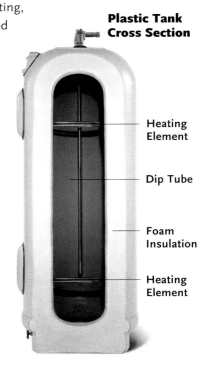

Plastic Tank Cross Section

Heating Element

Dip Tube

Foam Insulation

Heating Element

WATER HEATER ANATOMY

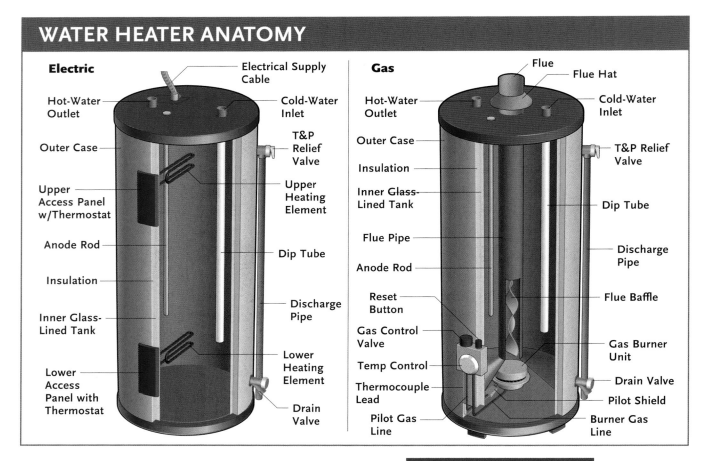

Electric

- Electrical Supply Cable
- Hot-Water Outlet
- Cold-Water Inlet
- Outer Case
- T&P Relief Valve
- Upper Access Panel w/Thermostat
- Upper Heating Element
- Anode Rod
- Dip Tube
- Insulation
- Discharge Pipe
- Inner Glass-Lined Tank
- Lower Access Panel with Thermostat
- Lower Heating Element
- Drain Valve

Gas

- Flue
- Flue Hat
- Hot-Water Outlet
- Cold-Water Inlet
- Outer Case
- T&P Relief Valve
- Insulation
- Inner Glass-Lined Tank
- Dip Tube
- Flue Pipe
- Anode Rod
- Discharge Pipe
- Reset Button
- Flue Baffle
- Gas Control Valve
- Gas Burner Unit
- Temp Control
- Thermocouple Lead
- Drain Valve
- Pilot Shield
- Pilot Gas Line
- Burner Gas Line

ment can be a problem in plastic heaters, but it's more manageable: they have rounded bottoms with large, centered drain plugs for easy draining. Plastic tanks are also highly insulated.

Gas-Fired Water Heaters

A gas-fired water heater is like an electric unit in many respects (glazed tank, anode rod, dip tube, relief valve), but its open-flame heating components require design differences. A flue tube runs through the center of the tank, from bottom to top. Viewed from above, the tank looks like a donut. To capture latent exhaust-gas heat, a wavy steel damper is suspended in the flue like a ribbon. The bottom of the tank is convex, which helps send sediment to the outer edges. At the bottom of the heater is the circular burner, and at the top, the exhaust-gas flue hat.

Gas goes to the burner through a thermostatically triggered control valve mounted on the front of the heater. The burner is joined to the control valve by three tubes. The largest of the tubes is the main gas feed. The mid-size tube is the pilot-gas feed, and the smallest is the thermocouple lead, which holds the gas control valve open.

GREEN SOLUTION

POWER-VENTING

A POWER-VENTED, or direct vent, heater has a sealed burner. A fan expels its gases, so the vent pipe does not need to be vertical (photo below). You can run the vent pipe horizontally down a long joist space and have it exit at a rim joist, for example. Power-vented units have a price tag two to three times that of a conventional gas-fired water heater and higher repair costs; however, they are a little more efficient.

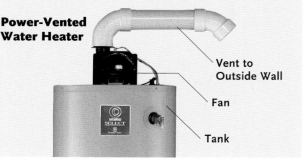

Power-Vented Water Heater

- Vent to Outside Wall
- Fan
- Tank

COMMON WATER-HEATER PROBLEMS

You will generally come across just a few main problems with a water heater: a faulty T&P relief valve, accumulation of sediment in the bottom of the tank, and a corroded anode rod. You'll learn how to deal with other more-specific problems on pages 216 to 223.

Faulty T&P Relief Valves

A temperature-and-pressure (T&P) relief valve is a water heater's primary safety device. Should a thermostat stick and the heater not shut off, the resulting increase in heat and pressure would be relieved through the T&P valve. Otherwise, the heater could explode.

The problem with T&P valves is that you can't always tell when they are no longer working. A leaky valve may signal a defect, but just as often, it indicates that the valve is working just as it should. A temporary pressure surge elsewhere in the system may have been relieved through the T&P valve.

smart tip

TOP-MOUNTED AND REPLACEMENT T&P VALVES

IF YOUR WATER HEATER'S T&P RELIEF VALVE IS THREADED INTO THE TOP OF THE UNIT, YOU MAY NEED TO CUT THE OVERFLOW PIPE TO REMOVE IT. AFTER YOU'VE INSTALLED THE NEW VALVE, RECONNECT THE OLD OVERFLOW PIPE, USING A SOLDERED COUPLING OR COMPRESSION COUPLING TO REPAIR THE CUT. IF YOUR OLD VALVE WAS WITHOUT AN OVERFLOW PIPE, INSTALL ONE. RUN IT AT LEAST TO THE TOP OF THE OVERFLOW PAN LIP.

WHEN BUYING A REPLACEMENT T&P VALVE, BE SURE THAT IT HAS A PRESSURE RATING LOWER THAN THAT OF THE WATER HEATER. IF YOUR HEATER IS RATED AT 175 PSI, AS NOTED ON ITS SERVICE TAG, BUY A VALVE THAT IS RATED AT 150 PSI. LOCAL CODES MAY SPECIFY A PRESSURE RATING.

REPLACING A T&P RELIEF VALVE

IT'S GOOD PRACTICE to test your water heater's T&P valve every six months or so. Just lift the test lever and let it snap back (inset photo right). This should produce a blast of hot water through the valve's overflow tube. If no water appears, or if the lever won't budge, replace the valve immediately. If water does appear, but you notice that the valve now drips steadily, open and close the valve several times. If it still leaks, tap lightly on the lever pin with a hammer and then retest. If it still leaks, replace the valve.

To replace a T&P relief valve, shut off the water and power, and let the water cool for at least a few hours. Open an upstairs faucet and the tank's drain valve. You won't need to empty the entire tank, just drain it to a point below the valve fitting. Remove the overflow pipe from its outlet. Then use a pipe wrench to unscrew the old valve from the tank (photo above right). Coat the new valve's threads with pipe-thread sealing tape, and tighten it into the heater (photo bottom). Stop when the valve's outlet points straight down. Screw the overflow pipe into the valve's outlet, and turn the water back on. Bleed all air from the tank through an upstairs faucet, and turn on the power.

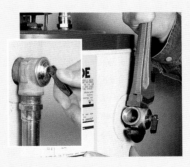

Test the T&P valve periodically (inset), and remove it if it seizes up or doesn't seem to work properly.

Wrap pipe-thread sealing tape around the threads of the new valve, and tighten it into the opening in the water heater.

Sediment Accumulation

If you know you have a problem with sand, rust, or scale in your water system, your water heater is likely to experience a buildup of sediment over time. If you have a gas-fired water heater and it makes a steady, rumbling noise each time it cycles on, but otherwise seems to work fine, then you'll know that it has accumulated several inches of sediment. The rumbling occurs when heat from the bottom of the tank percolates through the sediment.

Remedy this problem by flushing as much sediment from the tank as possible. Start by turning the gas control to "Pilot" or turning off the electricity at the electrical panel. Then shut off the water above the heater and con-nect a hose to the drain valve. Lay the other end on the floor, near a floor drain, or in a bucket. Open the valve, and drain all water from the tank. A good deal of sediment should flow out with the water. With the drain still open, turn the water back on for a moment. This should flush more sediment from the tank. Repeat the procedure several times. Finally, close the drain and fill the tank half full; then drain it through the hose once more.

Scale is produced at a faster rate when the water temperature is above 140° F. To limit further accumulations of scale, turn the heater down to 130° F. If the sediment is mostly sand, install a sediment filter early in the cold-water trunk line.

REPLACING A SEDIMENT-CLOGGED DRAIN VALVE

IF YOU FIND THAT THE DRAIN FITTING DRIPS, especially after clearing out the water heater, sediment is probably blocking the stop mechanism. You may be able to get rid of the sediment by operating the valve under pressure by opening and closing it a few times. If not, replace the drain valve. Start by turning the gas control to "Pilot" or turning off the electricity at the electrical panel and letting the water cool until it cools down. Turn off the water, and drain the heater. If the drain valve is a standard hose bib-cock, just back it from the tank with a wrench (bottom left photo). If it's a round plastic fitting, removal can be tricky. Start by rotating the valve counterclockwise while pulling back. After five or six turns, you'll cease to make progress. At this point, rotate the valve clockwise while pulling back.

With the drain valve removed, you may want to flush the tank one more time. Sometimes, getting the drain out of the way makes a difference. When you've finished, install a brass hose bibcock in the opening, using pipe joint compound or pipe-thread sealing tape (bottom right photo).

Remove the leaky old drain valve using large groove-joint pliers.

Install a new valve, turning it clockwise. Use pipe-thread sealing tape or pipe joint compound on the threads.

CHANGING AN ANODE ROD

Even though steel water-heater tanks are lined with a vitrified porcelain finish, the coating process is far from perfect. In fact, every glass lining has dozens of tiny pinholes where rust can start. The tank could rust through were it not for its anode rod. An anode works by sacrificing itself to corrosion, thereby preventing the tank from corroding. Most metals corrode, but they do so at different rates. Anode rods are typically made of magnesium because it corrodes at a faster rate than iron. It usually takes 4-5 years for an anode rod to fail.

TOOLS & MATERIALS

▪ New anode rod ▪ Pipe joint compound
▪ Breaker bar with 1¹⁄₁₆-in. socket
▪ Torch (as necessary)

1 Anode rods have a built-in nut at the top that threads into an opening in the tank. Usually, this nut requires a 1¹⁄₁₆-in. socket wrench. To loosen it, install a long breaker bar on the socket to increase the mechanical advantage.

2 This 5-year-old anode rod has exceeded its service life and must be replaced. Over half the rod is gone. Once the nut is loose, carefully lift the rod out of the tank to avoid breaking it and dropping part of it back into the tank.

3 Once the old rod is gone, clean up the threads at the top of the tank. Then feed a new rod carefully into the opening. Apply pipe joint compound to the nut threads, and turn the rod into the tank in a clockwise direction.

SERVICING GAS-FIRED WATER HEATERS

Gas-fired water heaters have service needs that are different from those for electric units. You may experience problems with the combustion process, the vent pipe, the burner itself, or the thermocouple and pilot.

Combustion-Air Problems

If your gas-fired water heater makes a puffing sound at the burner, it's not getting enough air. Air shortages are especially common in utility rooms, where a water heater, furnace, and clothes dryer all compete for air. Allow the water heater to run for a few minutes with the utility-room door closed. When it starts puffing, open the door and a nearby window. If the flame settles into smooth operation, inadequate combustion air was the culprit. Either cut a large vent through the utility-room wall or replace the solid door with a louvered door.

Vent-Pipe Problems

When a gas-fired water heater runs, it creates a thermal draft, in which air from the room is also drawn up the flue. If there is insufficient secondary air or if the flue is partially clogged, deadly carbon monoxide gas may spill into your living space. Because carbon monoxide is odorless and invisible, it pays to test the efficiency of the flue from time to time. You can call your gas company for a sophisticated test, but the following method works. If the utility room has a door, close it. Wait a few minutes for the heater to develop a good draft, and then hold a smoking match,

Use the smoke from a match or incense stick to test the flue.

incense stick, or candle about 1 inch from the flue hat at the top of the heater (photo above). If the flue draws in the smoke, all is well. If the smoke is not drawn in or is pushed away from the flue, there's a problem. To determine whether the problem is an air-starved room or a clogged flue, repeat the procedure with a door and window open. If the smoke is now drawn into the flue, make the corrections noted in "Combustion-Air Problems," at left. If the flue still does not pull the smoke, call in a professional to inspect it.

WATER-HAMMER ARRESTORS

IF YOU HEAR A POUNDING NOISE in your water system and see an occasional spill of water near your water heater, you have a water-hammer problem. (See page 32.) Water-hammer arrestors have internal rubber bladders that act as shock absorbers for high-pressure back-shocks. You typically install them between the water piping and fixture. For a clothes washer, screw the arrestors onto the stop valves and the hoses onto the arrestors.

To stop T&P valve leaks, install water-hammer arrestors on the washing machine shut-offs.

MAINTAINING THE BURN-ER AND THERMOCOUPLE

project

The job of the thermocouple is to hold the gas valve open so the gas can flow to the burner and pilot. If the pilot on your gas-fired water heater goes out and the heater goes out again after you re-light the pilot, the thermo-couple is probably misaligned or defective. Check to see whether the existing unit's sensor is positioned directly in the path of the pilot flame. If not, bend the fastener clip so that the pilot flame surrounds the sensor. Then light the pilot to see whether it keeps burning. If not, a new thermocouple is in order.

TOOLS & MATERIALS

- Adjustable wrench
- Wire
- Vacuum cleaner

1 Shut off the gas and electric to the water heater. Begin servicing the burner by removing the burner from the heater. Start by removing the access panel; then loosen the nuts that hold the gas and thermocouple lines to the gas valve. Pull the whole assembly out of the tank.

2 Clean the burner by reaming the individual jets with a piece of a stiff wire and removing any rust form the top and bottom with a wire brush. Vacuum up any dust and debris that has fallen onto the floor on the tank.

3 Replace the old thermocouple with a new unit, and secure it with a clip. If the wire is too long, coil it under the burner. Slide this assembly back into the tank, and attach all the tubes to the gas valve (inset). Reattach the access door.

217

SERVICING ELECTRIC WATER HEATERS

Electric water heaters are simple appliances. Diagnosis is easy using a volt-ohmmeter, which most plumbers carry. However, few homeowners own testing equipment, so a symptomatic is best.

If the water heater stops working, the trouble may be as simple as a tripped circuit breaker, blown fuse, or loose wire. Or it could be the thermostat. Do the easy things first. Check the electrical service panel and the thermostat reset button.

Thermostat. If a thermostat should stick in the "On" position, the high-limit switch (usually with the upper thermostat) will sense the added heat and open the circuit, putting the heater out of commission instantly. Press the reset button, and let the heater cycle. If the circuit opens again, investigate.

High-limit switches don't often fail, so suspect a faulty thermostat. But is it the upper or lower thermostat? Because the lower element won't come on until the upper element kicks off, this sequence offers a clue as to which thermostat has failed. With the power on, start by pressing all the reset buttons. Cover both thermostats with insulation. Then, listen for the expansion noises in the tank that signal a heat cycle. If the high-limit switch trips early in the 40-minute heating cycle, expect a faulty upper thermostat. If it trips much later, expect the lower thermostat.

Replacing a Heating Element

Faulty thermostats and heating elements can display some of the same symptoms (only testing will tell for sure), but elements are subject to more stress and generally fail more often than do thermostats. How will you know which element has failed? If the heater produces plenty of warm water, but no hot water, the upper element has probably failed. If you get a few gallons of hot water, followed by cold water, it's most likely the lower element.

To remove a defective element, shut off the power and drain the tank. If you need to replace an upper element, just drain the tank to that level. Remove the access panel, and loosen the two terminal screws securing the wires to the element. If your heater has a bolted flange, remove all four bolts, and pull the element out. If it's threaded-in, grip the wrenching surface with a pipe wrench, and back it out. If the sheet-metal cabinet keeps a pipe wrench from reaching the element, as it often does, use an element wrench.

To install the new element, slide the rubber gasket in place, and coat both sides with pipe joint compound (photo below, right). Then thread or bolt the element into the tank. Trim the stressed ends of each wire, and strip the insulation back about ⅝ inch. Insert the wires under the binding clips or around the screws. Tighten both screws, making sure the wires don't drift outward. Press the reset buttons, and replace the panels. Before turning the power back on, fill the tank and bleed all air through an upstairs faucet. The heater takes 45 minutes to recover.

REPLACING A HEATING ELEMENT

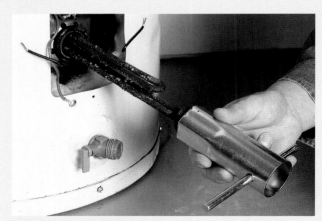

Use an element wrench to remove the old element from the heater. Some are held in place by four bolts.

Coat the element's threads and rubber washer with pipe joint compound, and screw the element in place.

TROUBLESHOOTING THE WIRING & THERMOSTAT

project

While tripped breakers often signal a defective heater component or a loose wire, the problem may be as simple as a momentary voltage spike. Reset the breaker or install a new fuse. If the breaker trips immediately, or in a day or two, look for a loose wire. If you find a defective thermostat, you can buy an inexpensive, universal replacement at most hardware stores. Just make sure that it has the same voltage and wattage specs. Remember, before you do any wiring work on a water heater, shut off the power to the heater at the service panel.

TOOLS & MATERIALS
▌ Insulated screwdriver
▌ Electrician's pliers

1 With the power off, remove the water-heater access door. Typically, just one or two screws hold it in place. Once the cover is off, push the tank insulation to one side to expose the thermostat and check for charred wires (inset).

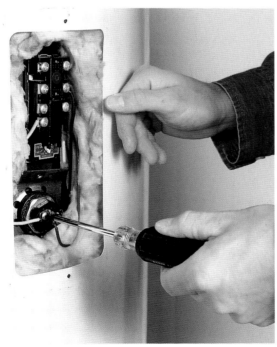

2 If you find a charred wire, remove it from its screw terminal, and trim away the burned area. Then strip back the insulation ⅝ in. from the end of the wire, and reattach it to the screw terminal, making sure it's securely tightened.

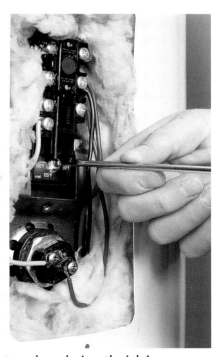

3 If the thermostat needs replacing, the job is simple. Loosen all the screw connections, and pull the old unit out through the insulation (inset). Then, screw the new one into place, and attach the wires to the heating element.

REPLACING A GAS-FIRED WATER HEATER

Start this job by shutting off the tank's water and gas supplies. Then, drain the tank with a hose and separate the gas line at the union next to the gas valve. Disconnect the vent at the top of the unit, and cut the copper water pipes that bring water to and from the appliance. Sometimes, in old houses, these pipes are made of galvanized steel and will have union fittings located directly above the tank. To free the tank, just separate these unions.

TOOLS & MATERIALS
- New heater, T&P valve ▪ Garden hose
- Self-tapping screws ▪ Pipe wrenches
- Power drill-driver ▪ Nut driver
- Pipe joint compound ▪ Tubing and cutter
- Solder, flux, torch

1 Begin this job by turning down the thermostat on the water heater, and shutting off the cold water supply. Drain the water by attaching a garden hose to the valve at the bottom of the tank and running the other end outdoors or into a laundry sink or floor drain.

4 Grip the heater by the supply tube stubs that are left on the top of the appliance. Then tip it from side to side as you pull it toward you, so the tank "walks" across the floor. Or have someone help you carry it. Once the water is drained out, the tank isn't hard to lift.

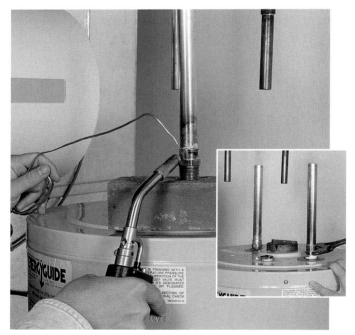

5 Slide the new tank into place, and cut pieces of tubing to fit between the existing tubes and the top of the tank. Solder the tubes to threaded male adapters. Once these joints are cool, wrap pipe-thread sealing tape on the adapter's threads, and turn the assemblies into the tank holes (inset).

2 Shut off the gas line that supplies the water heater by turning the valve just above the pipe union (inset). When the red handle is perpendicular to the pipe, the flow is stopped. Separate the gas line by unthreading the nut that holds the two sides of the union together.

3 Use a nut driver or a screwdriver to remove the screws that hold the flue in place. If it's easy to remove the entire flue section above the heater, do so. If not, hold up the flue with wire after it is disconnected. Then, cut the water pipes above the tank with a tubing cutter (inset).

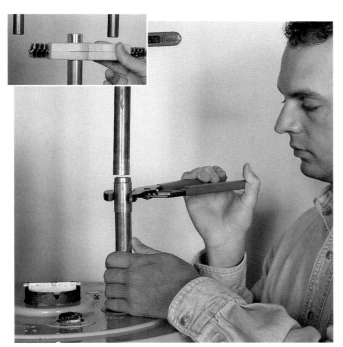

6 If a tank tube is too long, cut it using a tube cutter before trying to install the coupling (inset). The best way to join the supply and tank tubes is with slip couplings. These fittings are designed to slide over the end of one tube and then onto the second tube once the two are aligned.

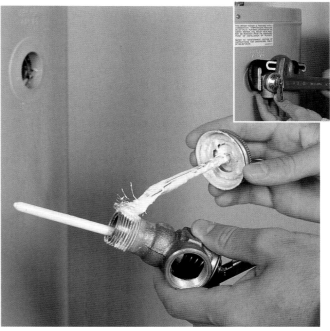

7 It's a good idea to install a new temperature-and-pressure relief valve (T&P valve) when you are putting in a new tank. To do this job, lightly coat the threads on the new valve with pipe joint compound. Then tighten the valve into the opening on the side of the tank with a wrench (inset). Continued on next page.

Continued from previous page.

8 Once you have threaded the T&P valve into place, attach a discharge pipe to the bottom of the valve. Cut a piece of copper tubing so it will extend from the valve at least to the top of the overflow pan lip. Then solder an adapter on the end of the tube, and thread this assembly into the valve.

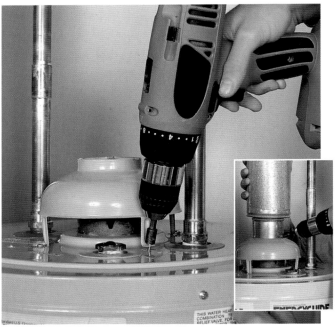

9 A flue hat comes with the heater. To install it, just screw it to the top of the heater with self-tapping hex-head screws. Once the hat is in place, cut the flue to length, and attach it to the hat (inset). Self-tapping screws eliminate the need for drilling pilot holes for the screws.

10 The gas valve is designed so the gas line will come into a side opening. The gas service assembly requires three things: a T-fitting for the pipe that goes to the valve; a drip tube at the bottom of the assembly (inset); and a union that joins the new piping to the old piping.

11 Once the gas service is attached, turn on the red valve handle, and light the pilot following the directions on the gas valve. A gas grill lighter usually works for this job. Once the pilot is on, install the access panel; make sure the drain valve is closed; and turn on the water.

INSTALLING AN ELECTRIC WATER HEATER

An electric water heater is easier to install than a gas-fired one. There are no flue or gas connections to make, and electric heaters usually come without legs, so any minor unevenness in the floor can be ignored. (A significant slope will still require shims, however.)

The big difference, of course, is the electrical connection. The heater will come with its own built-in electrical box, but you'll need to connect electricity to the box using a box connector and conduit. The conduit needs to run from the water heater to a disconnect box on the wall or into a joist space overhead. (Check with your building department.) Once in the joist space, the cable usually doesn't need to be in conduit. Use rigid, thin-wall conduit from the ceiling or flex conduit from a disconnect box on the wall.

Remove the knockout plug from the heater's box, and install a conduit-to-box connector (photo below, left). Bring the existing 240-volt cable through conduit and through the box connector. (If running new cable, make it a 3-wire with ground, usually 10-gauge wire with a 30-amp breaker.) Tighten the conduit in the connector. Then, using approved twist connectors, join the black and red circuit wires to the black and red (or black and black) lead wires in the water-heater box (photo below, right). Join the ground wire to the ground screw near the box. If you used

GREEN SOLUTION

smart tip

SAVING FUEL COSTS

TODAY'S WATER HEATERS ARE BETTER INSULATED THAN WERE THOSE OF EVEN A FEW YEARS AGO. STILL, YOU CAN BOOST EFFICIENCY BY INSTALLING AN AFTERMARKET INSULATION BLANKET. THE MORE HOURS YOU SPEND AWAY FROM THE HOUSE, THE MORE BENEFIT YOU'LL GET FROM THIS ADD-ON. CAUTION: BE CAREFUL NOT TO COVER THE ACCESS COVER, T&P RELIEF VALVE, OR CONTROL VALVE. YOU'LL ALSO NEED TO HOLD THE INSULATION AWAY FROM THE FLUE HAT BY SEVERAL INCHES ON GAS-FIRED HEATERS.

AN EVEN BETTER INVESTMENT IS TO INSULATE ALL HOT-WATER SUPPLY LINES. THEY SHED A GOOD DEAL OF HEAT, AND THE MORE YOU CAN DO TO SLOW HEAT LOSS, THE LOWER YOUR ENERGY BILLS. YOU'LL FIND SEVERAL KINDS OF INSULATION ON THE MARKET. THE BEST IS PRE-SLIT FOAM RUBBER.

rigid conduit, secure its upper end to the joists or a brace nailed between joists. Before turning on the power at the service panel, fill the tank with water and bleed all air from the tank through the faucets.

INSTALLING AN ELECTRIC WATER HEATER

Thread a conduit connector into the water heater, and run the wires in ½-in. EMT conduit.

Bond the grounding wire under the grounding screw, and join the like-colored wires in connectors.

9

sump pumps, filters & softeners

225 SUMP PUMPS

230 WATER TREATMENT SYSTEMS

MANY PLUMBING SYSTEMS function well with basic equipment, but problem situations may call for additional equipment. Sump pumps, filters, and water softeners are the most common residential problem solvers. Sump pumps remove unwanted ground water, gray water, or sewage. Filters strain contaminants from drinking water, and water softeners protect plumbing equipment by altering the chemical/electrical composition of water.

SUMP PUMPS

There are three types of sump installations:

■ **Conventional sumps** pump seasonally high groundwater from under a basement floor and require a below-floor drain tile system and sump pit. You'll find several types of pit liners, or plastic sump containers. Some have perforations around the top half, while others are solid. The perforated models are for groundwater installations. A typical groundwater setup includes a 12-to-14-inch-deep trench, which contains perforated plastic drain tile and coarse gravel, around the basement perimeter. Water drains into the pit liner and is pumped onto the lawn.

■ **Gray-water installations** are similar, but the water comes from plumbing fixtures and must be pumped into a plumbing stack.

■ **Ejector pumps** grind solid sewage, so they can drain a bathroom or an entire house.

Groundwater/gray water pumps come in two forms. Some are submersible, motor and all; others have the motor elevated above the pump on a pedestal. Pedestal models are less expensive, but they're a bit noisy.

Sump Pumps for Groundwater and Runoff

Many homes have groundwater and runoff problems. In the worst cases, a system made up of a sump pump and drainage tile (or nowadays, more likely perforated plastic pipe) can mean the difference between a dry, usable basement and a damp, crumbling foundation. If you plan to build a home and seasonally high groundwater is even a remote possibility, install a drainage system and a sump pit. You may never need to buy and install the actual sump pump, but doing this preparation work in the open stages of construction is a fraction of the cost of a retrofit. Each site is different, of course, but if a new-construction drainage system costs $350 (without a pump), you could easily pay $3,500 for a retrofit installation.

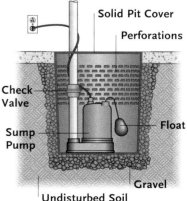

Submersible Sump

Solid Pit Cover

Perforations

Check Valve

Sump Pump

Float

Gravel

Undisturbed Soil

225

Installing a Submersible Sump Pump

Once you've worked out a perimeter drainage strategy to direct water to the sump pump (either installing a perimeter drain in new construction or using a rented concrete saw and jackhammer to cut out 18 inches of concrete along the basement walls and dig in a trench, gravel, and tile in an existing house with drainage problems), you need to install the pump and attach the discharge piping.

Install the Pit Liner. You'll find two types of sump-pit liners on the market. One has perforations around its upper half; the other doesn't. Use the perforated liner for groundwater sumps and the non-perforated type for gray-water pits. (See "Gray-Water and Ejector Sumps," page 230.) Dig a hole for the pit liner so that its top rim is flush with the top of the concrete floor.

If your drainage piping installation was made on the outside of the footing, use a tile spade to tunnel under the footing. Slide a length of pipe through the tunnel, and join it to the perimeter perforated pipe with a T-fitting. Then cut an opening in the pit liner (usually at its midpoint), and run the pipe through this hole several inches. Backfill the pit below the pipe connection with soil, tamping it in 4-inch lifts. Fill the upper half of the excavation with coarse gravel, packing as much as possible into the footing tunnel. Finally, backfill the exterior, and pour concrete for the basement floor.

SUMP PUMPS AND RADON GAS

High levels of radon gas in basements and crawl spaces have been linked to cancer and can be the equivalent to smoking several packs of cigarettes a day. Radon gas level test kits are readily available at your local home center.

A radon gas–approved sump pump lid with space for a vent pipe.

Sump pits liners have universal ribbed edges, but radon gas–approved lids have specific gaskets for the discharge pipe, electrical wires, and the required vent pipe. In addition, the edges of the concrete where to liner fits into the hole in the floor need to be sealed with caulking.

The discharge pipe can be installed to discharge to the outside air through an appropriate roof boot for passive systems or connected to a vacuum pump on the outside of the foundation connected to an exhaust pipe that allows the exhaust to correctly enter the outside air at the roof line.

project

INSTALLING A SUBMERSIBLE SUMP PUMP

While installing a pit liner, sump pump, and drainage pipe is not an easy job, it should take less than a day. If you are really efficient, it could take just a couple of hours. But that doesn't include the job preparation, which means digging the hole for the pit liner. This usually involves renting a rotary hammer to cut a hole through the concrete and then using a shovel and (often) a steel bar to loosen the soil so it can be removed. A good way to carry out this soil is in empty 5-gallon drywall joint compound buckets.

TOOLS & MATERIALS

▌ Sump pump kit ▌ Drill, saber saw
▌ PVC pipe, fittings ▌ Pliers, nut driver
▌ Primer, cement ▌ Hacksaw ▌ Pipe hangers

3 Once the adapter is in place, glue a length of pipe into the open end of the fitting. After the glue has cured, use the pipe to help lower the pump into the pit liner. You can cut this pipe to finished length once the pump is located.

1 Once the hole for the pit liner is excavated, slide the liner into this hole to make sure it fits properly. When satisfied with the fit, remove the liner, and cut holes for the drainage pipes that empty into it. Use a saber saw for this job.

2 A sump pump discharges water through a port that faces up on the side of the motor. This port needs to be outfitted with plastic (PVC) pipe. Start by threading a male adapter into the opening using pipe thread-sealing tape to seal the joint.

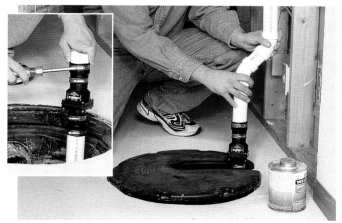

4 Install a check valve just above the pit liner lid, using a nut driver or a screwdriver (inset). In most installations, you'll have to offset the riser that comes out of the check valve by using a couple of 45-deg. elbows to make the turn.

5 Choose the shortest possible route to the outside of the house for the discharge pipe. Cut the pipe in convenient lengths, and join these with couplings to make installation easier. Cut as few holes as possible in the framing members.

smart tip

RUNOFF DRAINAGE SOLUTIONS

MOST WET BASEMENTS CAN BE FIXED ABOVEGROUND, WITHOUT A SUMP PUMP, BY IMPROVING THE GUTTER SYSTEM AND SLOPING THE GROUND AWAY FROM THE HOUSE. IF YOUR GUTTERS OVERFLOW, CLEAN THEM. IF THEY LEAK, REPLACE THEM. IF DOWNSPOUTS EMPTY NEAR THE FOUNDATION, INSTALL DOWNSPOUT EXTENSIONS OR SPLASH BLOCKS. IF A RAISED FLOWER TERRACE NEXT TO THE HOUSE CAN'T BE PROPERLY DRAINED, REMOVE IT. IF A CONCRETE PATIO SLOPES TOWARD THE HOUSE, REPLACE IT. AND IF THE SOIL AROUND THE FOUNDATION DOES NOT SLOPE AWAY FROM THE HOUSE FOR A DISTANCE OF AT LEAST 4 FEET (PAST THE PERMANENTLY ABSORBENT BACKFILL), REGRADE THE AREA.

INSTALLING AN ABOVE-FLOOR SUMP PUMP BOX

project

You'd like to add a basement sink, but you're not up to breaking into a concrete floor to install a pit liner. Consider installing an above-floor sump box available at most plumbing supply stores. The box receives the water from the sink drain and pumps it away like a sump pump installed below the floor. You'll have to tie the discharge side of the box to the house drainage system. And you'll need to install a vent pipe that extends above the roof.

TOOLS & MATERIALS

- Utility sink and fittings ▮ Primer, cement
- PVC drainpipe and fittings
- Groove-joint pliers ▮ Nut driver, screw
- PVC primer, cement ▮ 120-V receptacle
- Sump box kit driver, wrench, and cable

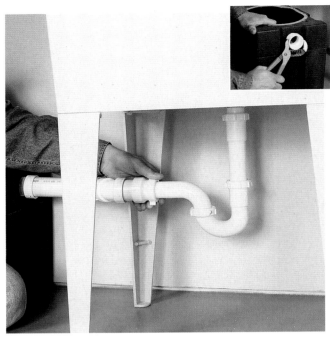

1 Place the box on the side of the sink where the discharge connection to the house waste system is the most convenient. Begin by installing a male adapter in the side of the box (inset). Then attach a trap to the bottom of the sink, and connect the two with pipe and fittings.

4 Install a threaded coupling on top of the threaded nipple. Again, use thread-sealing tape to seal the joint. Then thread a check valve into the top of the coupling, and tighten it in place with groove-joint pliers. This valve prevents wastewater from backing up in the sink.

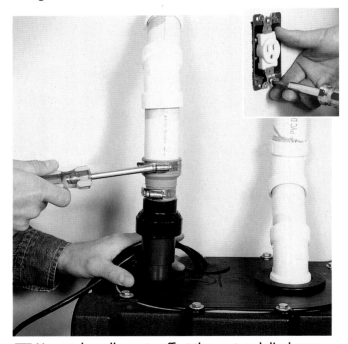

5 Use 45-deg. elbows to offset the vent and discharge pipes toward the wall. When you have completed the discharge assembly, tighten the banded adapter. To provide power, install a single-outlet 120-volt receptacle near the box (inset). Connect the receptacle to its own 15-amp GFCI-protected circuit.

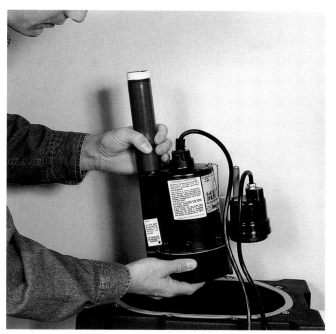

2 Attach a threaded plastic pipe nipple into the top of the sump unit. Use thread-sealing tape to make sure the joints don't leak. Then lower the unit into the box, positioning it so the pipe nipple can extend through the hole provided for it in the box lid.

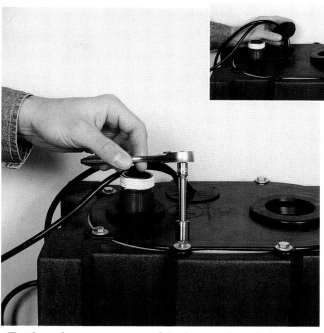

3 Place the cover on top of the box, and push the rubber gaskets into the holes for the cords and pipes (inset). Use the bolts or screws provided in the kit to install the lid. Tighten these fasteners progressively, turning each about one full turn before moving onto the next.

6 Usually, the most convenient place to splice in the sump box discharge pipe is in the main waste stack. To do this, just cut out a section of pipe, and replace it with a 3-in. Y-fitting with a 1½-in. hub on the side. Add pipe nipples to the Y, and install it with banded couplings.

7 The easiest way to vent the sump box is to connect its vent port to an existing dry vent line. To do this, install a T- or Y-fitting in the vent piping, and join the hub on this fitting to the adapter that's installed in the sump box with a combination of pipe sections and fittings.

GRAY-WATER AND EJECTOR SUMPS

IF YOUR HOUSE DRAIN exits a basement above floor level, the only way you can have a sink or laundry in the basement is to install a gray-water sump. The only way to have a full bath in the basement is to drain it through an ejector sump.

The difference between a groundwater and gray-water sump installation is in where the water must go. Groundwater is pumped outdoors, while gray water must be pumped into the plumbing system. Gray-water installations are usually low volume. The fixture drainpipe can enter the pit through the lid (below) or through the side of the liner, below the floor. A below-floor installation looks neater and tends to be less noisy than the alternative.

An ejector sump differs in that it handles—grinds and pumps—solid sewage. (It cannot handle other solids, like cigarette filters or tampons.) The inlet drainpipe always enters the pit liner under the floor through a watertight fitting. Because raw sewage is involved, the pit must be airtight and vented. Both the pit-liner lid and the soil-pipe connection have gaskets and are bolted in place.

Ejector pumps are expensive, but they are quite reliable. Like all mechanical things, however, they will eventually fail. When this happens, it can make quite a mess. For this reason, some manufacturers use two pumps that kick on alternately. When one of the pumps fails, it triggers an alarm.

A sump-box kit, left, or ejector pump, right, won't handle a lot of volume. The former works with a sink; the latter with sink and toilet.

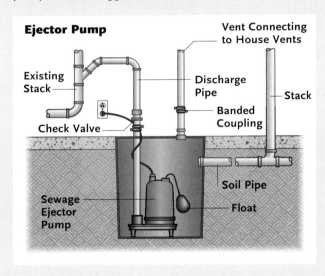

Ejector Pump — Vent Connecting to House Vents — Existing Stack — Discharge Pipe — Stack — Check Valve — Banded Coupling — Sewage Ejector Pump — Soil Pipe — Float

WATER TREATMENT SYSTEMS

The quality of groundwater across the country varies a great deal. There are two basic problem water categories: the unhealthy and the unpleasant. Unhealthy water might contain nitrates, heavy metals (like lead and mercury), disease-causing bacteria, organic and inorganic compounds (arsenic, atrazine, benzene, and the like), and dissolved gases such as radon. Unpleasant water might be high in iron, sulfur, manganese, sediments, non-hazardous bacteria, and so on. Municipal water is rarely unhealthy, but it may be unpleasant; private well water may be unhealthy, unpleasant, or both.

Common Water Problems. Private well water may be high in mineral salts, sediments, and chemicals. It may contain soluble iron, which leaves fixtures black or rusty. A harmless but nasty-looking bacteria, called bacterial iron, can feed on iron particles in well water and coat pipes and fixtures with an orange slime. (If so, you should have the well chlorinated.) Water may have a high sulfur content, making it taste and smell awful. It may contain Giardia, a common microorganism that causes flu-like symptoms. And finally, it can contain nitrates, disease carrying bacteria, lead, and cadmium. Most public water is monitored and treated, but treatment facilities can't do everything. They can't soften water, and they can't prevent aging piping from adding traces of lead, rust, copper, or asbestos.

Filter Types

The first line of defense is usually a filter or series of filters. If you think you need a series of filters (to treat a variety of problems) you should get professional help in deciding which filters to use and in which order.

Sediment Filters. If your only problem is sediment, a single sediment filter will do. These in-line filters consist of a filter body, a reservoir, and a filter medium. In most models, the filter must be replaced regularly, but expensive models have a filtering screen and reservoir. These have no filter cartridge to replace. Instead, a drain valve allows you to backflush the sediment from the filter.

Carbon Filters. If tests reveal high levels of organic or inorganic contaminants, granular carbon filters may work. (The porous surface of activated carbon traps particles.) High-volume carbon filters are installed in supply lines; low-volume carbon filters serve a single fixture.

Carbon and other media filters have a serious drawback. There's no foolproof way of knowing when the media is saturated with contaminants. (Some filter systems have flow meters or built-in timers to remind you to change them.) If you don't change the filter in time, it can dump contaminants (nitrates, chemical compounds, and so on) into the water in high amounts. If you stay on top of the maintenance, activated carbon filters will do a reasonably good job on contaminants in gas, liquid, and particle form.

Reverse-Osmosis Units. Reverse-osmosis (RO) units are reliable in dealing with nitrates and chemical contaminants. They should not be used with water high in dissolved minerals (hard water), and they can't remove biological contaminants. In an RO unit, water is forced

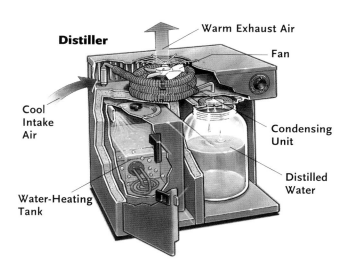

Distiller — Warm Exhaust Air — Fan — Cool Intake Air — Condensing Unit — Distilled Water — Water-Heating Tank

Reverse-Osmosis Filter System

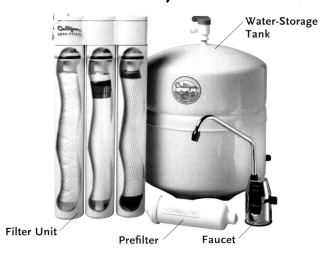

Water-Storage Tank — Filter Unit — Prefilter — Faucet

through a permeable membrane. Though the water makes it through, the contaminants do not. The membrane is dense, so RO units have an extremely low output, and they can't purify an entire plumbing system. They use a 2-to-3-gallon storage tank. Most are installed under kitchen sinks in series with sediment and carbon filters.

Disinfection Units. When water contains only biological contaminants, it's a good choice to send it through a disinfection chamber. The chamber can be treated with chlorine, bromine, or iodine, but it might also be bombarded with ultraviolet light or saturated with ozone. These units should be professionally installed.

Distillers. One of the most effective ways to purify water is through distillation. In a distiller, water is heated until it steams, or vaporizes. The vapor rises into a cooling chamber, where it condenses back into liquid. Heavy metals, trace chemicals, and minerals are left behind in the heating chamber. Biological contaminants are killed in the heating process. (See the illustration above.)

Distillers do an excellent job on a variety of contaminants, including nitrates, lead, mercury, radium, iron, calcium, magnesium, copper oxide, copper sulfide, coliform bacteria, Giardia, and more. No other system does as much, as simply. However, distillers have limited capacity; they use quite a bit of electricity; the larger units are fairly expensive; and the heating chambers must be cleaned of minerals and contaminants frequently.

Most residential distillers are counter-top models. Larger units are usually piped directly, while smaller models are either piped through a saddle-tap valve (like an ice maker) or have no connection at all: you pour tap water into the unit's reservoir.

Installing In-Line Filters

In-line filters must be installed so that the body is vertical. They usually require a shutoff valve on both sides, so cartridge changes can be made without draining the entire system.

Filter bodies normally come with threaded connections. When joining copper or plastic male adapters to threaded ports, solder or solvent-cement the adapters to pipe stubs first. Then thread the adapters into the filter.

In a horizontal installation, the piping connection is a simple splice in the water pipe (below). In the case of a vertical pipe, you'll need to create a horizontal offset so that you can install the filter with its reservoir vertical.

Assuming a horizontal installation in copper, begin by cutting out a section of pipe about 14 inches long, or long enough to accommodate the top of the filter body, the soldered stubs, and the male adapters. Then install two full-flow shutoff valves. Valves with compression nuts eliminate the need for soldering. Finally, if your copper or galvanized-steel piping system serves as a partial ground for the electrical system, install a jumper wire. Bolt an approved grounding clamp onto the pipe on each side of the filter, and join them with 6-gauge copper wire.

Installing Reverse-Osmosis Units

Reverse-osmosis units usually consist of the filters and, because the units operate slowly, a 2 or 3-gallon storage tank so that fresh water is always available for use. Identify and label all the color-coded tubes attached to the filter unit. Then hang the unit on the side or back wall of the

cabinet, using the screws provided. Set the storage tank in another corner of the cabinet. Most codes require a backflow-preventing air gap for the drain tube, with the gap at countertop level. (Faucets with a built-in air-gap connection are typically sold with RO units.) The faucet will have two barbed fittings next to the faucet shank. Insert the faucet through the fourth sink hole, and tighten the jamb nut onto the shank.

INSTALLING AN IN-LINE FILTER

Thread male adapters, with attached stubs, into the filter head.

Splice the filter into the water supply line with shutoff valves.

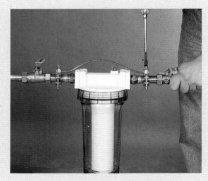

Install a jumper wire to restore the electrical path to ground.

INSTALLING A REVERSE-OSMOSIS UNIT

project

Reverse-osmosis filtration units do a very good job of removing impurities from all kinds of water. But they are more complicated than simple charcoal filters and require routine maintenance to work properly. This is not difficult, but it can be made difficult if you locate the components in difficult-to-reach spots. The manufacturers instructions should explain the best arrangement. But if they don't, the product literature will explain what parts have to be replaced at what intervals. Just install everything so you can make the switch.

Air gap fitting code updates may affect your installation or repairs (see page 292). Before replacing or repairing any section, consult your local building authority for applicable enforcement.

TOOLS & MATERIALS

▌ RO filter kit
▌ Drill and bit ▌ Adjustable wrench
▌ Screwdriver

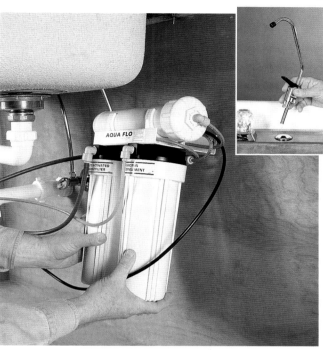

1 Different units can be installed different ways. But they usually have a similar approach. First, you attach the filter unit to the side of the sink cabinet. Then you attach the water lines to the filter and the faucet to the sink (inset).

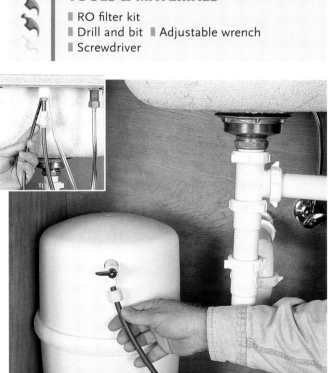

2 Attach the supply tube from the filter to the separate filter faucet (inset). Then set the storage tank in the corner of the cabinet, and attach the hose from the filter unit. These hose fittings are usually designed to be finger-tightened.

3 Attach the waste line from the filter to the waste pipe under the sink, using a saddle valve. First drill a hole in the waste pipe; then attach the saddle valve so the waste tube opening falls over the hole in the waste pipe.

INSTALLING A WATER SOFTENER

Hard water contains high levels of calcium and magnesium, which form a crusty scale on metallic surfaces. Detergents and soaps don't work well in hard water because they react with these dissolved metal ions. Most water has some hardness, measured in grains per gallon. The surest way to tell if you have hard water is to test it. Generally, anything above about ten grains of hardness may pose a problem. Most hardness problems can be solved by installing a water softener, like the one shown here.

Air gap fitting code updates may affect your installation or repairs (see page 292). Before replacing or repairing any section, consult your local building authority for applicable enforcement.

TOOLS & MATERIALS

▮ Copper tubing ▮ Flux, solder, torch
▮ Water softener unit ▮ Hose clamps
▮ Bonding cable ▮ Adjustable wrench
▮ Pipe joint compound ▮ Groove-joint pliers

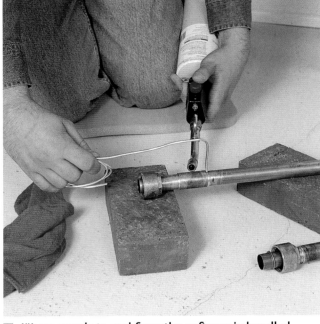

1 Water supply to and from the softener is handled by two copper tubes. These tubes are joined to the softener by two short stubs with a union nut on each. To make risers, slip the union nut over the tube stub, and solder it to a section of tube using flux, solder, and a torch.

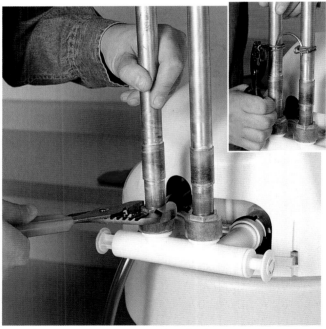

4 The riser tubes are joined to the bypass valve with the union nuts shown in photo 1. Tighten these until they are snug using groove-joint pliers. Some units require an electrical bonding wire between the tubes (inset).

5 Slide the softener unit into place next to the wall. Then route the purge line into an adjacent laundry sink, waste line, or floor drain. While the softener does require regular attention, chose a location that doesn't interrupt the work flow in the rest of the room.

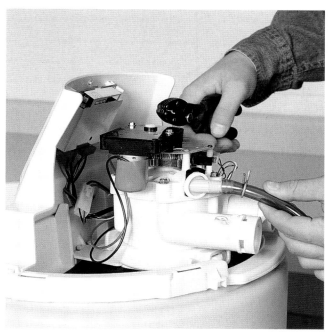

2 Every softener requires a discharge tube that provides a means for purging the unwanted minerals from the softener. This hose is usually attached with a simple hose clamp. This tube can be emptied into an adjacent laundry sink, drain waste line, or floor drain.

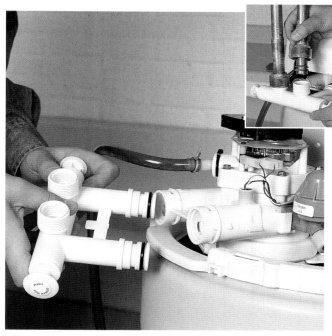

3 Every softener requires a bypass valve to allow you to isolate the water softener for service without shutting off the water supply to the rest of the house. Most softeners come with a factory designed bypass valve. This unit fits into the side of the softener head.

6 To bring water to the softener, cut into the cold water line so that one tube supplies hard water and the other carries away treated water (inset). Also install a T-fitting in the hard water line to supply a sillcock. It's wasteful to use treated water for watering plants or washing the car.

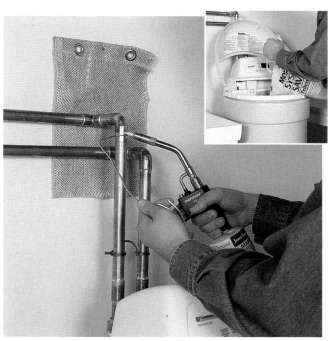

7 Solder all water-line connections using lead-free solder, and protect combustible surfaces with a flame shield. After disinfecting the unit, pour several bags of special water-softener salt into the reservoir (inset). The salt is available as sodium chloride or potassium chloride.

Model # WR60.0LC

10

septic systems, wells & lawn sprinklers

THOSE WHO LIVE beyond the reaches of public water and sewer systems rely on private water wells for fresh water and septic systems for sewage and wastewater disposal. Keeping these systems up and running is not difficult or complicated, but they require some maintenance. And knowing how these systems work is important because you'll be less likely to push them beyond their limits.

There are three types of code-approved waste-disposal systems in use today:

- **Anaerobic (Septic) System.** The septic system is the most common private waste system, consisting of a buried septic tank, a distribution box, and a gravel-lined leach, or drainage, field.
- **Aerobic Lagoon System.** The lagoon system, the second most popular waste-disposal design, is a precisely constructed open-air sewage pond. Lagoons are also known as stabilization ponds.
- **Hybrid Aerobic/Anaerobic System.** The least popular (and little used) system is a mechanical version of the buried tank and leach field. In this system, electrically powered equipment in the tank stirs and aerates the sewage, thereby speeding the digestion process.

While the tank is aerobic, the leach field is anaerobic. All three processes use naturally occurring bacteria to break down and consume solids. In an aerobic, or aseptic, system, a form of bacteria that thrives on the oxygen present in air does the work. Anaerobic systems use bacteria that cannot tolerate significant quantities of air.

SEPTIC SYSTEMS

In a septic system, sewage flows from the house by force of gravity through a 4-inch soil pipe that spills into the inlet baffle of the tank. The tank and distribution box may be made of steel, concrete, ABS plastic, or fiberglass. Once in the tank, the sewage solids are digested by naturally occurring anaerobic bacteria in human waste. The bacteria also produce heat, which keeps the tank and leach field from freezing. The digestion process yields two tank-bound by-products. Insoluble particles settle to the bottom of the tank to form sludge, while grease and soap float to the top to form a surface scum. The sludge and scum must be pumped out periodically to keep the system functioning properly.

Distribution. Every time wastewater enters the tank, an equal volume exits through the outlet baffle. This water, called effluent, flows into the distribution box and then into the leach field. Ideally, effluent contains no solids, but is rich in urea and organic nitrogen compounds. Once it's in the leach field, another digestion process begins. Naturally-occurring microbes that inhabit the top 2 feet of soil convert the waste in the water to compounds that can be used by plants. The water is either evaporated (60 percent by volume) or absorbed by soil and plants.

Avoiding Permanent Failures. If scum is allowed to accumulate for long, it will eventually become so deep that it extends below the outlet baffle, flows into the leach field,

SEPTIC SYSTEM ANATOMY

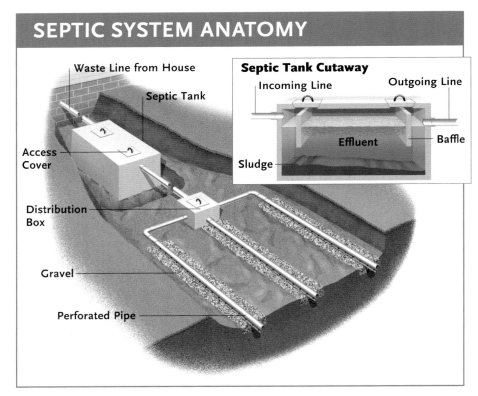

Waste Line from House

Septic Tank

Access Cover

Distribution Box

Gravel

Perforated Pipe

Septic Tank Cutaway

Incoming Line

Outgoing Line

Effluent

Baffle

Sludge

Minimizing Temporary Failures. Temporary failures, which often occur in the spring, may be due to soil saturation. Spring snow melt and heavy rains can saturate the soil above and around the leach field. Saturated soil not only fails to absorb water but also stops— even reverses—the nitrification process. The very term "septic" implies a lack of oxygen. When oxygen-rich surface water enters a field, the soil shifts from a septic to an oxygenated state. At this point, the leach field is merely a storage chamber. It does almost no work. As you add more sewage to the system, it's almost certain to back up.

In a well-designed, properly-sized system, these temporary setbacks are accommodated, but in marginal systems, sewage backups are an annual occurrence. To help a seasonally stressed septic system, try to keep the field dry.

and plugs up the trench. (Sludge contributes to the problem by reducing, from the bottom, the volume of water the tank can hold.) This reduces storage capacity and halts most of the absorption and nitrification. At this point, the leach field is ruined. The tank and distribution box still work, but the leach field must be replaced. *For this reason, it's important that you have the septic tank pumped every three to four years.* You'll find septic-pumping companies in the Yellow Pages.

- **Change the roof-gutter** downspout locations, or extend them with underground diversion pipes if they send water in the direction of the field.
- **Move some soil** to divert runoff, if necessary.
- **Use less water** during the wettest weeks of the year. You might take your clothes to a coin-operated laundry, shorten the length of your showers, wash the dishes by hand, and flush your toilets less often.

Other Septic-System Considerations. A well-designed septic system will handle a range of household materials, from toilet paper to food waste. (White toilet paper is best.) However, septic systems can't tolerate large amounts of cleaning solvents or furniture strippers, strong acids, paint thinners, petroleum-based lubricants, antifreeze, and some chemicals used in photography.

Septic-System Additives. According to health department officials and other experts, "enzyme" additives available in supermarkets and from other sources won't hurt your system and may even help a bit, but they're not needed. While it takes several months for a new system to develop a full-strength colony of bacteria, the official consensus seems to be that additives don't do much to speed up the process. And of course, once the system is up and running, it will regulate itself.

smart tip

CLEANING THE SEPTIC TANK

HAVING THE SEPTIC TANK PUMPED OUT PERIODICALLY IS VITALLY IMPORTANT FOR LONG-TERM, TROUBLE-FREE OPERATION OF A SEPTIC SYSTEM. SOME HOMEOWNERS FIND IT HELPFUL TO USE A WELL-RECOGNIZED RECURRING EVENT AS A REMINDER TO HAVE THE JOB DONE. FOR EXAMPLE, YOU MIGHT WANT TO TIME YOUR PUMPING RITUAL WITH THE PRESIDENTIAL ELECTION. THAT WAY, WHEN ELECTION TIME COMES, YOU KNOW IT'S TIME TO PUMP OUT THE SEPTIC TANK.

AEROBIC LAGOON SYSTEMS

WHEN SOIL DRAINAGE IS EXTREMELY POOR, lagoon systems are an alternative. Lagoons, or stabilization ponds, are conically shaped earthen enclosures designed for sewage disposal. The wastewater is always exposed to air, so treatment is accomplished with aerobic bacteria, algae, and surface evaporation. Lagoons must be built to exacting specifications if they are to last. When properly constructed, they work efficiently and odorlessly for decades with little maintenance.

The perimeter of the lagoon must be fenced with wire mesh, and because surface breezes must sweep a lagoon for it to work properly, no tall vegetation should be allowed to grow near it.

As for the digestion process, heavy solids settle to the bottom and are consumed by bacteria. Suspended bacteria break down lighter particles, giving off carbon dioxide. In the presence of nitrogen, phosphate, carbon dioxide, and light, algae flourish. The algae produce oxygen, which in turn feeds the bacteria. As in a septic system, the digestion process keeps the water from freezing. Unlike a septic system, however, a lagoon and its digestive process are all but odorless.

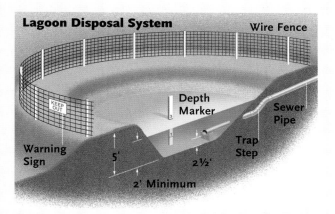

Lagoon Disposal System — Wire Fence, Depth Marker, Sewer Pipe, Trap Step, Warning Sign, KEEP OUT, 5', 2½', 2' Minimum

PRIVATE WATER WELLS

Well types range from the tireless windmill wells of generations past to shallow-well jet pumps to deep-well submersible pumps (most common). Most deliver water to a pressure tank located in a well pit or basement.

The Pressure Tank. Pressure tanks maintain a steady line pressure and keep the pump from kicking on with every small draw of water. A pressure switch installed in the tank piping controls the electric well pump. The switch turns the pump on and off at preset pressures.

The most common pressure-tank problems are waterlogged tanks and pressure switches that have drifted off their settings or have simply worn out. Some pressure tanks (especially older ones) are simply hollow, galvanized containers that have a top-mounted snifter, or air, valve. The snifter, sometimes called a Schrader valve, looks and acts like a valve stem in a car tire. In these tanks, a quantity of air is held in the top of the tank as a pressure buffer and is in direct contact with the water in the tank. It's not uncommon for them to have waterlogging problems. Most modern tanks have an air-filled rubber bladder inside at the top. It has a snifter valve as well, but it's always charged at the factory and usually requires no on-site adjustment on your part,

making them nearly maintenance free.

Minor Well Problems and Solutions. If your system stops working abruptly, look first for tripped breakers and blown fuses. Pump motors require roughly ten times the power to start than to keep running, so occasionally they will trip.

The tank pressure must be drawn down for the pump to kick in when you turn the system back on, so open a few faucets to test

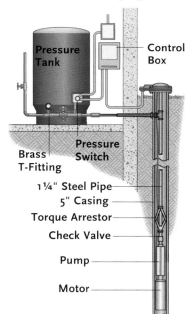

Submersible Pump System — Pressure Tank, Control Box, Brass T-Fitting, Pressure Switch, 1¼" Steel Pipe, 5" Casing, Torque Arrestor, Check Valve, Pump, Motor

your work. If the breaker trips or the fuse blows when the pump comes on, call a professional.

If the pump runs for a time before tripping the breaker, or shuts off normally but seems to run much longer than it used to, you may have a pressure-switch problem.

FIXING A PRESSURE SWITCH

TO GAIN ACCESS to the pressure switch, shut off the power; loosen the captive nut on the top of the switch cover; and remove the cover. You'll see several wire connections and two spring-loaded pressure sensors: one long and one short. The short spring will control the cut-out pressure (the range), which is usually the culprit, while the larger spring will control the cut-in (low) pressure while maintaining the cut-in/cut-out pressure differential. Both springs can be adjusted with the range nuts on top of the springs. Adjust the larger spring nut first. For a higher cut-in pressure, rotate the nut downward. For a lower cut-in pressure, rotate the nut upward. Follow with similar adjustments for the smaller cut-out spring. If these corrections don't help, or if the pump cycles erratically, replace the pressure switch.

If your pump seems to kick on every time you use a little water, expect a tank with too little air. Test the tank pressure with a tire tester. If the high pressure is not as high as the tank specifications require (usually about 28 pounds), add some air.

If these simple procedures don't put your well back in working order, call a professional.

To set the cut-in pressure while maintaining the pressure differential, adjust the nut on the long spring.

To set the cut-out pressure only, adjust the nut on the short spring.

HOW TO REPLACE A LEAKY PRESSURE TANK

project

The pressure tank on a well keeps the water pressure within the home at a constant level. It also prevents the well's pump from switching on every time a small amount of water is used in the home. To replace a leaky or damaged tank, start by turning off the well pump, shutting down the electric power, and draining the system. A professionally installed tank should include a boiler drain placed near the tank for just this purpose. Pressure tanks are connected to the system through a T-fitting.

TOOLS & MATERIALS

▌ Pipe wrenches ▌ Adjustable wrench
▌ Screwdriver ▌ Wire-cutting tool
▌ Thread-sealing tape ▌ Pressure switch
▌ Pressure-relief valve

3 Mount a new pressure switch on a ⅜-in.-dia. threaded nipple, and thread the nipple into the tank T-fitting. Make sure the pressure switch is compatible with the pumping system you have.

1 With the power and water turned off, drain the system through the boiler drain next to the shutoff. Then disconnect the wires, and undo the box connector (inset). Loosen the unions on all sides of the pressure tank T-fitting. Slide the old tank out.

2 Replace the valves and fittings with new ones. Tighten a new pressure-relief valve into one of the openings in the tank T-fitting.

4 With the T-fitting assembled, install it on the pressure tank. Use the half union mounted on the tank for attachment.

5 Attach the tank to the piping using unions with thread-sealing tape on the mating surfaces. Then reconnect the wiring (inset), and recharge the system. Adjust the pressure switch so that the pump cuts in and out at the pressures stated by the manufacturer.

HOW TO INSTALL AN UNDERGROUND SPRINKLER SYSTEM

As you might expect, planning an effective layout is a big part of installing a lawn sprinkler system. Manufacturers know this, so they provide help. Most have design personnel on staff who will create a layout for you.

The company can provide a detailed drawing with all the lines sized and color coded with a complete parts list, including components not made by your company. (No company makes all the pieces.) Our Rainbird plan and materials list came with an engineer's stamp for instant approval at our building codes office.

An Overview

The typical system consists of the water piping, a control panel, a vacuum breaker, zone valves, and the sprinkler heads. You can buy all the components at your home center or shop the professional sprinkler shops. In warm climates, you can use PVC pipe (not CPVC) throughout, but PVC may crack due to freezing, so black polyethylene is a better choice in colder climates. A good plan is to run PVC from the house to the main distribution box. Use type M copper between the internal T-fitting at the water meter and the vacuum breaker. If your water supply pipe is only ¾ inch, immediately upsize to 1 inch for the sprinkler system piping.

Every sprinkler system needs a vacuum breaker to keep the underground piping from contaminating the household water. There are two types: surface level and reduced pressure. A surface-level vacuum breaker must be installed 1 foot above the highest point of the property, so with a banked yard, a riser for a standard vacuum breaker can turn into a flagpole. In this case, a reduced-pressure vacuum breaker is a better idea. It can be installed indoors in a closet, garage, or basement.

The number and locations of the zone valves will depend on your layout, but many yards will need six to eight. For example, the main PVC distribution line could feed the first control box, which might contain three electric zone valves, serving three separate lines.

When the system is connected to the control panel, each zone will kick on, run its prescribed length of time, and shut off, at which time the next valve will kick on. This sequencing will continue until the entire yard is watered. All the zone valves can be set for the same duration, or each can be programmed differently.

TIME-SAVING VALVE

SELF-TAPPING SADDLE VALVES let you skip the drilling associated with standard inline taps. (The product shown here is called the "Blazing Saddle.") The saddle snaps onto the polyethylene pipe. You then screw the valve home using finger pressure to create a self-sealing hole. You can also find self-tapping valves for PVC pipe. Such valves are another good reason to use plastic pipe for sprinkler systems.

To make an in-line tap, spread the saddle valve's collar over the pipe and snap it together.

Thread the tapping point all the way down, until it punctures the pipe.

LAYING THE SPRINKLER PIPING

project

Even though sprinkler plumbing is buried relatively shallow, always contact your local Digger's Hotline to have buried utility lines located and flagged. This is usually required by law, and it just makes good common sense. You can dig the trenches by hand, but the work will go much quicker if you use a turf plow. The blade on the plow will knife through the turf, shimmying the tubing into the ground. The blade can cut through tree roots up to 3 inches thick.

TOOLS & MATERIALS

▌ Water pressure gauge
▌ Groove-joint pliers ▌ Spade
▌ Polyethylene or PVC tubing
▌ Pipe puller or turf plow

1 Attach an inexpensive water pressure gauge to an outdoor sillcock to test your home's water pressure. Measure the flow rate by running water into a 5-gal. bucket and timing how long it takes to fill the bucket. Flow rates are measured in gallons per minute.

2 Connect the turf machine's friction grip to your coil of polyethylene. If using PVC, glue several lengths of tubing together before pulling. When running to zone valves, be sure to wrap the low-voltage cable around the tubing so that you can install both at once.

3 Start the machine forward, and let the vibrating blade shimmy the tubing into the ground. You will need to dig 8 to 10 in. deep. Avoid tree roots if possible, but the knife can cut through those that are up to 3 in. thick.

CONNECTING SPRINKLER COMPONENTS

project

In addition to the sprinkler plumbing, you will need to install the zone valves, drain fittings, and sprinkler heads. The layout of your system will determine where each component is located. Make as many connections aboveground as possible, but for most of the components, you will need to excavate shallow depressions to make the necessary connections. Use cement when working with PVC; use barb fittings for polyethylene tubing.

TOOLS & MATERIALS

- Adjustable wrenches ▮ Zone valves
- Low-voltage cable ▮ Pipe cutters
- Sprinkler tubing ▮ 90-deg. elbows
- Brass drain fittings ▮ Saddle valves
- Silicone sealant

1 Following the layout of your system, assemble the zone valve piping aboveground so that you will have room to work. Thread the PVC adapters into the valves, and then cement the drainpipes into the adapters. Place the assembly in a hole on a bed of sand.

4 Use a scissor-style tube cutter to slice through polyethylene tubing. To make connections, use barb fittings for T-fittings, couplings, and elbows. Slide crimp rings on each pipe; push the tubing into the fittings until they bottom out on the barbs; slide the rings forward; and crimp them in place.

5 Use a 90-deg. elbow when a sprinkler head is installed at the end of a line and drainage isn't a concern. This type has a barb fitting on one end and a female thread on the other to hold the sprinkler head.

2 On the downstream side of each valve, install a drain-down fitting. Connect the assembly to the buried poly tubing using barb fittings secured with staness-steel crimp rings. Connect the low-voltage wires to the valves with twist connectors. Install an access box over the valves and backfill.

3 Install a drain valve at the main feed line's lowest points. Attach a T-fitting with a threaded branch, and screw a brass drain into the branch. Cement the T-fitting to the main line. Cant the drain downward about 45 deg.; dig a depression for the drain; and fill the depression with sand.

6 When water will be trapped at the end of the line, use a self-draining elbow to connect the sprinkler head. This allows you to connect the head using a threaded riser.

7 Swing pipe allows you to move the head where you need it so that you can make last-minute adjustments. It doesn't need crimp rings. To add an in-line head, install a saddle valve. (See "Time Saving Valve," page 242.) Backfill heads about 4 in. at a time, tamping with a rubber mallet.

PLUMBING AND CONTROL-PANEL CONNECTIONS

project

To connect the water system to the outdoor sprinklers, drill through the house's siding and rim joist near where you will tap into the home's supply line. The hole must accommodate a 1-inch copper feed line and the wiring for the sprinkler system. Insert the tube in the hole, and solder on a 90-degree elbow. If using a surface vacuum breaker, turn the elbow facing up, and install a riser and the vacuum breaker 1 foot above the highest sprinkler head. Run PVC to the first zone valve location.

TOOLS & MATERIALS

- Adjustable wrenches ▪ T-valve
- Ball valve ▪ Vacuum breaker
- Control panel ▪ Copper tubing
- Hose bibcock

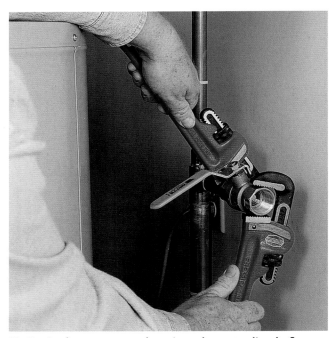

1 Drain the system, and cut into the water line before it branches to any fixtures. Solder in a T-fitting. Add a few inches of copper tube and a male adapter. Backhold the male adapter, and thread a ball valve onto it. This will be the primary shutoff valve, and it allows you to turn the water back on to the rest of the house during installation.

4 Pull the plugs from the sprinkler heads, and run the system briefly to flush any soil from the lines.

5 When the water runs clear, the lines should be clean. Install the grit strainers and nozzles on the sprinkler heads, and adjust the heads for the best coverage. Make adjustments based on manufacturers directions. Each sprinkler head's field of coverage should slightly overlap the next one.

2 Install a brass nipple in the ball valve, followed by the reduced-pressure vacuum breaker if you are using one. Install a hose bibcock for cold-weather draining. Attach the line to the copper tube coming in through the rim joist. Install the vacuum breaker's air-gap drain, and run a tube to a nearby floor drain.

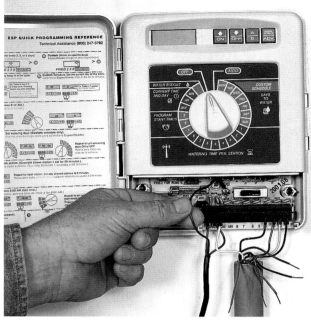

3 Place the control panel on a wall in an accessible location near a 120-volt receptacle. Bring the low-voltage cable into the house, and connect the wires to the control panel. The panel is coded for installation.

TUNNELING UNDER OBSTACLES

MOST YARDS WILL HAVE A FEW OBSTACLES under which you may need to tunnel, including sidewalks and tree roots. If you've rented a turf plow, it will come with a built in boring device. If not, one approach is to dig a hole on each side and tunnel directly through with a tile spade.

Another is to dig a hole on each side and use a sluice pipe to burrow across. This method works well for longer distances. Just attach a pointed spray nozzle to one end of a PVC pipe and a hose fitting to the other. Connect the hose, and turn the water on. Then push the pipe across in bursts, pushing forward and pulling back. The sluice water will carry the soil toward you as the pipe progresses across the span.

Make a sluice pipe for boring under sidewalks and other obstructions. The brass nozzle clears away the soil.

PART II:
Fundamentals

plumbing basics

11

A HOME'S PLUMBING SYSTEM can seem daunting and unapproachable. But it is comprehensible if you take a few minutes to organize it into manageable segments according to function. Even if you go only so far as to separate the incoming water lines from the outgoing drainage pipes, most of the battle is won. Separating these two systems is easy because the piping sizes are different. Drainpipes are larger than water lines, ranging between 1¼ and 4 inches in diameter. Water lines are generally ½ and ¾ inch in diameter. These are typically inside diameters, so exterior measurements will be a little larger.

Pipes versus Tubes. Although most homeowners don't realize it, there is a technical difference between a "pipe" and a "tube," and it relates to how they are sized by diameter. The diameter of a pipe is measured inside, while the diameter of a tube is measured outside. That difference notwithstanding, most industry professionals refer to household copper lines, as well as any lines that don't need 90- or 45-degree fittings, such as PEX lines and ⅜-inch chromed water supply lines, as tubing.

WATER SUPPLY SYSTEM

Each home has a water service line that travels underground from its supply source—either a private well and pump or public water main. Service lines are buried between 1 and 6 feet deep, depending on climate. The colder the climate, the deeper the lines are buried. City water services typically have three or four shutoff valves, offering several points at which to interrupt service. Only the last two valves are readily accessible to homeowners, however.

DIGGING IN PUBLIC PROPERTY

ALTHOUGH YOU'RE CHARGED WITH maintaining the sidewalk in front of your house and the space, if any, between the sidewalk and street, this area is really public property. It's sometimes known as the public parking, and any qualified person needing to work on buried pipes can dig in this ground, even if it seems to be part of your lawn. You have a right to dig across the street to make installations and repairs, and your neighbor from across the street has the right to dig on your side when he or she needs to make repairs. There is no need to tear up an area of more than about 4 by 8 feet. Whoever does the digging, whether by hand or machine, should also try to restore the grass.

Plumbing Basics

Municipal Supply Lines

The first shutoff in the water line is actually part of the water-main tap mechanism, in which a motorized tapping machine has bored a self-tapping valve directly into an iron or plastic water main. The tap/valve, once in place, is called a corporation stop, or more commonly, a corpcock. You can reach it only by excavating the soil above it.

The second shutoff is also underground, usually in the public area between the street and sidewalk. The term for this in-line valve is a curb stop. It's only accessible with a long street key—which plumbers carry—and has a pipelike sleeve that reaches to grade level. This vertical extension is called a stopbox. If you've been mowing around a protruding stopbox for years, you should know that it is adjustable, up or down. Call a plumber to adjust it.

A final pair of valves connects directly to the water meter, although in some houses there is only one valve. The valve on the street side of the meter is called a meter valve, and the one on the house side, a house valve. These are the valves you'll use to shut down the system for emergencies and repairs, so it pays to know where your meter is located and how to reach both valves.

WHERE ARE THE WATER AND SEWER MAINS HIDDEN?

WATER AND SEWER MAINS are usually located near the street curb, with the sewer on one side of the street and the water on the other. Plumbers usually have to bore under the street with a mechanical boring machine to reach a main. Both mains need to be unearthed to make the service connections. Water mains are usually 3 to 6 feet deep, while sewers are between 3 and 15 feet deep, with most in the 8- to 10-foot range.

In older neighborhoods, city mains may be located under the street. To gain access for installations or repairs, you must cut out a section of street. Street repair costs generally accrue to the homeowner. In the oldest neighborhoods, sewers are often located in alleyways. Your town's public works department should be able to tell you exactly where your sewer and water mains lie.

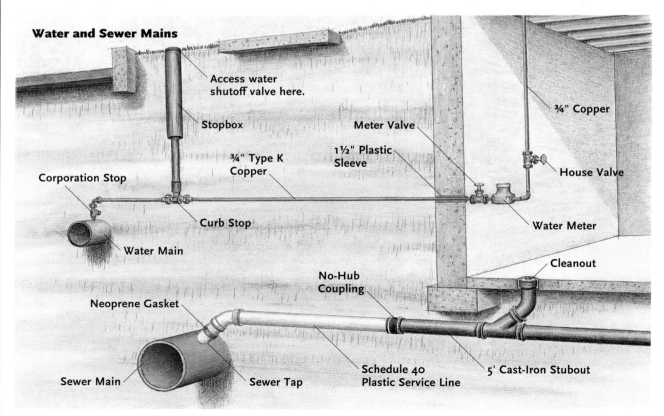

Water and Sewer Mains

Access water shutoff valve here.

Stopbox

¾" Copper

Meter Valve

1½" Plastic Sleeve

House Valve

¾" Type K Copper

Corporation Stop

Curb Stop

Water Meter

Water Main

Cleanout

No-Hub Coupling

Neoprene Gasket

Sewer Main

Sewer Tap

Schedule 40 Plastic Service Line

5' Cast-Iron Stubout

Water Meters

The water meter marks the end of the service line and the beginning of the in-house plumbing system. In some cases, the city owns the meters, while in others, users buy and maintain them at their own expense. Meters are usually inside the house, in a utility room, basement, or crawlspace, but some are outside, buried in meter pits. Meter pits are usually near the street, on either side of the sidewalk. If the meter is in a pit, the system rarely has a curb stop. In this case, the system has an additional valve just inside or outside the house, depending on the climate.

Types of Service Line

Service Lines in Older Homes. In homes built before World War II, the original service line is usually galvanized steel with a ¾-inch inside diameter. Pipes predating

1930 may terminate in a lead pipe loop. To identify a lead loop, look for a blackish gray pipe that curves between the meter valve and the galvanized-steel service line. You should see a bulge at each end where the loop joins the pipe and meter valve. These hand-formed splices are called wiped-lead joints and are now a lost art. Plumbers made them with candle wax and molten lead, sweeping the lead repeatedly around the joint with the aid of a heat-resistant glove.

Plumbers also used lead underground at the water main because it was soft enough to accommodate seasonal ground movement. If you see a lead loop at the meter-end of your service line, expect one at the other end as well. While clearly obsolete, millions of lead loops are still in place, even though lead is now known to be extremely toxic. In most cases, mineral deposits sealed the water from the lead long ago.

Service Lines in Newer Homes. Homes built after 1950 have soft-copper service lines. Again, the predominate size was ¾ inch, though larger diameters—up to 1¼ inch—were sometimes used to compensate for low pressure. Today, 1-inch services are common, especially when plans include underground sprinkler systems. If your system draws water from a recently installed private well, expect to see plastic lines in place, 1 to 1¼ inches in diameter. Most plumbers place a shutoff valve on the well side of the pressure tank. (Never shut off this valve without also shutting off power to the pump.)

Lead Pipe Loop

Galvanized-Steel Service Line

Meter Valve

Wiped-Lead Joint

THE MIDCENTURY BENCHMARK

The 1950s were watershed years in home building, with methods and materials changing dramatically. A house built in the 1950s not only looks different from its 1940s predecessor, it is in fact built differently. In the 1950s, balloon framing was out; platform framing was in. Plaster was out; drywall was in. Knob-and-tube wiring was out; sheathed cable was in. Galvanized-steel piping was out; copper tubing was in. And so on.

The end of World War II had a great deal to do with these changes. Thousands of returning young soldiers sought affordable housing through the GI Bill

as they settled down to start families. Builders began to use mass-production building methods to meet the demand for housing, and once-rationed materials of every description were readily available. Not surprisingly, some of the 1950s tract homes, furrowed in neat rows, bore a noticeable resemblance to military-base housing. In any case, this midcentury benchmark is useful when assessing the best approach for repairs and upgrades. Different methods and materials require different approaches.

Plumbing Basics

Interior Water Supply Systems

A home's in-house water system starts with the water meter or, in the case of a private well, a pressure tank. If your home has an outdoor meter pit, consider the first full-size shutoff valve in the house as the starting point. From here, a single supply line—the cold-water trunk line—travels to a central location where a T-fitting splits it into two lines. One of these enters the top of a water heater while the other continues to feed cold water to the house. A third line exits the heater and becomes the system's hot-water trunk line. The hot-and-cold-water trunk lines usually run side by side along the center beam of the house, branching to serve fixtures along the way.

Smaller houses generally have ¾-inch-diameter trunk lines that run under the floor joists. Branch lines will be reduced to ½ inch at some point. Larger homes with more fixtures or especially high-volume fixtures, such as body-massage showers, will require larger pipes— at least at the beginning. You may need to run 1- or 1¼-inch pipe from the meter to the first fixture group and water heater, even if the meter and service lines are ¾ inch. Check with your local code office for specific size requirements.

NO BASEMENT?

WHERE ARE THE WATER PIPES if you live in a house without a basement? It depends on the type of home. If your ground floor spans a crawl space, the sytem is likely to resemble that of a basement installation, except that the water heater may be located on the main floor in a utility closet.

If yours is a slab-on-grade home with a concrete floor, expect to find soft-copper water tubing buried under the concrete slab. The water service in this case will usually enter a utility room through the floor. In most cases, the fixture supply lines will also be run under the slab, surfacing in the utility room near the meter on one end, and near each fixture or group of fixtures on the other. In the extreme southern reaches of the country, where hard freezes are unlikely, copper or plastic water lines may be run in the attic, with branch lines dropping into plumbing walls.

WHERE ARE THE WATER AND SEWER MAINS HIDDEN?

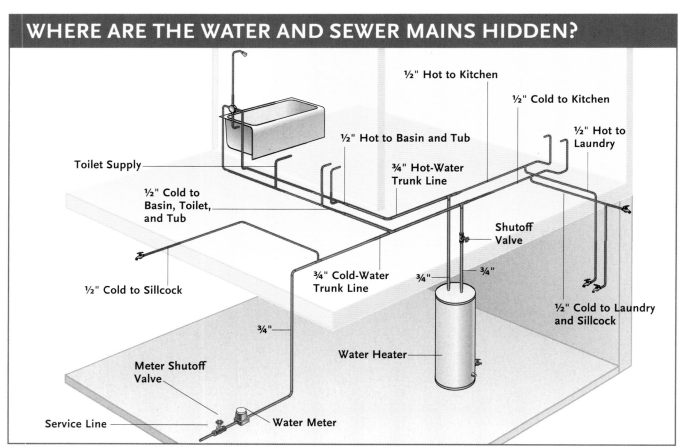

½" Hot to Kitchen

½" Cold to Kitchen

½" Hot to Basin and Tub

½" Hot to Laundry

Toilet Supply

¾" Hot-Water Trunk Line

½" Cold to Basin, Toilet, and Tub

Shutoff Valve

¾" Cold-Water Trunk Line

¾"

¾"

½" Cold to Sillcock

½" Cold to Laundry and Sillcock

¾"

Water Heater

Meter Shutoff Valve

Service Line

Water Meter

TYPES OF PIPING AND TUBING

	Characteristics	Cutting Tools	Joining Method
	PVC Pipe. Polyvinyl chloride plastic pipe is the preferred drain and vent piping for houses. It can't corrode, and it's easy to assemble.	Wheel cutter, hacksaw, scissor cutter	PVC solvent cement
	ABS Pipe. Acrylonitrile butadiene styrene is a black plastic used in the same applications as PVC. It is not as rigid as PVC.	Wheel cutter, hacksaw	ABS or PVC solvent cement
	Cast-Iron Pipe. Once used in drain and vent systems, this durable but brittle metal pipe has been largely replaced by PVC and ABS plastic.	Snap cutter, chisel	Banded neoprene couplings or rubber gaskets
	Rigid Copper Tube. The dominant water-delivery material today, copper tubing is usually joined with soldered (sweat) fittings.	Wheel cutter, hacksaw	Sweat or compression fittings
	Soft Copper Tube. Used primarily for natural gas and propane but also for water, this tubing is allowed under concrete.	Wheel cutter, hacksaw	Compression, solder, or flare fittings
	Chromed Copper Tube. This flexible tubing is used as fixture water-supply tubes between fixtures and permanent water lines.	Wheel cutter, hacksaw	Compression fittings
	Flexible Braided-Steel Supply Line. This flexible tubing, often used as fixture supply tubing, is easier to use than chromed copper tubing.	Fixed length	Factory-installed fittings
	Chromed Ribbed Copper Tube. Available only as fixture supply tubing, the ribbed section of this tube makes it easy to bend.	Can't cut	Compression fittings
	CPVC Pipe. Chlorinated polyvinyl chloride plastic water piping was created to replace rigid copper. Does not meet all local codes.	Wheel cutter, hacksaw, or scissor cutter	PVC cement or compression or crimp ring fittings
	Pex Tube. Cross-linked polyethylene plastic tubing is a flexible material gaining acceptance for in-house water systems. It requires few fittings.	Scissor tool, hacksaw	Several brands of proprietary fittings
	Galvanized-Steel Pipe. Once used for in-house water systems, steel pipe is now used mostly in repair situations.	Wheel cutter, and threading dies	Threaded fittings
	Black Steel Pipe. Steel pipe was once used for in-house gas piping, though it's fast losing ground to soft copper and CSST.	Wheel cutter and threading dies	Threaded fittings
	CSST. Corrugated stainless-steel tubing is a flexible, plastic-coated tube made of stainless steel for in-house natural gas and propane.	Hacksaw or wheel cutter	Proprietary compression fittings

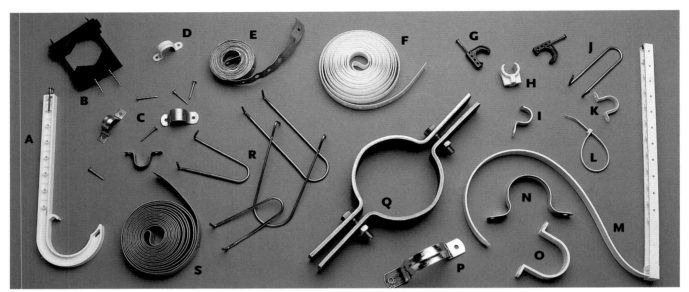

Hangers and Straps: A—plastic waste-pipe hanger, **B**—plastic water-tube support, **C**—copper tubing straps, **D**—plastic pipe strap, **E**—copper hole strap, **F**—plastic hole strap, **G**—plastic pipe hook, **H**—plastic pipe clamp, **I**—conduit strap, **J**—copper tube hook, **K**—plastic pipe strap, **L**—zip tie, **M**—plastic drainpipe support, **N**—galvanized drainpipe strap, **O**—plastic drainpipe strap, **P**—galvanized drainpipe strap, **Q**—stack clamp, **R**—wire pipe hooks, **S**—plastic hole strap

Copper or Plastic? While copper tubing still dominates the market and likely will for a long time, plastic is gaining ground, especially in warmer climates. In some areas, including parts of the Desert Southwest, acids in the soil attack copper tubing, so plastic really is a better choice. And where codes allow plastic underground, you will usually find it approved for aboveground use as well. Do-it-yourselfers have long favored chlorinated polyvinyl chloride (CPVC) plastic water pipe because it requires little in the way of special skills or tools. Newer cross-linked polyethylene (PEX) plastic tubing is gaining in popularity with DIYers because of its ease of use. This white, red, or blue tubing is flexible and has little coil memory, so you can unwind it easily. You stretch it over rigid brass or plastic fittings to make a seal. Although it may require a special fitting tool and is usually connected to copper stubs at fixtures, PEX tubing saves labor. It can also take a freeze better than copper or CPVC and is impervious to corrosive water or soil.

Connecting Fittings. Fittings for water lines are limited. They include male and female threaded adapters, 90-degree elbows, 45-degree elbows, T-fittings, couplings, and unions. Copper and brass fittings that are made to be soldered are called sweat fittings.

Elbows, as the name implies, create right-angle turns. Couplings join pipes end to end, and Ts split one pipe run into two separate lines. All of these fittings are available as copper or brass sweat types for copper tubing; threaded steel or brass types, which are threaded onto steel or copper lines, respectively; and plastic types, which are cemented to plastic pipes. Push-fit and barbed fittings may be used with some plastic pipe.

You'll use a special coupling called a *dielectric union* when you need to connect copper to iron. This fitting prevents electrolytic corrosion, which is caused by electrochemical reactions that take place when the dissimilar metals like iron and copper come in direct contact in the presence of water. Corrosion rates vary widely by locale, so some codes require dielectric unions and others don't.

Fixture and Appliance Connections

Most fixture, faucet, and appliance connections are mechanical, meaning that they can be taken apart. (Two exceptions are those on water heaters and tub/shower faucets.) Sinks, waste-disposal units, laundry hookups, dishwashers, toilets, and bidets all have mechanical connections, which make them easy to service and replace.

The transition fittings that join permanent pipes to fixture-supply tubes take several forms. Prior to the 1950s, most of these connections were made with cone-shaped rubber washers, called *friction washers*. Plumbers simply slid a friction nut and cone washer onto a ¼- or ⅜-inch supply tube and inserted the tube into the larger ½-inch riser tube. They then threaded the nut onto the tube, forc-

Fittings: A—¾ x ½ x ½-in. T-fitting, **B**—drop-eared elbow, **C**—brass union, **D**—90-deg. street elbow, **E**—dielectric union, **F**—copper sweat cap, **G**—sweat × female adapter, **H**—tubing strap, **I**—repair coupling, **J**—coupling, **K**—reducing coupling, **L**—¾ x ½ x ¾-in. T-fitting, **M**—¾-in. T-fitting, **N**—½ x ½ x ¾-in.T-fitting, **O**—¾ x ¾ x ½-in.T-fitting, **P**—45-deg. street elbow, **Q**—sweat × male adapter, **R**—90-deg. elbow, **S**—45-deg. elbow

ing the tapered washer between the water line and supply tube to make the seal.

Since the 1960s, compression fittings—both adapters and shutoff valves—have almost completely replaced cone washers. A compression fitting consists of a brass body with a tapered seat at top; a beveled compression ring, or *ferrule*; and a compression nut with a tapered inner rim. You slide the nut onto the supply tube, and follow it with the ferrule. When you insert the tube and tighten the nut, the pressure drives the ferrule into its tapered seats

until it compresses around the supply tube and locks into place.

Compression adapters and shutoffs—stops—are available to fit both copper tubing and threaded steel pipe risers. Those made for ½-inch steel pipe have female threads at their lower ends, while those for copper have a second, larger compression nut and ferrule that grips the copper. While adapters do a good job of connecting risers to supply tubes, many codes now require shutoffs. With shutoffs under every fixture.

FRICTION AND COMPRESSION FITTINGS

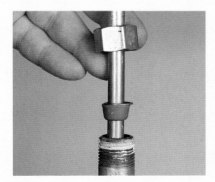

Cone washers were common before compression fittings.

Modern installations use brass compression valves and fittings.

Compression valves are also available with pipe threads.

Plumbing Basics

Supply Tubes. Supply tubes take several forms. The oldest and simplest is the copper ball-head tube, which is usually chrome plated. It's a simple ¼-inch flexible copper tube with a formed ball at the top. To make the connection, you place the ball against the tapered inner rim of the faucet shank and tighten a coupling nut over it. As the nut draws down, the formed-copper ball crushes against the shank's taper, making the seal. You trim the lower end to length, and join it to the supply riser with a compression adapter or shutoff valve.

A new user-friendly version of the supply tube is the stainless-steel-enmeshed polymer tube (or flexible braided-steel tube), which is made for both sinks and toilets. Available in several lengths, it comes with compression and coupling nuts already installed. Homeowners much prefer this braided-steel tube because they don't have to trim it to fit, bend it into shape, or deal with separate fitting components. They just buy the lengths they need and tighten the coupling nuts in place. These tubes cost a few dollars more than their traditional counterparts, but they are usually worth it. Braided nylon and polybutylene versions are also available, but again, code compliance may be a problem depending on local codes.

A simple extension tube with a compression coupling is sometimes useful. Some modern faucets don't have threaded brass or plastic shanks. Instead, they have ¼-inch copper supply tubes brazed directly to the faucet body. These are usually about 18 inches long, but when the water pipes come through the floor, extension tubes are in order.

DRAINAGE & VENT SYSTEMS

A plumbing drainage system has three basic segments, each with its own function: drainpipes, vent pipes, and fixture traps.

- **Drainpipes** direct wastewater away from the fixtures and house to the sewer, septic tank, or cesspool.
- **Vent pipes** allow air to enter the drainage system to equalize air pressure, allowing the wastewater to flow freely, and prevent suction.
- **Fixture traps** hold a small amount of water at fixtures to prevent the passage of sewer gas and vermin from the drainpipes into a living area.

Unlike potable-water systems, which flow under pressure, waste systems operate by gravity. Consequently, designers of these fittings emphasize gradual flow patterns and broad sweeps instead of abrupt turns. The abrupt geometry of water fittings is too severe for drain fittings. It's a difference you can see.

Sizes and Materials

As noted earlier, drainpipes are larger than water lines, ranging in diameter from 1¼ to 4 inches. A pipe 1½ inches in diameter or larger is likely to be a drainpipe or vent pipe; most 1¼-inch pipes are drainpipes only. Galvanized steel, copper, cast iron, and plastic are all common drainage and vent-pipe materials. Cast-iron stacks were standard for decades, usually with galvanized-steel or copper

TYPES OF SUPPLY TUBES

The ball head of this water supply tube crushes against the faucet shank.

Flexible braided-steel water supply tubes come equipped with factory-installed fittings.

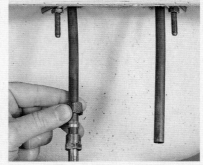

Compression coupling tubes are often used for faucets that have built-in supply tubes.

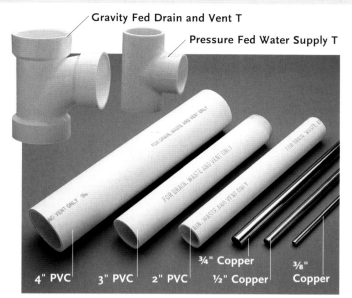

Gravity Fed Drain and Vent T

Pressure Fed Water Supply T

4" PVC | 3" PVC | 2" PVC | ¾" Copper | ½" Copper | ⅜" Copper

The plastic pipes carry wastewater; the copper tubes, fresh water. These are the most common sizes for modern homes. Note the gradual flow pattern in the drain T-fitting.

The Drainpipe System

The in-house waste drainage system starts with a main pipe several feet outside an exterior wall, which passes under the house footing and into the basement or crawl-space. This below-floor drainpipe, usually 4 inches in diameter, is called the *main soil pipe*. (Any pipe carrying solid waste is technically a soil pipe.) The main soil pipe continues horizontally under the basement floor or slab (or along a wall), sometimes branching off—and reducing—to serve a laundry standpipe, maybe a floor drain or two, and possibly a basement bathroom and kitchen riser. Where the main soil pipe runs under the basement floor, it terminates in a 90-degree sweep bend through the floor and becomes the base for the primary vertical stack, or the main stack. Smaller drain stacks also come off of the main soil pipe and continue upward. These secondary drain and vent stacks may tie into the main stack above the highest fixture or pass through the roof independently.

Every home should have a full-size (3- or 4-inch-diameter) stack that travels from the soil pipe to roughly 12 inches above the point of exit through the roof. Think of this main stack as the trunk of a tree. At each floor, horizontal branch lines reach out to serve individual fixtures or fixture groups.

branch lines. Galvanized steel was used through the 1940s, and copper and brass emerged in the 1950s. The 1950s and 1960s also saw the installation of a good many copper and brass systems. After the mid-1970s, drain/vent systems usually tended to be plastic, either black acrylonitrile-butadiene-styrene (ABS) or white polyvinyl chloride (PVC).

DRAIN AND VENT LINES

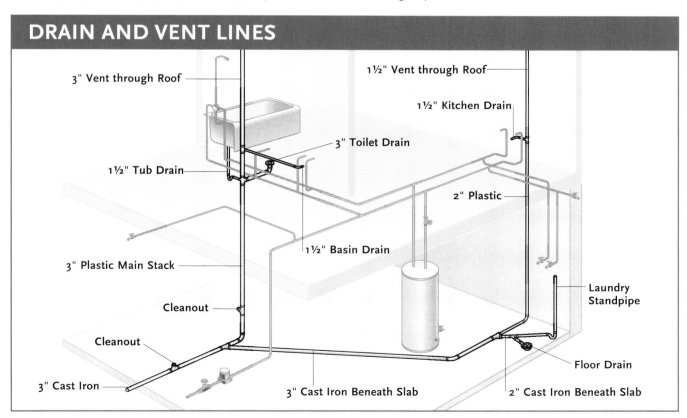

3" Vent through Roof

1½" Vent through Roof

1½" Kitchen Drain

3" Toilet Drain

1½" Tub Drain

2" Plastic

1½" Basin Drain

3" Plastic Main Stack

Cleanout

Laundry Standpipe

Cleanout

3" Cast Iron

Floor Drain

3" Cast Iron Beneath Slab

2" Cast Iron Beneath Slab

TYPES OF DRAIN TRAPS

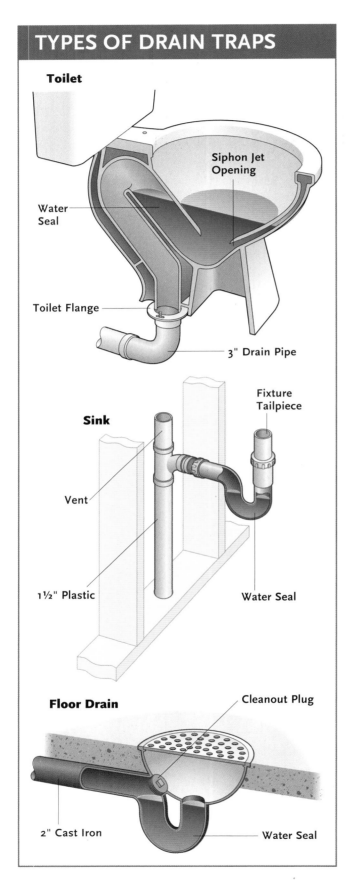

Toilet

Siphon Jet Opening

Water Seal

Toilet Flange

3" Drain Pipe

Sink

Fixture Tailpiece

Vent

Water Seal

1½" Plastic

Floor Drain

Cleanout Plug

2" Cast Iron

Water Seal

Drain Traps. Each fixture in the system is joined to its branch line via a water trap. These are critical components because they hold back the considerable volume of sewer gas present in every sewer system. While every fixture and drain must have a trap, not all traps are the same. Toilets, for example, have built-in, or *integral*, traps. The water you see standing in a toilet bowl is trapped there by an outlet passage that sweeps up before it sweeps down to its drain-pipe connection. Floor drains also have integral traps, as do bidets and some urinals.

Sinks have sharply curved external chrome or plastic tube traps (P-traps), which can be disassembled for service work. Tub, shower, and laundry traps, in contrast, are usually fixed one-piece units.

Vent Pipes. Vents are important to a drainage system because they allow traps and drains to function properly. When water flows from a fixture through a pipe, it displaces an amount of air equal to its own volume, creating negative pressure behind the flow. This localized suction can be quite strong, especially at bends in the pipe. A toilet flushing near a sink, for example, can easily pull water from the sink's P-trap, allowing poisonous sewer gas into the living quarters. In fact, without adequate venting, a toilet won't flush properly.

Every home needs a stack vent through the roof, of course, but that's not always enough. All sorts of common situations can choke a vent, so it's necessary to have auxiliary vents, called *re-vents*. The shape, size, and location of these vents are critically important. (See Chapter 13, starting on page 274, for an in-depth discussion of venting.)

Fixtures. The point of all this piping begins and ends with the fixtures: sinks, toilets, tubs, and shower stalls. Fixtures are not as permanent as they appear. They are designed to be taken up and put back with relative ease and at moderate expense. Even bathtubs, which can look as though they've grown right out of the structural timbers of a home, are not that difficult to replace. If an old or defective fixture has you mumbling to yourself with every use, don't be intimidated: tear it out and put in a new one.

Appliances. The list of plumbing-related appliances has grown over the years to include water heaters, dishwashers, water softeners, water purifiers, clothes washers (which are not really plumbed in), waste-disposal units, hot-water dispensers, whirlpool tubs—and even refrigerators, with their ice makers. However, only the water heater is an essential and code-required part of every home's plumbing system.

TOOLS

It's easy to get carried away when buying tools, so focus on the basics first. A good rule is to hold off buying specialty tools until you need them. And if you can't imagine needing a large, expensive tool more than once or twice, rent it if possible. You're likely to find many large tools, like chain wrenches and oversize pipe wrenches, in the local rental store. And finally, if you need to do special work on piping, many full-service hardware stores will cut, thread, or flare pipes for you in the store.

Of the specialty holding and turning tools shown here, the basin wrench and spud wrench are the real problem solvers. They will save you a lot of headaches, and they are not that expensive to buy.

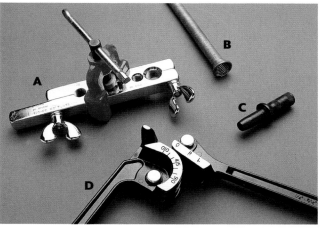

Forming Tools: A—clamp-type flaring tool, **B**—spring-type tubing bender, **C**—hammer-type flaring tool, **D**—mechanical tubing bender

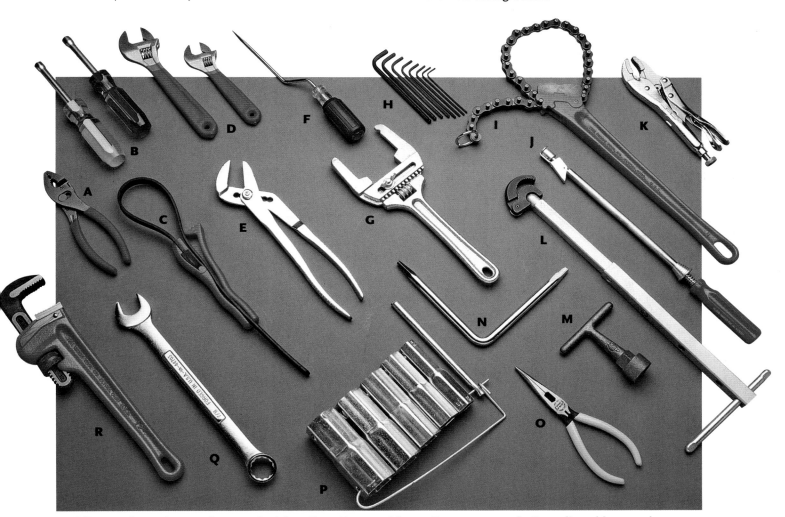

Holding and Turning Tools: A—slip-joint pliers, **B**—nut drivers, **C**—strap wrench, **D**—adjustable wrenches, **E**—groove-joint pliers, **F**—offset screwdriver, **G**—spud wrench, **H**—Allen wrenches, **I**—chain wrench, **J**—sink-clip (Hootie) wrench, **K**—locking pliers, **L**—basin wrench, **M**—stop-box wrench, **N**—faucet-seat wrench, **O**—needle-nose pliers, **P**—deep-set faucet sockets, **Q**—combination wrench, **R**—pipe wrench

Plumbing Basics

Shopping for Tools

The right tools really do make the work easier and, in the end, better. While you can certainly cut water and gas piping with a hacksaw, a wheel cutter is faster and leaves a cleaner pipe edge. This may seem insignificant, but a crooked or ragged pipe end can affect fit and create turbulence as water passes through it. Excess turbulence can erode the pipe wall. The little pipe reamer shown below serves a similar purpose with plastic water pipe.

Similarly, a high-quality saber saw with an auto-scroll feature cuts more accurately and is less damaging to countertop laminates than bargain saws. And finally, no one who remembers life before cordless drills will ever take them for granted. If you doubt their contribution to quality, count the number of screws used in older homes.

DRILL BITS

MOST DRILL BITS cut holes, but they provide differing results. The two self-feed bits shown here are for rough-in work, for chewing quick, crude holes through framing lumber. They work best on a right-angle drill. The hole-saw, in contrast, cuts a neat, precise hole, so it's better suited to drilling countertops and tub-shower surrounds, where chipping may be a problem. Speed bits are good starter bits. They bore a quick, moderately clean hole. Spiral bits work best when you need to drill a lot of holes; the spiral literally pulls the bit through. The high-speed bit is good for small holes in metal.

Cutting Tools: A—scissor-type tubing cutter, **B**—wheel cutter, **C**—plastic-pipe saw, **D**—reciprocating saw, **E**—rat-tail and slim tapered files, **F**—bastard-cut (flat) file, **G**—multi-tool, **H**—utility saw, **I**—hacksaw, **J**—utility knife, **K**—cold chisel, **L**—plastic-pipe reamer, **M**—faucet valve seat grinder, **N**—saber saw, **O**—miniature tubing cutter

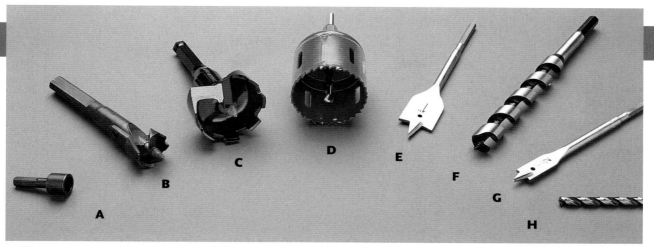

Drill Bits: A—nut driver, **B**—1-in. self-feed bit, **C**—2-in. self-feed bit, **D**—hole saw, **E**—speed bit, **F**—auger bit, **G**—speed bit, **H**—high-speed bit

Common General Tools: A—claw hammer, **B**—multitester, **C**—stud finder, **D**—flashlight, **E**—chalk-line box, **F**—caulking gun, **G**—aviation snips, **H**—cordless drill/driver, **I**—measuring tape, **J**—2-ft. level, **K**—small sledgehammer, **L**—flat-blade screwdriver, **M**—Phillips screwdriver, **N**—putty knife, **O**—torpedo level, **P**—drywall taping knife

Plumbing Tools

Most plumbing tools are very specialized, so they have to earn their keep on plumbing projects alone. But basic tools, like plungers, pipe wrenches, and drain augers, should be part of every homeowner's tool kit.

The snap-cutter for cast-iron pipe and threading die for galvanized or black steel piping probably occupy the far end of the spectrum. With such limited application, snap-cutters and dies are almost always rental tools. The rest of these tools are more house specific. If you plan to add a bathroom or do other major additions or rerouting to copper water tubing, then you need a soldering torch. If you have small children, you may find yourself needing a closet auger to retrieve small toys or other objects flushed down the toilet.

PLUMBING SUPPLIES

The products shown at right cover a variety of projects. Silicone and latex tub and tile caulk are essential for preventing leaks around sinks, tubs, and showers. Silicone is more durable and generally lasts longer before you must replace it. You'll need PVC primer and solvent cement when working with PVC drainpipes and CPVC water pipes. Both PVC cement and ABS cement work on ABS plastic pipe. You'll need pipe joint compound or pipe-thread tape for sealing threaded connections. For soldering copper, you'll need flux, solder, and abrasives to clean the tubing.

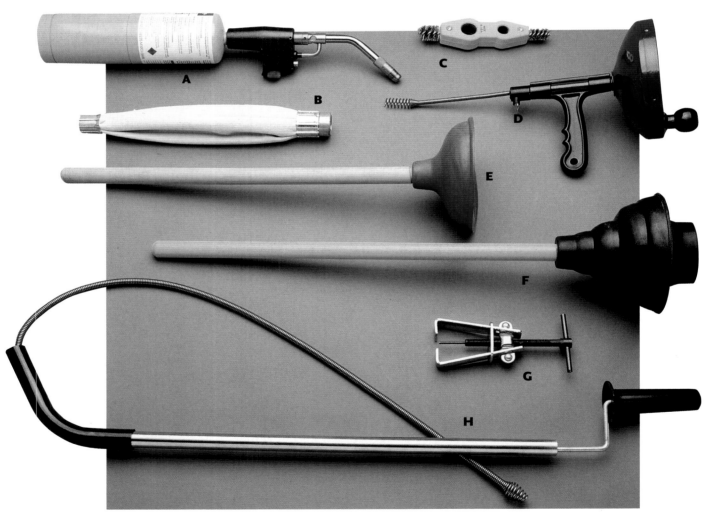

Common Plumbing Tools: A—soldering torch, **B**—blow bag, **C**—pipe and fitting cleaning tool, **D**—hand auger, **E**—standard plunger, **F**—combination plunger, **G**—handle puller, **H**—closet auger

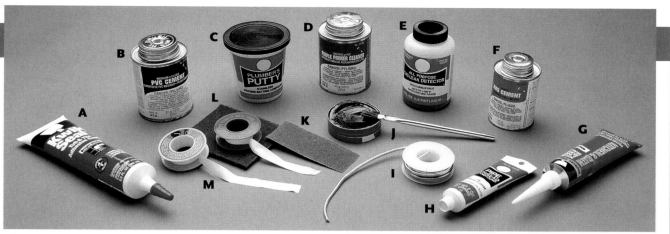

Plumbing Supplies: A—latex tub and tile caulk, **B**—PVC solvent cement, **C**—plumber's putty, **D**—PVC primer, **E**—leak-detection fluid, **F**—ABS solvent cement, **G**—silicone caulk, **H**—pipe joint compound, **I**—solder, **J**—flux, **K**—emery cloth, **L**—abrasive pad, **M**—pipe-thread sealing tape (yellow spool: gas, blue spool: water)

Rental Tools: A—power drain auger, **B**—steel-pipe cutter, **C**—steel pipe rachet threading die, **D**—right-angle drill, **E**—snap-cutter for cast-iron pipe, **F**—snap-cutter drive wrench

12
planning plumbing
changes &
additions

IF YOU ARE PLANNING to make improvements that require installation of plumbing fixtures or appliances where there are no water or drainage facilities, draw up a detailed plan to scale before proceeding. Once you have prepared the plan (along with a list of materials), have someone with experience doing similar projects check to see whether you have overlooked anything. The local plumbing-supply store may have trained consultants on staff who can provide this service. The two main tasks ahead of you will be, 1) to make sure your plans comply with code, and 2) to make accurate rough-in measurements.

BUILDING CODES

A building code is a collection of legal statutes that specify which building materials may be used and how these materials are to be assembled in the construction of residential and commercial buildings. A code office has legal authority within its jurisdiction, be it state, county, or municipality. A typical building code covers everything from framing to concrete installations to ventilation and insulation. To comply with code, you must secure a building permit from the local building department and allow the building inspector or inspectors access to the work for inspection and approval at various stages of completion.

Lack of Uniformity. Unlike electrical codes, which are written at the national level (the National Electrical Code, or NEC) and enforced (and sometimes modified) at the state, county, and city level, plumbing codes are often written and enforced only at the city or county level. So while electricians normally carry both state and county licenses, plumbers are usually licensed only at the city/county level. Plumbers who work in more than one county or city must often have more than one license.

Many small towns and rural areas remain entirely without plumbing codes. More and more codes are going on the books in these areas, but the conversion is far from complete and may never be. The most far-reaching enforcement bodies are state and county health departments, which have little to do with home construction but claim jurisdiction over private wells and septic systems. You should check with the local building department to see what plumbing codes, if any, are enforced.

Properties financed with the aid of government guarantees, such as those of the U.S. Department of Housing and Urban Development's Federal Housing Administration (FHA) and the U.S. Department of Veterans Affairs (VA), must meet the guidelines established by these agencies in addition to meeting any applicable codes. These federal guidelines are not considered codes and have more to do with home financing than home building.

Permits and Inspections

If you're planning to install a permanent appliance such as a water heater or to make a plumbing upgrade that will require piping changes or additions, visit the local building department and apply for a permit. Explain exactly what plumbing is in place and how you plan to expand or change it. If your plan is not likely to meet applicable codes, the code official will most likely be ready with suggestions to make it code-worthy.

Once you have worked out the details of your plan, you'll need to fill out a permit application. The form will probably have a list of fixtures and project descriptions. In most cases, you'll simply check the boxes next to each appropriate fixture or work category. The cost of the permit is usually determined by the number of boxes you check, but don't let this intimidate you. Permit fees are seldom expensive, especially for simple upgrades.

Inspections. You'll need to call for an inspection, usually a minimum of a day ahead of time, at the completion of each stage of work. Try to be present at the inspections. At the very least, arrange for a friend or family member to open the door for him or her.

Just how many plumbing inspections will you need? It depends. If your remodeling involves in-wall piping and fixture installations, you'll need two inspections—a rough-in inspection and a finish inspection. But if you're simply replacing a water heater, where all of your work will be visible at a glance, then only a final inspection will be necessary. (New-home construction usually requires four plumbing inspections: one for the underground sewer and water service lines and taps, one after the installation of any under-slab or basement floor piping, another after the installation of above-floor rough-in piping, and a final inspection when the work is complete.)

Remember that an inspector must approve all in-wall and underground work before you cover it up. A building inspector has the authority to make you remove drywall, soil, or even concrete if he or she suspects substandard work.

LAYOUT & FEASIBILITY

Before making any plumbing improvements, you'll need to assess feasibility. To do that, you'll need to know how much space each new fixture will require, both practically and as a matter of local codes. You'll also need to know the rough-in specifications for each new fixture. Most are standard, meaning that they'll work with any common brand of fixture. If you choose a specialty fixture with nonstandard dimensions, make certain that you get the rough-in measurements from the manufacturer's product literature or from the retailer before you start.

Standard Rough-In Measurements

Roughing-in piping for fixtures and appliances in exactly the right places is critical because finished walls and cabinets usually cover the piping by the time you install the fixtures. Going back to reposition connector fittings at the finishing stage is costly and time-consuming. The following sections give standard fixture and appliance rough-in measurements. Note: all measurements are taken from the center of the pipe or fitting.

smart tip

GRANDFATHER CLAUSE TO THE RESCUE

IF YOU'RE LIKE MANY HOMEOWNERS, WHAT KEEPS YOU FROM APPLYING FOR PERMITS AND CALLING FOR INSPECTIONS IS THE FEAR THAT ONCE AN INSPECTOR DARKENS YOUR DOOR, HE OR SHE WILL STORM THROUGH THE HOUSE CONDEMNING EVERYTHING IN SIGHT. IS THIS POSSIBLE? PERHAPS, BUT IT'S NOT LIKELY. UNLESS INSPECTORS SEE SOMETHING THAT IS CLEARLY A HEALTH HAZARD, THERE ISN'T MUCH THEY CAN DO.

IF YOU HAVE WATER RUNNING PROPERLY TO THE BASIC FIXTURES, REASONABLY FEW APPLIANCES PLUGGED INTO WORKING ELECTRICAL OUTLETS, NO MISSING STEPS, AND WALLS AND A ROOF THAT APPEAR AS IF THEY'LL STAY PUT, YOU ARE PRETTY MUCH COVERED BY THE GRANDFATHER CLAUSE. THE CLAUSE BASICALLY APPLIES COMMON SENSE BY STATING THAT NEW STANDARDS CANNOT BE APPLIED TO OLD WORK UNLESS THAT WORK NOW POSES A GENUINE HEALTH RISK TO INHABITANTS OR PASSERSBY. IF OLDER INSTALLATIONS MET THE STANDARDS OF THEIR DAY, THEY WILL DO UNTIL YOU DECIDE TO IMPROVE THEM. WHEN YOU MAKE THOSE IMPROVEMENTS, HOWEVER, CURRENT STANDARDS WILL APPLY.

PLUMBING 101: Do You Need a Permit?

IN AREAS WITH ENFORCED PLUMBING CODES, only those plumbing projects with pipe changes and permanent appliance installations generally require permits and inspections. If, for example, you plan to take up your old toilet and replace it with a new one, you usually don't need a permit or inspection. But if you plan to move the toilet from one wall to another, which would require changing the permanent piping, you'll need to apply for a permit and have your work inspected and approved. The reasoning here is simple: replacing a fixture does nothing to compromise the plumbing system, but piping changes can alter the balance between drain and vent segments; therefore, they deserve closer scrutiny.

Water heaters are another case in point. The issue, as always, is safety. Code officials need to know whether you've vented your gas water heater adequately or wired your electric heater properly. They'll also check the temperature-and-pressure-relief valve. If the relief-valve installation is wrong, the heater can explode when its thermostat fails. Too often, homeowners plug relief-valve openings with iron plugs or transfer the old water heater's nearly spent temperature-and-pressure-relief valve to the new heater. In other cases, homeowners install relief valves that have pressure ratings higher than those of the heaters they are meant to protect. Code officials, therefore, have several good reasons to check your water-heater installation.

Water-Heater Relief Valves: Permit Required

New Valve

Clogged Valve That Won't Pass Inspection

Toilets. Most toilets fit into a space that is 22 inches wide, but codes require a 30-inch opening. The toilet flange fitting should be centered in this opening, with at least 15 inches between the center of the toilet flange and the nearest side wall or cabinet. Spot the center of the flange 12½ inches out from the stud wall in the back, or 12 inches out from a finished wall. All standard toilets are set up for this 12-inch rough-in, but toilets with 10-inch and 14-inch rough-ins are available. These work fine, but they're always more expensive. You'd use one if you mismeasured or if a structural framing member kept you from roughing-in the flange at 12 inches. Some plumbers might accommodate the difference of an inch or so by using an offset flange. Offset flanges are not as sturdy as standard flanges, so avoid them if possible. A 12-inch rough-in is the best way to go, even if it means cutting floor joists and installing structural headers on each side of the cut.

When it comes to a toilet's water-supply rough-in, you'll have two choices. You can bring the supply riser through the floor or through the back wall. If you use a floor riser, center it 6 inches to the left of the drain outlet and 3 inches from the back wall. For a wall installation, spot the stub-out 6 inches to the left of the outlet and 6 inches off the floor. The toilet supply tube will handle minor variations.

Kitchen Sinks. The water piping for a kitchen sink should be centered on the eventual placement of the sink. Again, these pipes can enter from the wall or floor. Wall installations make it easier to set the cabinet, however. The standard height for water pipes exiting a wall is 18 inches. Kitchen sinks have 8-inch center spreads, so place each pipe (hot and cold water) 4 inches off center.

The drain piping should exit the wall at least 16 inches above the kitchen floor (not the cabinet floor). With a 36-inch cabinet base (and double-basin sink), position the drain 12 inches in from one or the other side of the cabinet wall. The side from which you'll measure depends on whether you plan to install a waste-disposal unit. If so, position the unit for convenience, and install the drain on the remaining side. With a single-basin sink, locate the drain 2 inches off center in either direction.

STANDARD ROUGH-IN MEASUREMENTS

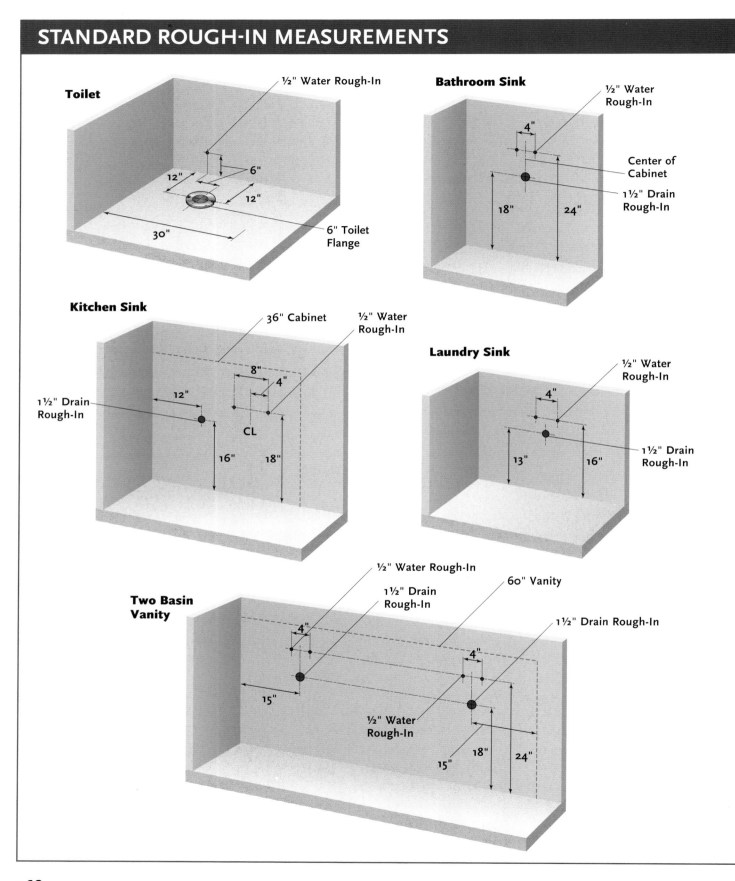

Toilet
- ½" Water Rough-In
- 6"
- 12"
- 12"
- 30"
- 6" Toilet Flange

Bathroom Sink
- ½" Water Rough-In
- 4"
- Center of Cabinet
- 1½" Drain Rough-In
- 18"
- 24"

Kitchen Sink
- 36" Cabinet
- ½" Water Rough-In
- 8"
- 4"
- 12"
- 1½" Drain Rough-In
- CL
- 16"
- 18"

Laundry Sink
- ½" Water Rough-In
- 4"
- 1½" Drain Rough-In
- 13"
- 16"

Two Basin Vanity
- ½" Water Rough-In
- 1½" Drain Rough-In
- 60" Vanity
- 1½" Drain Rough-In
- 4"
- 4"
- 15"
- ½" Water Rough-In
- 15"
- 18"
- 24"

Bathtub

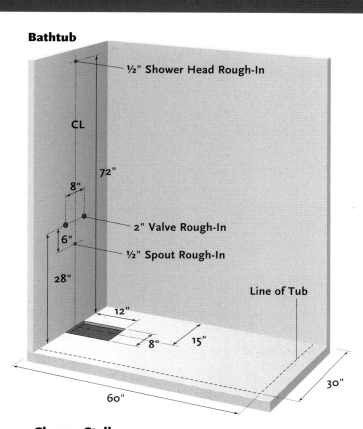

½" Shower Head Rough-In

CL

72"

8"

2" Valve Rough-In

6"

½" Spout Rough-In

28"

Line of Tub

12"

8"   15"

60"   30"

Shower Stall

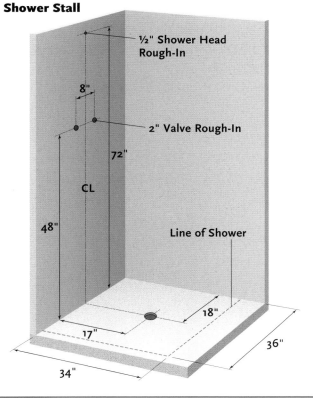

½" Shower Head Rough-In

8"

2" Valve Rough-In

72"

CL

48"

Line of Shower

18"

17"

34"   36"

Bathroom Sink. If you are using a single bathroom sink (sometimes called a lavatory basin), position the water piping a distance of 2 inches from each side of the centerline. For two sinks in the same cabinet, install two sets of stub-outs and two drains. A workable height for in-wall water stub-outs is 24 inches above the floor.

You can center drains inside the vanity cabinet, 18 inches off the bathroom floor. When installing two basins in a larger cabinet, divide the width of the cabinet into four equal parts, and center the two drains on the second and third dividing lines. Pedestal lavatories are always a tight fit, so check the manufacturer's specifications before running any pipe. When installing a wall-hung lavatory, locate the rim of the fixture 32 inches off the floor. The rough-in placements are the same as those for a single-basin vanity.

Bathtubs. Because bathtubs have bulky waste-and-overflow drain assemblies, you must cut a relatively large hole in the bathroom floor, under the drain, to accommodate them. Make the cut before setting the tub. Standard tubs are 30 inches wide, while some fiberglass models and most whirlpool tubs are 32 inches wide. In any case, measure 15 or 16 inches from the back wall to establish the drain center. Then cut an 8 × 12-inch opening in the floor. This will allow you to set the tub, install the waste assembly, and attach the P-trap, in that order.

Standard height for a tub-faucet valve is 28 inches off the floor. That places the spout pipe at 6 inches below the center of the valve and the shower head 72 inches above the floor. While these placements work well, there's nothing particularly special about them. If local codes specify other placements, use those. And if you'd like to raise the faucet and shower head slightly, feel free to do so. Many people prefer a 76-inch position for the shower head because it places the connection above the tops of manufactured tub surrounds.

Shower Stalls. Standard shower-valve height is 48 inches, with the shower head 72 inches above the floor.

Be sure to read the product specifications. In most cases, the drain is centered in the pan. For example, for a 34 × 36-inch pan, measure 18 inches in from the side walls and 17 inches in from the back wall. As with a bathtub, you'll cut the floor opening—which should be at least 6 inches in diameter—and install the pan or one-piece stall before connecting the P-trap. The only case when a trap needs to be piped first is when you are setting a shower on concrete.

Laundry Sinks. Laundry sinks have rough-in placements different from those of other sinks. Plan for a drain outlet no more than 13 inches above the floor and water stub-outs 16 inches above the floor. The drain can be centered or up to 2 inches off center. Laundry sinks, made of fiberglass and ABS plastic, are available in stand-alone, wall-hung, and floor-supported models and in drop-in models for cabinet installations.

Laundry Standpipes. A laundry standpipe (the riser drainpipe) should reach 36 inches above the floor. For in-wall installations, a recessed laundry box is a good idea. In this case, the water lines and standpipe terminate inside the box.

A laundry standpipe must be 2 inches in diameter and have a fixed, 2-inch trap. In basements and slab-on-grade homes, the trap can be below the floor and may not need to be vented. In basements, mount the shutoff valves about 42 inches above the floor. Secure all pipes to the wall to counter the back-shock, or water hammer, caused by a washer's electric solenoid valve.

Designing Efficient Plumbing Installations

The best plumbing configuration for a larger home is the back-to-back bathroom grouping shown on the opposite page. In this two-bathroom arrangement, both tubs, both sinks, and both toilets are stack-vented. Only the first floor kitchen line travels any distance, and so only that line is vented separately. A back-to-back layout also holds water piping to an absolute minimum. Less water piping means less heat loss through pipe walls and, therefore, less wasted energy. Having to drain the cooled water from a long hot-water piping run in order to get the hot water behind it is enormously wasteful and perpetually annoying. In a small home, another way to save piping is to back the kitchen up to the bathroom. If you don't intend to install a dishwasher, you can plan on stack-venting all of the first-floor fixtures.

LAUNDRY STANDPIPES

Cold Water Supply

Hot Water Supply

1½" Vent

Laundry Box

24"

36"

2" Trap

2" Drain

Water Line

smart tip

FREEING UP FLOOR SPACE

If you have to accommodate the cramped space of an extremely small bathroom where a conventional toilet will get in the way, you can use a corner toilet. This kind of toilet has a wedge-shaped tank to fit an inside corner and free up some floor space. The rough-in for this toilet is centered 12½ inches from the rough framing of each of the walls that form the corner.

This corner toilet is shaped to save floor space, but you plumb it as you would any other toilet.

BACK-TO-BACK BATHROOM LAYOUT

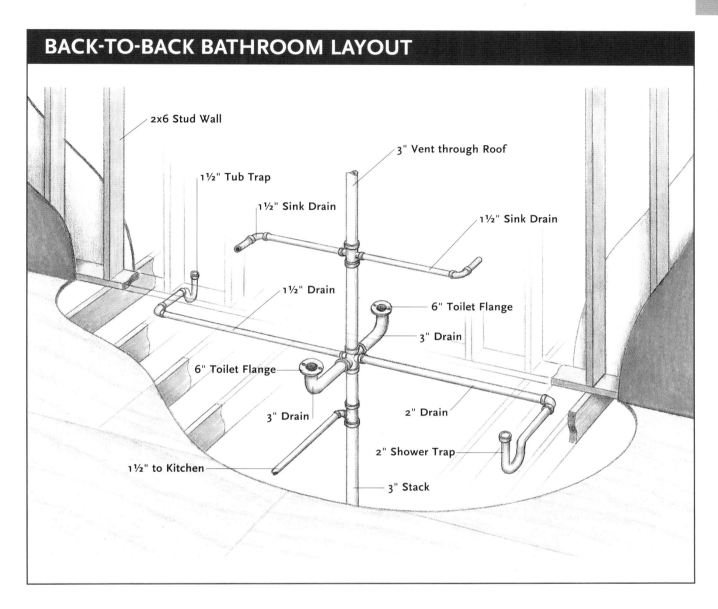

- 2x6 Stud Wall
- 1½" Tub Trap
- 1½" Sink Drain
- 3" Vent through Roof
- 1½" Sink Drain
- 1½" Drain
- 6" Toilet Flange
- 3" Drain
- 6" Toilet Flange
- 2" Drain
- 3" Drain
- 2" Shower Trap
- 1½" to Kitchen
- 3" Stack

When it comes to fixture layout within a bathroom, the most economical and easy-to-vent approach is to place the toilet between the sink and tub/shower. This is a perfect small-bathroom configuration: it saves water pipe, drain-pipe, and fittings. Older homes, in which the sink and tub share a common floor-mounted drum trap, have an older layout. In these cases, the sink occupies the space between the toilet and tub. Whenever possible, remove a trouble-some drum trap, and trap each fixture independently.

Planning for Access

In a few instances, you can install new plumbing without opening walls and ceilings, but these situations are lim-ited. In many cases, you'll need to open at least one side of a wall so that you can install and secure the new piping.

This is especially true when working in multistory houses.

Don't let the notion of tearing into walls and ceilings scare you. Cutting out drywall or plaster is easy, and if you don't feel up to repairing these openings yourself, you shouldn't have a problem finding a drywall or plaster con-tractor to do it for you. In any case, don't let a few dollars' worth of drywall or plaster repair keep you from having the bath, kitchen, bar, laundry, or whirlpool you've always wanted.

If you're simply remodeling a bathroom or kitchen, most of the piping from the old bath or kitchen will do. Al-though you may need to replace or move some of it, there will always be existing plumbing with which to work. But if you hope to bring plumbing to a new area of the house, then planning the pipe route is an important first step.

smart tip

CLUSTER FOR EFFICIENCY

AS A GENERAL RULE, IT'S LESS EXPENSIVE AND LESS COMPLICATED TO CLUSTER PLUMBING GROUPS NEAR ONE ANOTHER AND CLOSE TO THE MAIN STACK. THE FARTHER ONE BATH IS FROM ANOTHER, OR THE FARTHER ANY FIXTURE IS FROM THE MAIN STACK AND WATER HEATER, THE MORE PIPING IS REQUIRED. WHILE THIS MAY SEEM OBVIOUS, THERE'S MORE TO IT THAN MERE DISTANCE. FIXTURES NEAR THE MAIN STACK CAN OFTEN BE STACK-VENTED. WHEN THEY ARE SPREAD OUT, THESE SAME FIXTURES WILL NEED SECONDARY VENTS, OR RE-VENTS, WHICH CAN MORE THAN DOUBLE THE AMOUNT OF PIPING NEEDED FOR THE JOB. (SEE CHAPTER 13, BEGINNING ON PAGE 274, FOR MORE ON VENTING.)

Planning New Vents. If you have a crawl space or unfinished basement, chances are you'll be able to run water and drainage piping easily to any area of the house from existing lines. The trickiest part will be venting your new fixtures. The new vent will need to extend into the attic, where it must either join an existing vent stack or exit the roof directly. If you are working in a two-story home and the plumbing to be installed is in the basement or on the first floor, you may need to open a wall or build an enclosed chimneylike pipe

JOINING VENT PIPES

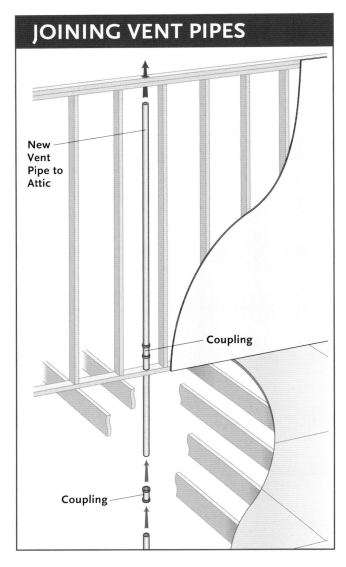

New Vent Pipe to Attic

Coupling

Coupling

ACCOMMODATING VENT PIPES

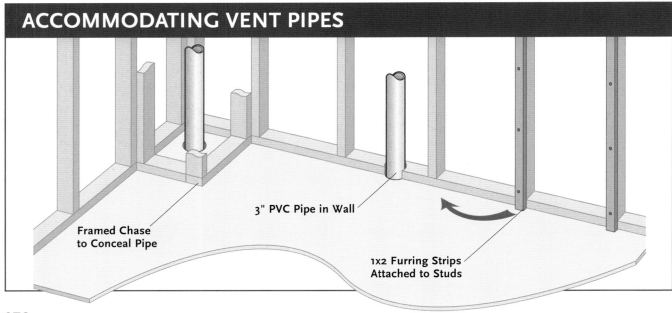

Framed Chase to Conceal Pipe

3" PVC Pipe in Wall

1x2 Furring Strips Attached to Studs

chase to cover the pipes. Some codes may allow an exterior vent as long as it terminates above the roof, away from attic windows or dormers.

If your upgrade involves a single, low-volume fixture—a bar sink, for example—you might consider an automatic vent device or a loop vent. Neither requires open-air access. Both vents are discussed in Chapter 13, starting on page 274.

If you live in a single-story home, venting basement plumbing is relatively easy. You can usually drill the top and bottom plates of an interior wall, and then slide pipe from the basement into the attic. It takes careful measuring, but it's not difficult. If that's not workable, consider running the vent pipe through a first-floor closet. You can then enclose the pipe with studs and drywall or caulk it where it passes through the closet floor and ceiling.

The size of the vent is another consideration. An isolated sink, laundry, or shower will only need a 1½-inch vent. A bath group, consisting of a tub/shower, sink, and toilet, can all be vented with a 2-inch vent as long as your house already has a 3- or 4-inch stack extending through the roof.

If you have a two-story home, you'll likely need to cut access openings in finished walls and ceilings. If you need to extend a 3-inch stack from the basement to the attic—to add an upstairs bath, for example—a conventional wall will be too narrow. You'll need to build a pipe chase or widen the entire plumbing wall about an inch. To widen the wall, nail 1×2 furring strips to the existing studs, and then screw and glue new drywall to these strips.

Before

After

Often, you can leave the fixtures in place when you remodel a bathroom or kitchen. That way you don't have to worry about rerouting plumbing.

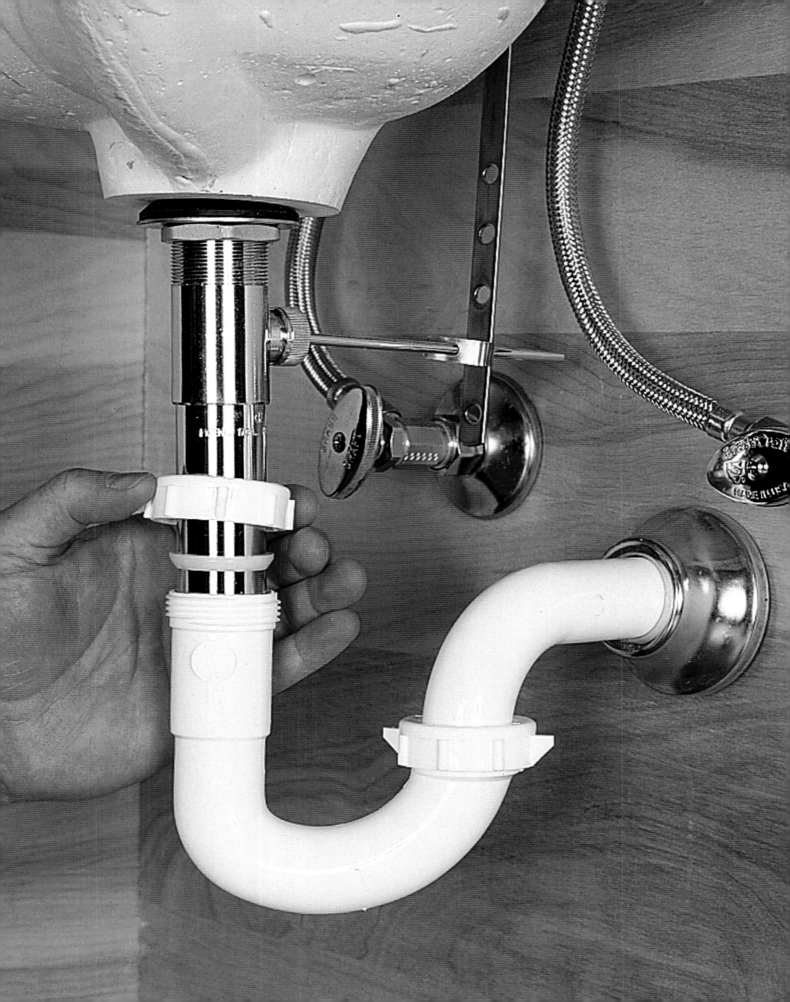

13

drains, vents & traps

DRAINPIPES HAVE THE IMPORTANT JOB of carrying all the wastewater a household generates away from the house and into a sewer or septic system.

The sewer or septic system creates a poisonously gaseous environment, however, and all that stands between you and that environment is the water in your fixture traps. If you lose the water in one of these traps (by evaporation, suction, or some other means), noxious sewer gases will rise into your living quarters, possibly causing headaches and even respiratory illnesses in those exposed to the gases. This is why traps are so important in your drainage system.

A properly designed venting system is the only way to ensure that you maintain adequate water flow in drainpipes and water levels in fixture traps, so vents are also vitally important.

DRAINS

Establishing an easy, gradual flow of wastewater is the prime consideration with drains. (Similarly, airflow is the critical, overriding objective of venting, which is essential to proper drainage.) Therefore, every drainpipe, whether buried under soil or threaded through walls, must be sloped just right. When installing drainpipes, shoot for a slope of ¼ inch per foot. If structural barriers force a compromise, try to maintain at least ¹⁄₁₆ inch per foot.

Too Much Slope. A drain must slope downward if

water is to flow by gravity, but you may not know that it's possible to have too much slope. When water moves through a horizontal pipe too quickly and that water is carrying solids, the water can outrun the solids, creating the possibility of clogs.

To prevent accumulation problems, limit the fall of any drain line to no more than ¼ inch per foot. If structural barriers force a slope greater than ¼ inch per foot or if the line will slope a distance greater than its own diameter along its length, use fittings to step up the line. Hold the line before and after the step at ⅛ to ¼ inch per foot. You must re-vent any line that slopes, from start to finish, more than its own diameter.

Fitting and pipe selection is also important. In a waste system, use only code-approved sanitary fittings. Do not use T-fittings in drainpipes except to drain and stack-vent the top fixture or the top bath group on a stack. A Y-connector, used in combination with a 45-degree elbow, offers a much more gradual flow, which is less likely to clog. This is true whether the stack is vertical or horizontal.

When you install a 90-degree elbow at the base of a vertical stack, use a long-sweep L configuration, or install two 45-degree elbows. Standard short-sweep elbows will do fine elsewhere in the system, but water falling vertically from a height of 8 to 24 feet needs a buffer as it changes direction. A more gradual turn will keep water from filling the elbow completely and prevent a momentary loss of vent.

TYPICAL DRAINAGE SYSTEM

WORKING BACKWARD—that is, starting with where the wastewater exits the house—the drainage system begins with the 4-inch soil pipe, which enters the house under an exterior footing or slab. (In some cases the soil pipe may enter the house through the basement or crawl space wall, especially in a septic-system arrangement.) Just after the cleanout, a 4 x 2-inch Y-fitting splits off to drain the laundry in the basement and the kitchen on the main floor. Codes often require that this takeoff be downstream of the larger toilet-branch line.

The next in-line fitting is a 4-inch Y-fitting that serves the basement bath group—toilet, tub, and sink. The toilet line is re-vented because it is a lower floor installation; it does double duty as a wet vent for the shower and sink.

Before the soil pipe sweeps up to become the primary 3-inch vertical stack, an unvented 2-inch Y-fitting serves a trapped floor drain. On the stack, a 3-inch T-fitting with a 1½-inch side inlet serves the toilet and shower. The fitting allows both fixtures to enter the stack at the same level and, therefore, allows both to be stack-vented.

Approximately 16 inches above the floor, a second T-fitting drains the sink basin, which is also stack-vented. From there, the stack continues through the roof.

Whole-House Drain and Vent System

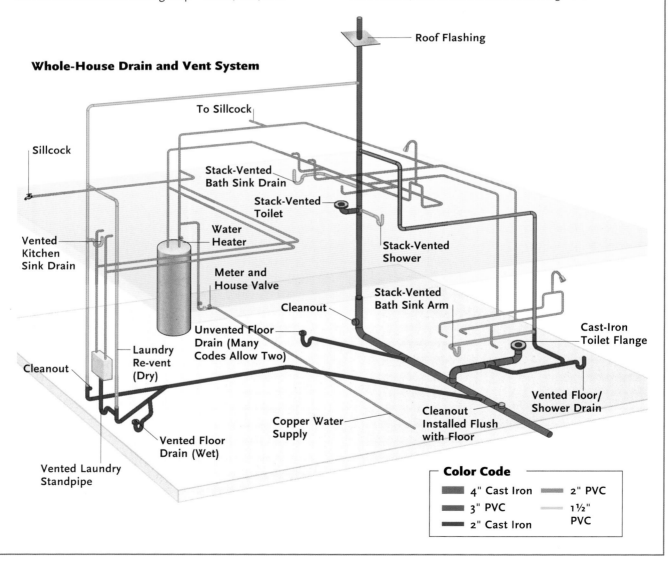

Roof Flashing

To Sillcock

Sillcock

Stack-Vented Bath Sink Drain

Stack-Vented Toilet

Water Heater

Vented Kitchen Sink Drain

Stack-Vented Shower

Meter and House Valve

Cleanout

Stack-Vented Bath Sink Arm

Cast-Iron Toilet Flange

Unvented Floor Drain (Many Codes Allow Two)

Laundry Re-vent (Dry)

Cleanout

Copper Water Supply

Cleanout Installed Flush with Floor

Vented Floor/ Shower Drain

Vented Floor Drain (Wet)

Vented Laundry Standpipe

Color Code

▬ 4" Cast Iron	▬ 2" PVC
▬ 3" PVC	▬ 1½" PVC
▬ 2" Cast Iron	

INSTALLING PIPE THROUGH JOISTS

Drill four holes if you don't have a large bit or hole saw.

Use a saber saw to cut out the lumber between the holes.

Fit short pieces of pipe between joists, and couple them together.

Supporting Pipe. When you run pipe through stud walls or through floor and ceiling joists, always drill the pipe holes slightly larger than the outside diameter of the pipe. Plastic pipe expands when warm water passes through it, and if the pipe fits too tightly, you'll hear a steady ticking sound when warm water is used. This annoying sound is the pipe rubbing against the wood as the plastic expands and contracts.

When you hang plastic drainpipes and vent pipes under floor joists, support the pipe with hole strapping or the appropriate pipe hangers. (See "Hangers and Straps" photo-graph, page 254.) Support plastic waste piping every 4 feet.

Cutting Structural Timbers. You'll rarely need to cut into load-bearing timbers to install drainpipes. When a drain line needs to travel with the joists toward an outside wall, you can usually tuck it up between the joists.

If you do need to run pipes through a few joists, drill only the center one-third of each joist, and where possible, stay within a few feet of a support wall. Never notch the bottom of a joist, because the bottom carries a disproportionate share of the load. Break these rules, and you could threaten the floor's load-bearing capacity.

Wire hangers, which you just hammer into joists to support pipes, are quick and affordable. Several other types of hangers are available.

Use a right-angle drill and a large self-feed bit to bore pipe holes in studs. Make the holes at least ⅛ in. over-size.

Drains, Vents & Traps

VENTS

As mentioned in Chapter 2 (pages 46 to 61), vents play a vital part in a home's drainage system. A faulty venting system will not only cause aggravation when the drainage system malfunctions, it could also make you and your family ill if it compromises the seal the traps are supposed to maintain against sewer gases. It is important to understand the various venting options possible, as well as size and installation basics, if you plan to alter or add to your current system.

Vent Types and Terms

The most common vent installations are illustrated and discussed in the following pages. If your planned system varies from those described here, draw a picture of it and ask the local building (or plumbing) inspector to check it before you do the work.

■ **Vent Terminal.** A *vent terminal* is roughly equivalent to the distance between floors, but because of layout considerations, it often starts a foot below one floor and ends a foot below the next.

■ **Broken Vent.** The term *broken vent* refers to an improperly installed vent that, consequently, is ineffective. A broken vent is little better than no vent at all. Often the problem stems from the location and elevation of the vent's takeoff fitting. The illustrations in "4 Common Broken Vents," opposite, show the situations you're most likely to face.

■ **Stack Vent.** Vertical drainage stacks are also considered vents because, in addition to carrying water, they pull relief air from above the roof. Short horizontal branch lines extending from a vertical stack do not require additional venting and, therefore, are said to be stack-vented.

■ **Vent Stack.** A *vent stack* is a vertical stack installed for the sole purpose of providing relief air. It does not carry water. A vent stack can be tied into the stack vent above the highest fixture or exit the roof independently.

You may have seen plumbing diagrams in the past with vent stacks starting below the lowest fixture on the stack and continuing into the attic. All fixtures are shown connected to this stack vent. However, dedicated system-wide vent stacks are rarely found in residential installations.

STACK-VENTED BATH GROUP

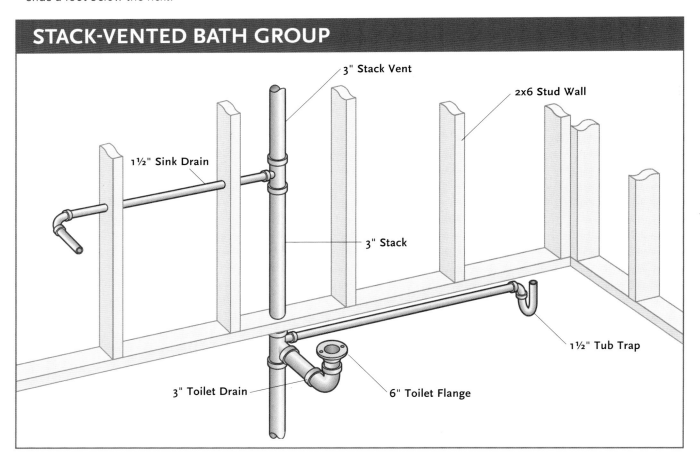

3" Stack Vent

2x6 Stud Wall

1½" Sink Drain

3" Stack

1½" Tub Trap

3" Toilet Drain

6" Toilet Flange

Drains, Vents & Traps

4 COMMON BROKEN VENTS

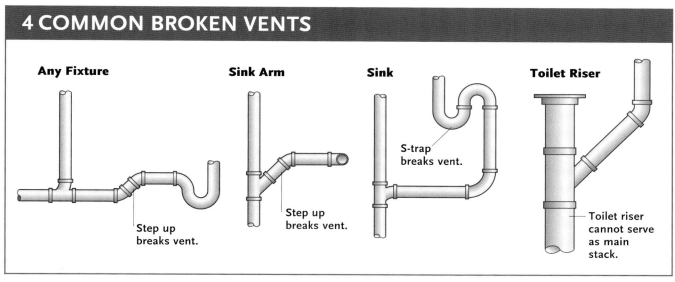

Any Fixture

Step up breaks vent.

Sink Arm

Step up breaks vent.

Sink

S-trap breaks vent.

Toilet Riser

Toilet riser cannot serve as main stack.

VENT STACK

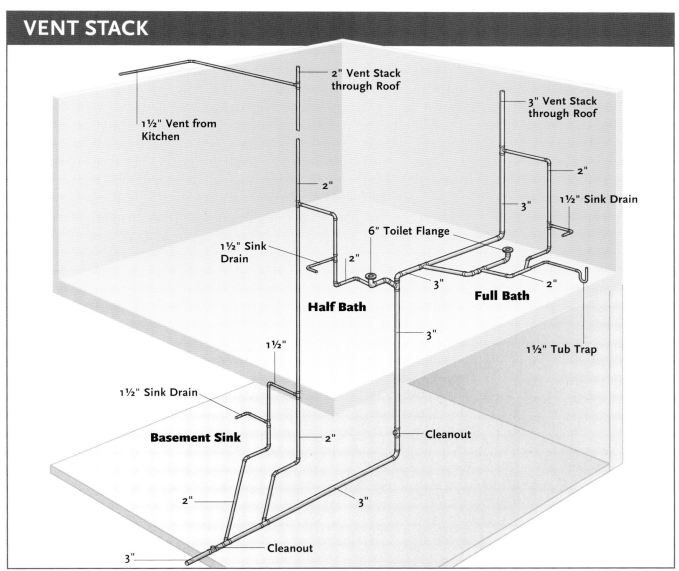

2" Vent Stack through Roof

1½" Vent from Kitchen

3" Vent Stack through Roof

2"

2"

3"

1½" Sink Drain

1½" Sink Drain

6" Toilet Flange

2"

3"

2"

Half Bath

Full Bath

3"

1½" Tub Trap

1½"

1½" Sink Drain

Basement Sink

2"

Cleanout

2"

3"

Cleanout

3"

Drains, Vents & Traps

■ **Stack-Vented Branch Arm.** In a properly designed system, fixture drain lines are sized so that they can never run more than one-half full. (See "Sizing Fixture Drains and Vents," page 283.) If a horizontal branch pipe is not too long, the unused top half of the pipe can serve as a vent. It feeds air from the stack to the back of the flow, preserving the trap seal. But such a stack-vented branch arm is limited.

- **The stack fitting** must be a T-fitting and not a Y-fitting.
- **The branch must serve** the highest fixture or fixture combination on the stack. This "highest fixture" rule applies to each individual stack, even when the highest fixture on a second stack is a floor or two above or below that of the primary stack, providing that all stacks extend to the open air above the roof. However, there are always exceptions: this rule does not apply to a basement bath group, because most of that piping is horizontal.
- **The length of the branch** should not exceed the limits established by code. The larger the pipe diameter, the longer the allowed length. For example, most codes specify that a 1½-inch-diameter branch arm can extend 5 feet from the stack to the trap, although many further limit that to 3½ feet. Most codes allow 8 feet for a 2-inch line and 10 feet for a 3-inch line. You must re-vent anything longer. (See "Maximum Stack-Vented Branch Arm Lengths," below.)

MAXIMUM STACK-VENTED BRANCH ARM LENGTHS

Fixture	Pipe Size (inches)	Max. Length of Branch Arm (feet)
Toilet	3"	10'
Toilet	4"	12'
Sink	1½"	5'
Lavatory	1½"	5'
2 Lavatories	2"	8'
Tub	1½"	5'
Shower	2"	8'
Laundry	2"	8'

All horizontal branch arms longer than noted must be re-vented.

STACK-VENT EXCEPTION

THERE IS ONE EXCEPTION to the "highest fixture" rule regarding stack-vented branch arms, and it's an important one. When the uppermost fixture is part of a bathroom group, you can also join the toilet piping to the stack with a T-fitting, even though it's technically the second-highest branch on the stack. Making the most of the rule that says two fixtures can be stack-vented if their pipes enter the stack at the same level, manufacturers now sell 3-inch T-fittings with 1½- or 2-inch side inlet fittings. A T-fitting with a side inlet allows a nearby tub or shower to enter the stack at the same level as the toilet. What this means, of course, is that you can stack-vent the entire bath group. You can handle the toilet and tub/ shower with a 3-inch side inlet T, and the lav with a standard T-fitting. (See "Side-Inlet Ports," pages 285–286.)

■ **Re-vent.** A re-vent is a vent added to a fixture drain when the primary vent is broken in some way. When a branch arm is too long, for example, or when it must step up to accommodate structural barriers, you must install a re-vent.

■ **Wet and Dry Vents.** A wet vent carries water as well as air through some part of its piping. A basement bathroom provides a good example because the vent for the toilet line also carries water from the tub and lavatory. While wet vents are handy and efficient, they have their limits. Because air must share space with water, wet vents are low-volume vents. They cannot vent many fixtures. A dry vent does not carry water, so it has greater fixture capacity. (See "Vent/Drain Sizes per Fixture/Appliance" and "Assigned Fixture Units per Fixture," page 283.)

■ **Common Vent.** A common vent is a single vent pipe that provides relief air to two or more fixtures on a single branch line. When you are venting two fixtures, you must place a common vent between the drain outlets. When you are venting more than two fixtures, install the vent between the last and second-to-last fixture outlets.

■ **Loop Vent.** When you want to install a sink in a kitchen island, a conventional vent will not work. In this situation, a loop vent is appropriate. Loop vents easily handle the low-volume fixtures (such as kitchen sinks) that are installed in island cabinets.

RE-VENTED BRANCH ARM

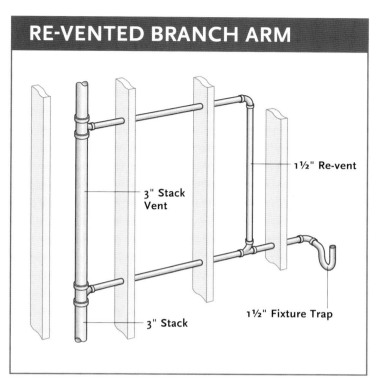

1½" Re-vent

3" Stack Vent

1½" Fixture Trap

3" Stack

COMMON VENT

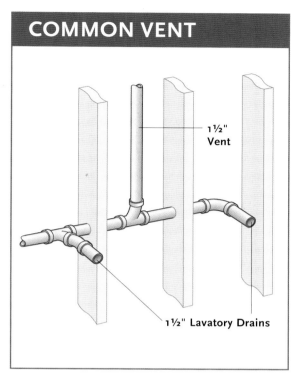

1½" Vent

1½" Lavatory Drains

TOILET, TUB, AND SINK SERVED BY A WET VENT

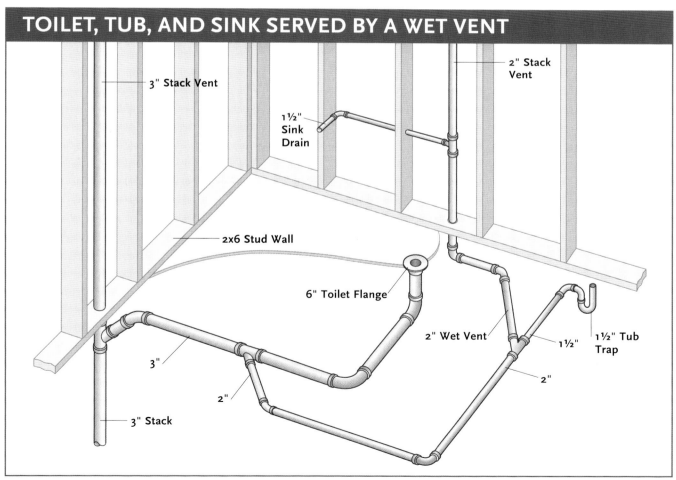

3" Stack Vent

2" Stack Vent

1½" Sink Drain

2x6 Stud Wall

6" Toilet Flange

2" Wet Vent

1½"

1½" Tub Trap

2"

3"

2"

3" Stack

UNDERCOUNTER LOOP VENT

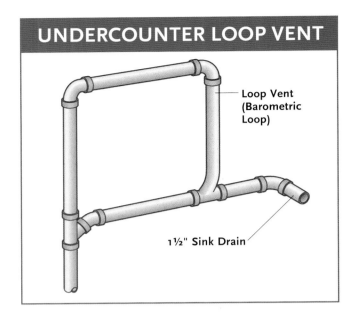

Loop Vent (Barometric Loop)

1½" Sink Drain

AUTOMATIC VENT DEVICE

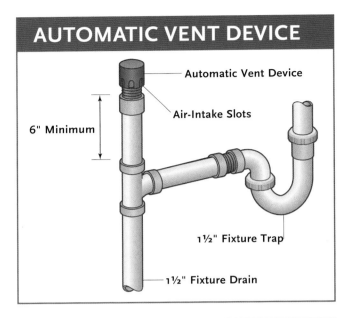

Automatic Vent Device

Air-Intake Slots

6" Minimum

1½" Fixture Trap

1½" Fixture Drain

Loop vents do not extend to open air or tie into stacks or vent stacks. They have no direct, fresh-air connections. Instead, the volume of air in the piping loop provides the needed pressure relief. For this reason, loop vents are also known as *barometric vents*.

■ **Automatic Vent Device.** An *automatic vent device* can take the place of a loop vent. It consists of an inverted cup with a spring-loaded diaphragm. The spring allows the piping system to pull air from the cabinet space but keeps gas from escaping. It's basically a kind of check valve, allowing air to flow in only one direction. While automatic vent devices can be system savers, do not use them where it's possible to install conventional vents. Because automatic vents have limited capacity, don't use them to vent high-volume drains or to vent more than one fixture. Install them so that they always remain accessible because they can wear out and may need replacing.

Basic Drain and Vent Considerations

Codes stipulate that you can stack-vent only branch lines that stem from a vertically positioned T-fitting and that Ts can only be used in drainpipes when the fixture to be drained is the highest fixture on a vertical stack. When two uppermost fixtures are stack-vented, their drain lines must enter the stack at the same level. Therefore, to connect two lines to the stack you will need a *side-inlet fitting* or a *cross-fitting* (a cross fitting is basically a double T-fitting). (See the photo on page 286.)

T-FITTING WITH SIDE INLET

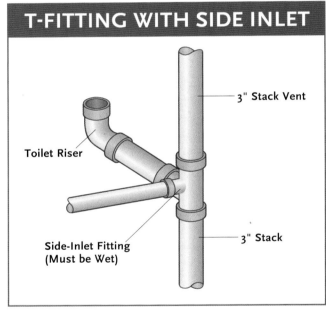

3" Stack Vent

Toilet Riser

Side-Inlet Fitting (Must be Wet)

3" Stack

Re-venting. Some codes state that any time a horizontal branch line falls more than its own diameter along its length, it too must be re-vented. Y-fittings, by their very shape, constitute what's called in the industry a step-up and cause the line to fall by more than its diameter, so all branch lines connected to a stack via a Y-fitting should also be re-vented. (Some codes waive this rule for 1½ inch lines, so you may want to check with your local building department.)

Whenever you install a vertical stack that jogs horizontally, you must re-vent all fixture lines attached to the horizontal section and all fixtures below the horizontal section.

And finally, you must re-vent even an uppermost branch arm if it's too long, as defined by local codes.

Some codes allow one or two unvented traps below the basement floor. If these exceptions are granted, plan on re-venting every other branch line in the system—except stack-vented branches. Place these vents near the fixtures, and always install the takeoff fitting on the highest horizontal plane of a branch line. If the drain line steps up, install the vent between the step and the fixture trap.

Secondary Stacks. Some codes require that homes have at least one full-size stack, but if you need additional stacks to serve bath groups elsewhere in the house, you can reduce the vent portions of the other stacks to 2 inches in diameter. Bring these vents through the roof, or tie them into the primary stack. Tie all of the secondary vents into a stack at least 6 inches above the flood plane of the highest fixture on that stack. For a 36-inch-high kitchen sink, for example, you would connect the re-vent to the stack no lower than 42 inches above the floor.

Sizing Fixture Drains and Vents

Code officials use a formula known as the fixture-unit measurement to determine the minimum allowable pipe size for each fixture or group of fixtures. This is a fairly complicated formula. You won't need to know it in detail, but the results—as organized in the tables at right—are important. Use the tables to customize your plans and solve specific problems.

In brief, engineers on standard-setting boards have assigned each household fixture and appliance a numerical value based on the maximum flow rate that each is likely to produce. (See "Assigned Fixture Units per Fixture," at right.) The goal is to size each drainpipe so that it will never run more than one-half full, leaving the remaining half of the pipe open to carry vent air. When a single line serves multiple fixtures, the total number of fixture units it will carry dictates the line size. Because drains and vents work together, fixture units are used to size both drain and vent lines.

Installing Vents and Vent Fittings

Vent fittings don't normally carry water, so they don't need the gradual sweeps and curves that drainage fittings use. Their turns are abrupt and short, like water fittings. Special vent fittings are seldom used at the residential level, however. This makes them difficult to find in retail outlets.

Most plumbers, and nearly all weekend plumbers, use

VENT/DRAIN SIZES PER FIXTURE/APPLIANCE

Fixture	Drain Size	Vent Size
Sink	1½"	1½"
Lavatory	1½"	1½"
Tub	1½"	1½"
Shower	2"	1½"
Laundry	2"	1½"
Floor Drain	2"	1½"
Toilet	3–4"	1½"

Vents for below-grade traps may be waived.

ASSIGNED FIXTURE UNITS PER FIXTURE

Fixture/Appliance	FU
Clothes washer (2" trap)	3
Bath group (toilet, lav & tub)	6
Bathtub, w/wo shower (1½" trap)	3
Dishwasher	2
Floor drain (2" trap)	3
Kitchen sink (1½" trap)	2
Kitchen sink w/disposer & dishwasher	3
Kitchen w/dishwasher	3
Lavatory (1½" trap)	1
Laundry sink (1½" trap)	2
Shower stall (2" trap)	2
Toilet (3.5 gal./1.6 gal.)	4/3

drainage, or sanitary, fittings in both the drain and vent segments of the system. Because you must use drainage fittings for wet vents, this is a reasonable approach. Moreover, because drainage fittings are readily available, sold everywhere in quantity, they cost less.

While you must connect most drainage lines with Y-fittings, you can often use T-fittings in vent connections. When a re-vent takes off vertically from a horizontal branch line (which is wet), you can use a T-fitting on its back—that is, with its in-line openings positioned horizontally and its branch inlet facing up. You can also use a T-fitting with its branch inlet facing down or to either side when a horizontal vent is dry. These positions are common when tying several vents together in an attic.

DESIGNING A CODE-APPROVED DRAIN AND VENT

HERE'S HOW TO USE the fixture-unit and pipe-size tables on pages 280, 283, and 285 to ensure a code-worthy installation. Let's say that you would like to stack-vent a horizontal branch arm that serves a bathtub and a freestanding shower. Assume that the tub will be 6 feet from the stack and the shower 9 feet from the stack in the same direction. According to the table "Maximum Stack-Vented Branch Arm Lengths," page 280, a shower requires a 2-inch trap and pipe diameter, and the maximum length of its branch arm is 8 feet. In this case, you can't stack-vent the shower because it is too far away (9 feet), so there's little point in stack-venting the tub. You should still drain the tub and shower through a common line, however, because that's the most economical way, and you may have limited room to fit the piping. But where should you place the re-vent, and what size must it be to pass inspection?

As described under "Common Vent" (page 280), if one vent serves two fixtures, you must place it between them. In this case, you can vent the shower through the Y-branch serving the tub. This will make a short length of the tub branch a wet vent, in that it will both drain and vent the tub while also venting the shower.

What size will this branch need to be? If you check the table "Assigned Fixture Units Per Fixture," page 283, you will see that a tub and shower have a combined value of four fixture units. Refer to the table "Maximum Fixture Units Per Vent Size," opposite, and you will see that a 1½-inch wet vent accommodates only two fixture units, but a 2-inch wet vent will carry four fixture units. This means that from a 2-inch drain line, you will need to branch off with a 2-inch Y-fitting to serve the tub, and you'll need to continue this size until the pipes branch again. You can reduce to 1½ inches out of the front in-line opening of this second Y-fitting because a tub requires only a 1½-inch trap and drain, and this section serves only the tub drain. The vent from this point on will be dry and will extend upward from the branch of the Y-fitting.

To determine the size of the dry vent you will need to continue to the open air above the roof, see "Vent/Drain Sizes per Fixture/Appliance," page 283. The table shows 1½ inches for a shower and 1½ inches for a tub. But according to the table "Maximum Fixture Units Per Vent Size," a 1½-inch dry vent can handle only three fixture units, which is not enough. You have four, which means that you must maintain a 2-inch re-vent.

Code Approved: Stack-Vented Horizontal Branch Arm

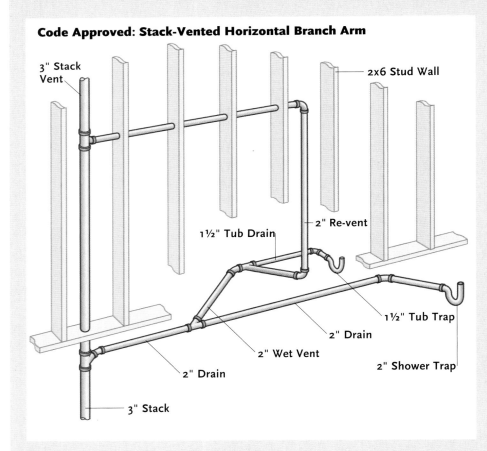

- 3" Stack Vent
- 2x6 Stud Wall
- 2" Re-vent
- 1½" Tub Drain
- 1½" Tub Trap
- 2" Drain
- 2" Wet Vent
- 2" Shower Trap
- 2" Drain
- 3" Stack

MAXIMUM FIXTURE UNITS PER VENT SIZE

Vent Size	FU/Wet	FU/Dry	Total for 2-3 Floors (Dry)*
1½" hor.	2	3	—
1½" vert.	2	3	4
2"	4	6	10
3"	—	20	48
4"	—	160	240

*A dry vent at its fixture unit limit on a lower floor is allowed additional capacity as it reaches second and third vent terminals.

INSTALLING FITTINGS

When you install drainpipe, make sure the flow pattern of T-fittings runs downward.

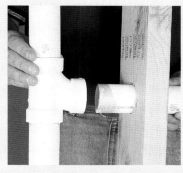

Invert the T-fittings for vent pipes, with the flow pattern running upward.

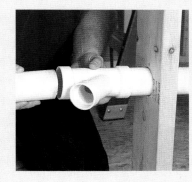

When you install a horizontal Y-fitting, rotate the branch upward slightly.

When you use a T-fitting to stack-vent a horizontal fixture drain, position the branch inlet so that the flow pattern sweeps downward, facilitating the flow of water. But do the opposite when connecting a horizontal dry vent to a vertical vent stack. In this case, you should install the T-fitting upside down, so the flow pattern of the branch sweeps upward, in keeping with the direction of airflow. Follow the same patterns when sloping piping that is connected to vent fittings. Just as you always position drainpipes to slope downward, you must always position horizontal vents to slope upward. A rise of 1/16 to 1/4 inch per foot is adequate.

When a wet re-vent takes off from a horizontal drain line, you must use a Y-fitting, of course, because a wet vent carries water through its lower section, and drainage rules take precedence. In this case, you must rotate the Y-fitting, used on its side, slightly upward. Canting the Y's inlet 1/8 inch above level is enough to ensure adequate flow and air intake. Because you often use Y-fittings in conjunction with 45-degree elbows, slightly cant the elbows, too.

■ **Cross-Fittings.** Codes stipulate that you can stack-vent two or more fixtures on the same level only when their drains enter the stack at the same level, so you'll need connectors with more than one inlet port to accomplish this. Cross-fittings, which are essentially double T-fittings, are examples of such connectors, as are Ts and crosses with side-inlet ports. All of these fittings expand the practical limits of stack-venting. Crosses work well when you splice them into vertical stacks, but do not work in horizontal piping runs. Additionally, some code authorities don't allow crosses on small drains because drain-auger cables tend to pass through them horizon-tally instead of dropping into the stack.

When two sinks are in the same wall, you can use a cross-fitting, with one sink on each side of the stack. You can also use cross-fittings to drain and stack-vent two toilets, again, with one toilet on each side of the stack, as is the case with back-to-back bathrooms. As always, codes allow Ts and crosses only on uppermost fixtures or fixture groups and on dry vents.

■ **Side-Inlet Ports.** T-fittings and cross-fittings with side-inlet ports are doubly useful in top-floor bathroom plumbing. You can drain and stack-vent a toilet with a 3-inch T-fitting, but you can also drain and vent a bathtub

TYPES OF DRAIN AND VENT FITTINGS

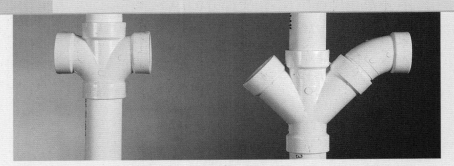

Comparing a cross-fitting with a double Y-connector, the Y has a more gentle sweep and requires two 45-deg. elbows, but takes up more space.

A 3- or 4-in. cross-fitting serves two toilets in back-to-back baths.

with a T that has a 1½ inch side inlet. In like manner, you can use a 3-inch cross-fitting to drain and vent two toilets, but a cross with two inlets also allows you to drain and vent a shower on one side and a bathtub on the other. When you use a 3 × 1½-inch cross-fitting above the toilet cross, it can drain and vent the two bath sinks if the bathrooms share a common wall. With stack-venting, you can plumb back-to-back bathrooms efficiently.

■ **Heel-Outlet 90-degree Elbows.** Heel-outlet elbow fittings can also save piping. The savings here are not in length but in diameter. The typical use for a heel-outlet 90-degree elbow is to vent a single toilet that is on a secondary stack. You bring a 3-inch vertical stack up to the basement ceiling and use the heel-outlet elbow (with the 2-inch outlet facing upward) to change from vertical to horizontal. A standard 90-degree elbow returns vertically into the toilet flange. In this case, the heel-outlet not only facilitates a smaller vent but allows the vent to be placed in a wall with standard 2×4s. You would have to fur out the wall at least an inch if you used a 3-inch vent. You can also cut lesser, nearby fixtures into the stack below the heel-outlet elbow, using a Y-fitting, and tie the vent serving these fixtures into the toilet vent 42 inches above the floor. Again, you can do all this within a standard wall. Heel-outlet elbows work well in these situations, but code does not allow the small port to carry water. While you can use a side-inlet port wet—washed by at least one fixture—you must always keep a heel-outlet port dry. If you need to drain and vent a toilet-and-sink combination, use a 3-inch elbow with a 2-inch side-inlet to drain the bathroom sink and create a vent. Install the elbow right below the toilet, and drain the sink through the inlet. Continue the inlet piping into the attic as a vent.

Roof Vent Considerations

Plumbing-stack flashing comes in three forms: lead and the more conventional neoprene rubber and sheet metal with neoprene inserts. You can also get a special sheet-metal flashing designed to prevent frost problems in northern climates.

Lead Flashing. Lead flashing was the standard for many years, but sheet metal and neoprene are better for today's modern asphalt shingles. Roofers still use lead when installing a wood-shake shingle roof, however, because lead flashing has a larger apron, so it offers greater leak protection with shake shingles.

Conventional Flashing. Neoprene flashing and sheet-metal flashing with neoprene inserts are easy to install. In new construction, simply press the flashing over the vent stack. When installing a new vent through an existing roof, you'll need to work harder for a watertight fit. Start by considering whether a new roof hole is even necessary. It's often easier and cheaper to tie the new vent into an existing stack in the attic, avoiding the roof entirely. If this isn't possible, proceed from the attic. Bring the new stack through the attic floor, and suspend a plumb line above the vent to determine the exact location of the roof cut. If the vent falls too near an existing structure (roof window, roof vent, or the like), offset it a couple of feet with 22-, 45-, or 90-degree elbows. When you've determined the best location, mark a circular opening on the roof sheathing about 6 inches in diameter. Drill ¼-inch holes in at least four equidistant spots along the cut line. Then go on top of the roof, and use a reciprocating saw or saber saw to cut through the sheathing and shingles by connecting the drill holes. You can also cut the shingles first using a utility knife.

Side-inlets allow a cross-fitting to serve toilets, plus shower and tub.

A heel-outlet elbow's 2-in. fitting, used vertically, can vent a toilet.

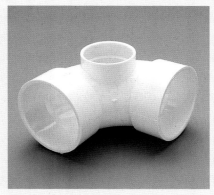

A side-inlet elbow's 2-in. inlet can vent a toilet and drain a sink.

Installations vary, but it is often easier to splice the new flashing into the existing shingles before bringing the vent through the roof. Slide the top half of the flashing under the shingles above the opening, and allow the bottom half to lie on top of the shingles below the opening. This will require removing a few nails from the shingles above the opening. Slide a hacksaw blade under the appropriate shingles, and cut through the nails. Then slide the flashing under these shingles so that the opening in the flashing is directly over the vent opening. You may also need to trim the top shingle flaps a bit to accommodate the raised insert.

With the flashing in place, go into the attic to install the last section of pipe, feeding it through the flashing opening from below. Return to the roof, and apply roofing compound—fiber-reinforced tar in a caulk tube—under the exposed outside edges of the flashing. Lastly, nail through the bottom corners of the flashing and apply a dab of compound to each nailhead.

This method works well, but you can also install the vent pipe before installing the flashing. In this case, press the flashing over the vent, turn it sideways, and rotate it under the upper shingles. Again, you may need to remove a nail or two.

Frost-Proof Flashing. In northern climates, the warm, damp air flowing up from a stack can freeze when it reaches the cold roof area, forming frost that blocks the vent pipe. Some codes call for pipes 3 inches in diameter.

There is also frost-inhibiting roof-vent flashing. This flashing has double sheet-metal walls with an air space between them. The air space serves as an insulator between the warm stack and the cold surrounding roof surface.

TYPES OF VENT-PIPE FLASHING

Lead vent-pipe flashing can be made watertight by folding its excess length into the stack.

Sheet-metal flashing with a neoprene insert can sometimes be rotated under existing shingles.

TRAPS

Traps vary in size and shape, according to the volume of water they handle. Traps that serve high-volume appliances, such as washing machines, are usually 2 inches in diameter. Toilet traps are even larger. A bathroom sink trap, in contrast, usually has a 1¼-inch trap.

Volume and trap size are not the only considerations, however. Modern traps have a precise geometry that permits them to allow wastewater to pass yet retain a reservoir of water as a seal, even under slight negative pressure. Because of this precision in design, you should always use manufactured traps; do not make your own from conventional fittings.

Trap Variations

Older houses can have a variety of fixture and drain traps, some good and some not so good. Two traps that were popular earlier in the last century, the S-trap and drum trap, for example, are no longer permitted by codes, although hundreds of thousands of them may still be in use.

■ **Drum Trap.** Used almost exclusively in bathrooms, the drum trap was installed in the floor, usually near the toilet. This trap has a juglike shape, with two inlets and one outlet. The top, usually capped with a brass cleanout plug, is usually visible in a bare floor. The inlet pipes enter the drum at a point lower than the outlet,

which creates the water seal. (See the illustration below.) The two inlet pipes typically drain the bathtub and sink. You can determine whether an older home has a drum trap in any of several ways. The easiest is to look for the cleanout fitting in the bathroom floor, although a floor covering may now conceal it. If your home was built before the 1940s and the bath has not been substantially remodeled, you'll likely find a drum trap. Layout can be another indicator. In drum trap installations, the sink is usually between the tub and toilet, while in modern installations the toilet is often between the sink and the tub. Slow drains also yield good information. A sink that backs up into the tub may signal a drum trap. For more

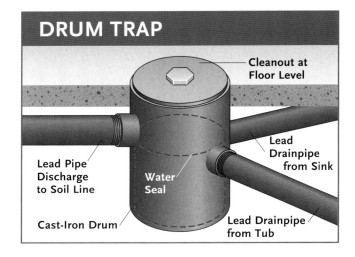

DRUM TRAP

Cleanout at Floor Level

Lead Drainpipe from Sink

Water Seal

Lead Drainpipe from Tub

Cast-Iron Drum

Lead Pipe Discharge to Soil Line

CONNECTING TRAPS

Join chrome traps using flat rubber friction washers.

S-traps connect to drainpipes at floor level. Try to avoid them.

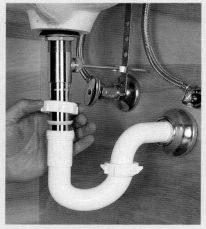

Join plastic P-traps using nylon compression washers.

on replacing a drum trap, see "Replacing an Old Drum-Trap System," pages 96–97.

- **S-Trap.** Do not install them unless you are directly replacing a leaking or broken one. Building codes do not allow using them for new construction or remodels.
- **P-Trap.** The only code-worthy external trap is the P-trap. You can easily take these traps apart. They consist of a U-shaped bend and trap arm, plus several nuts and washers. P-traps come in plastic and chrome-plated brass. Plastic lasts longer, but chrome looks better.

Chrome trap connections have hex nuts and flat rubber friction washers. Plastic traps have beveled joints and tapered nylon compression washers with nuts that you can tighten by hand. In contrast, you tighten chrome traps using a pipe wrench. Use pipe joint compound on chrome traps but not on plastic traps.

You will find the second type of P-trap on built-in fixtures, such as bathtubs and showers, and on laundry stands. These traps may come in two pieces, but once assembled, they are not meant to be taken apart.

- **Running Traps.** A running trap is a variation on the P-trap. Plumbers use this specialty fitting only when a permanent P-trap is not workable—such as in situations where structural framing members are in the way or where a P-trap could freeze. A drain at the bottom of an external stairwell gives a good example: you install the drain basin outdoors but move the trap indoors where the water stays warm. Codes require that the trap be within 2 feet of the fixture outlet or floor drain that it serves. Greater distances can cause the trap to siphon. (For more on this, see "Siphoning Action," page 290.)
- **Built-in Traps.** Some fixtures, such as toilets and bidets, have built-in traps. These fixtures do not need, and cannot have, additional traps. In fact, double trapping of any fixture is prohibited by code. House traps, as described below, are the only exception. Two traps in close proximity to each other can be a problem; the second trap can siphon the first—and be depleted itself—by momentum.
- **House Traps.** Some local codes require the use of a house trap in a dwelling. This trap is in the soil pipe beneath the basement floor and is the same size as the soil pipe, usually 4 inches in diameter. If a house trap is not required in your area, don't use one.
- **Area Drains.** Garage floors often have built-in drains called area drains. Builders frequently stand a square clay-tile flue liner on end to make the drain reservoir. Factory-made plastic drains are also now available.

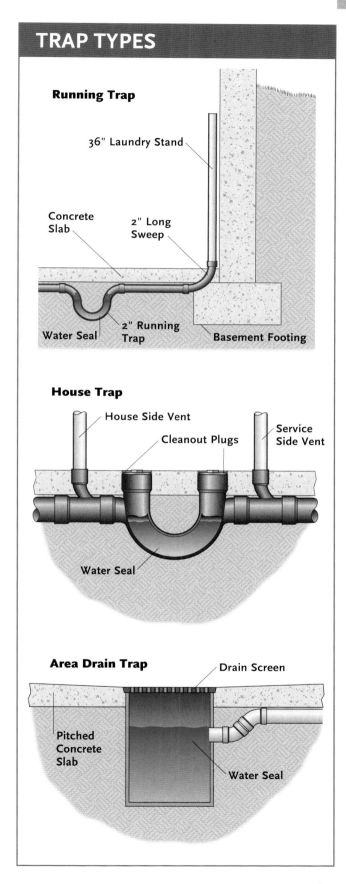

TRAP TYPES

Running Trap

36" Laundry Stand

Concrete Slab

2" Long Sweep

Water Seal

2" Running Trap

Basement Footing

House Trap

House Side Vent

Cleanout Plugs

Service Side Vent

Water Seal

Area Drain Trap

Drain Screen

Pitched Concrete Slab

Water Seal

Drains, Vents & Traps

Broken Trap Seals

A broken trap seal occurs when the water level in the trap recedes enough to allow sewer gas to escape. Broken seals may be caused by aspiration, capillary attraction, evaporation, momentum, or siphoning action.

Aspiration. Aspiration occurs when a large volume of water flows near the trap and creates negative pressure, which pulls standing water from the trap.

There are two major symptoms of aspiration. One is the possible odor of sewer gases, although most of the poisonous gases are odorless. The other is the familiar choke and gurgle of unvented low-volume drains. If you have a fixture drain in your house that goes "glug-glug-glug" when a nearby fixture or appliance is used, that's the unmistakable sound of aspiration resulting from improper venting. You should call in a plumber.

Every fixture with an S-trap is subject to aspiration because you can't vent S-traps. Short of replacement or installing an automatic vent device (page 282), you can do little about S-trap problems.

Momentum. The momentum of water passing through traps can also empty them. This occurs when the vertical distance between the fixture outlet and the trap is too great; limit it to around 12 inches. Most codes prohibit a vertical distance greater than 24 inches, except for laundry standpipes.

Capillary Attraction. Capillary attraction lifts water the way a wick lifts lamp oil. All it takes is a few strands of fabric caught by a rough edge in the trap. The threads absorb water from the bottom of the trap and lift it into the drainpipe. This process takes a few days, so it's most noticeable in drains that are used only once or twice a week. Laundry drains are particularly prone to broken seals caused by capillary attraction.

Evaporation. If you've ever returned from a long vacation to find that your house has a foul smell, chances are that evaporation opened a trap to sewer gas. Floor drains are especially prone to evaporation because they are so seldom used. Your basement floor drain should have a float ball, a lightweight plastic ball that drops to seal the drain if the water evaporates. When water flows into the drain, the ball floats, allowing the water to escape through the drain. If your drain does not have one, pour a pint of water into the drain every week, or install a float ball.

Siphoning Action. Siphoning, which can deplete the water in an S-trap, occurs as follows: when a rapid flow of water runs through an S-trap, the water fills the entire trap. The resulting water pressure is not relieved by vent-supplied air at the top of the trap bend, as it is in a P-trap, and a siphon occurs, in which the pressure created at the front of the flow pulls too much of the back of the flow through the trap, leaving too-little water behind.

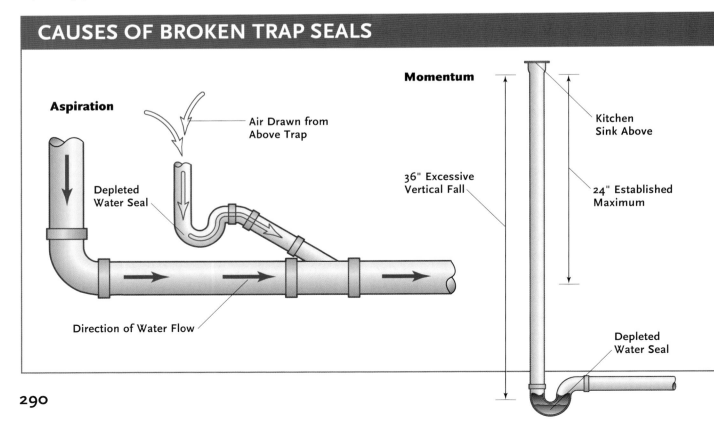

CAUSES OF BROKEN TRAP SEALS

Aspiration

Air Drawn from Above Trap

Depleted Water Seal

Direction of Water Flow

Momentum

Kitchen Sink Above

36" Excessive Vertical Fall

24" Established Maximum

Depleted Water Seal

PLUMBING 101: How to Avoid Aspiration

WHEN WATER FLUSHES through a drainpipe, it displaces a quantity of air equal to its own volume. The water pulls air in behind it to fill the void as it moves through the pipe. If air does not move into the pipe, a partial air lock occurs, slowing the water flow.

While flow through a drainage system never completely stops, local airflow problems are common. Unless a plumbing system is well designed with careful attention to venting, it may be subject to areas of isolated pressure when water moves through it in volume. Water from a high-volume fixture rushing past a smaller branch line that serves a bathroom sink or tub is more likely to pull air from that branch than it is to pull from the open air above the roof stack, which can be 20 to 30 feet away. When air is pulled from an improperly vented branch line, it pulls water from the trap at the end of that line.

But if you drilled a hole in the top of that branch line, a short distance from the trap, the rushing water moving past the branch would pull its air through that hole, leaving the water in the trap undisturbed and the seal intact. To get the same results, plumbers splice a T-fitting into the line instead of drilling a hole. They run piping up from the T until they can tie the pipe back into the main stack above the highest fixture or run it through the roof independently, creating a vent.

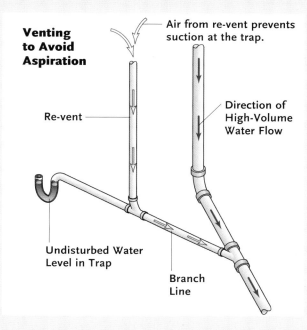

Venting to Avoid Aspiration

Air from re-vent prevents suction at the trap.

Re-vent

Direction of High-Volume Water Flow

Undisturbed Water Level in Trap

Branch Line

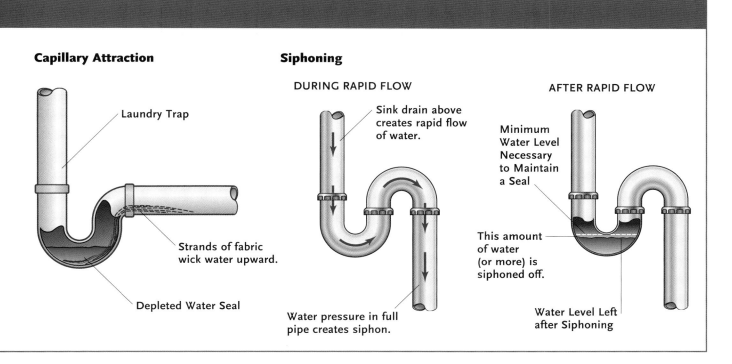

Capillary Attraction

Laundry Trap

Strands of fabric wick water upward.

Depleted Water Seal

Siphoning

DURING RAPID FLOW

Sink drain above creates rapid flow of water.

Water pressure in full pipe creates siphon.

AFTER RAPID FLOW

Minimum Water Level Necessary to Maintain a Seal

This amount of water (or more) is siphoned off.

Water Level Left after Siphoning

AIR GAP FITTING REQUIREMENTS

The 2021 Plumbing Code requires the use of air gap fittings for automatic dishwashers, water softeners, and water filtering systems such as reverse osmosis units. This requirement has been adapted to prevent the siphoning of traps attached to standpipes in the main wastewater piping of a residence. This also includes the risk of wastewater backups that can contaminate the drain lines, filters, and complete units attached to the house waste plumbing system.

Air gap fittings can be attached to the existing 1 ½-inch standpipe attached to a P-trap or on the sink deck flange in the kitchen. By doing so, the presence of a clog will become quickly evident as drain water will spill into the sink bowl or be present on the floor of a mechanical room, giving the homeowner sufficient notice that a clog or other issue has occurred.

Pictured here are the most common types of air gap fittings available (for various appliances and units) to comply with this code change.

EXAMPLES OF AIR GAP FITTINGS

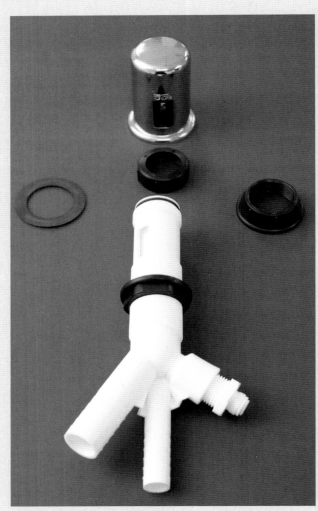

An example fitting for automatic dishwashers

An example fitting for reverse osmosis units

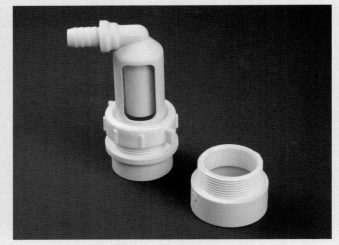

An example dual automatic dishwasher/reverse osmosis combination fitting

An example fitting for water softeners

Installation for Automatic Dishwashers

The automatic dishwasher installation method that was formerly acceptable was for the plumber or appliance technician to install the flexible drain hose for the dishwasher in a "high loop" method. The drain line would enter through a hole that was drilled through the bottom of the side wall of the sink cabinet and then be attached to either the garbage disposal inlet port or the drop tube connected to a ⅝-inch hose barb. Both attachments would use a stainless steel worm clamp or spring clamp to attach the drain hose. The upper loop of the dishwasher drain hose would then be clamped to the underside of the kitchen countertop or high on the back wall in the dishwasher cavity to prevent the loop from falling down.

Unfortunately, many dishwashers have been installed without the high loop in the drain line and with the excess hose length coiled inside the sink cabinet or behind the dishwasher. These methods can allow the sink water in the P-trap of the sink drain or the water that is in the bottom of the garbage disposal to siphon back into the automatic dishwasher, which will contaminate the dishwasher and create foul odors inside it. In addition, a clogged dishwasher drain line can allow water to back up inside the dishwasher, as it cannot properly drain to the garbage disposal or the kitchen sink drop tube that the hose is attached to.

To prevent these situations from happening, the 2021 plumbing code has adopted the use of an air gap fitting. This dishwasher air gap fitting is placed in a separate hole in the kitchen sink deck flange or mounted through a hole in the countertop directly next to the sink to allow backed-up overflow water to drain into the sink. This signals to the occupant that there is a plumbing issue that needs to be addressed immediately. Usually the problem will be in the form of a clog in the dishwasher drain line coming into the air gap fitting, the air gap fitting itself, the line running between the air gap fitting and the sink/garbage disposal, or the kitchen sink drain line.

One fitting design utilizes hose barb inlets and outlets and requires clamps that will connect to the dishwasher drain hose and then be connected to the inlet of the drop tube, where the water will safely drain into the kitchen sink P-trap or be connected to the inlet of the garbage disposal mounted under the sink (see the two images below).

AIR GAP FITTING FOR AUTOMATIC DISHWASHER

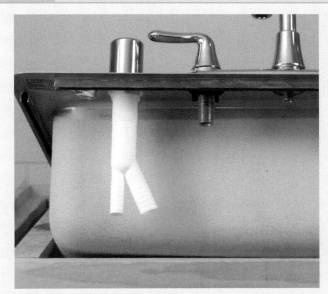

Here is a possible placement for the air gap fitting, in the sink deck flange. This one is installed where there was already a hole in the deck flange for, for example, a sprayer. A professional could punch a new hole for you.

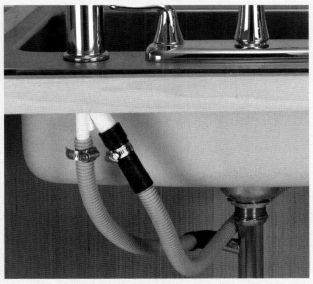

Here is an installed air gap fitting for a dishwasher where there is no garbage disposal, in its final setup leading directly into the sink tailpiece.

Installation for Reverse Osmosis Units

The reverse osmosis system installation method that was formerly acceptable was for the plumber or technician to install the flexible drain hose for the reverse osmosis system and have the drain line enter through a saddle drain port in the kitchen sink drain line, since most reverse osmosis systems are installed directly under the kitchen sink.

One issue that can occur from this setup is that the sink drain can be become clogged, thereby allowing the drain water to flow back into the filter cartridges, creating bacterial issues in your filtered drinking water. Another potential issue is that when the reverse osmosis system is draining but the sink drain stopper is in place, it has the potential to create negative pressure in the drain line, which will cause problems.

A dual-purpose air gap fitting design is available, which connects the dishwasher to the fitting inlet side, the fitting to the garbage disposal, and then can also include the drain line necessary for the reverse osmosis water filtering drinking system (see the two images below).

AIR GAP FITTING FOR REVERSE OSMOSIS UNIT

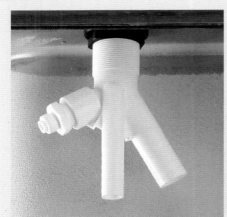

As shown on page 293, here is a possible placement for the air gap fitting, in the sink deck flange, but this time a combination dishwasher/reverse osmosis unit fitting has been installed.

Here is the finished installation including a garbage disposal. Have your installer connect everything as illustrated. The dishwasher line leads off to the right side; the clear tube leads to the reverse osmosis unit.

Installation for Water Softeners

The water softener installation method that was formerly acceptable was for the plumber or water softener technician to install the flexible drain hose for the water softener and have the drain line enter through a vertical 1½-inch standpipe attached to a P-trap in the drain line of the main sewer piping. The 1½-inch standpipe drain installation can allow the trap water to siphon back into the open end of the softener flexible drain line and contaminate the line (see the image below).

Even though every softener has a built-in check valve that only allows the discharge water to flow through the drain line, it still can cause problems if the check valve fails or if there are contamination issues from a water backup.

The use of a PVC air gap fitting that adapts to the main sewer 1½-inch standpipe above the P-trap, using the included compression washers and nuts, allows for the drain water to not siphon the trap water and flow back into the flexible drain line of the softener, thereby creating a safe situation for the drinking water of the home (see the image below).

AIR GAP FITTING FOR WATER SOFTENER

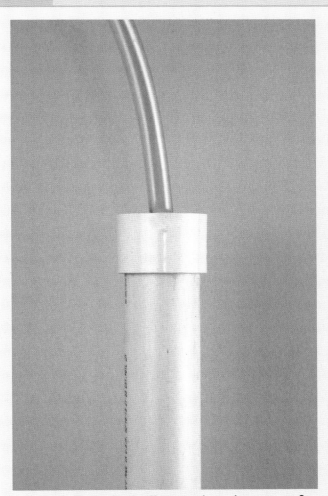

Here is a normal installation where the water softener drain line has been run directly into a 1½-inch riser, creating a possibility for contamination.

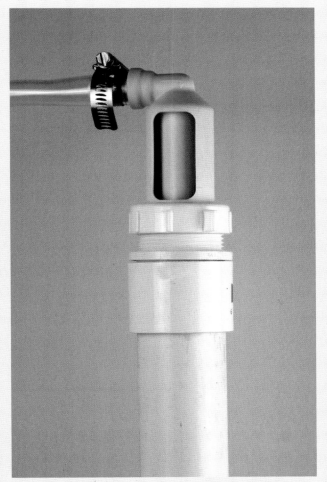

You can retrofit the existing 1½-inch riser to create a sealed connection with the approved air gap fitting. With this setup, contaminated water cannot escape back up the clear tube to the water softener.

This list of manufacturers and associations is meant to be a general guide to additional industry and product-related sources. It is not intended as a listing of products and manufacturers represented by the photographs in this book.

Air-Conditioning, Heating, and Refrigeration Institute (AHRI)
2311 Wilson Blvd., Ste. 400
Arlington, VA 22201
(703) 524-8800
www.ahrinet.org
Trade association that represents more than 300 manufacturers of air conditioning, heating, and commercial refrigeration equipment.

American Society of Plumbing Engineers (ASPE)
6400 Shafer Court., Ste. 350
Rosemont, IL 60018
(847) 296-0002
www.aspe.org
International organization for professionals skilled in the design, specification, and inspection of plumbing systems.

American Standard
865 Centennial Ave.
Piscataway, NJ 08854
(855) 815-0004
www.americanstandard.com
Manufactures products for the bath and kitchen, including faucets, fixtures, showers, and sinks.

American Water Works Association (AWWA)
6666 W. Quincy Ave.
Denver, CO 80235
(800) 926-7337
www.awwa.org
Dedicated to the promotion of public health and welfare in the provision of drinking water of unquestionable quality and sufficient quantity.

Delta Faucet Company
55 E. 111th St.
Indianapolis, IN 46280
(800) 345-3358
www.deltafaucet.com
Manufactures fixtures for the kitchen, bath, bar, and laundry.

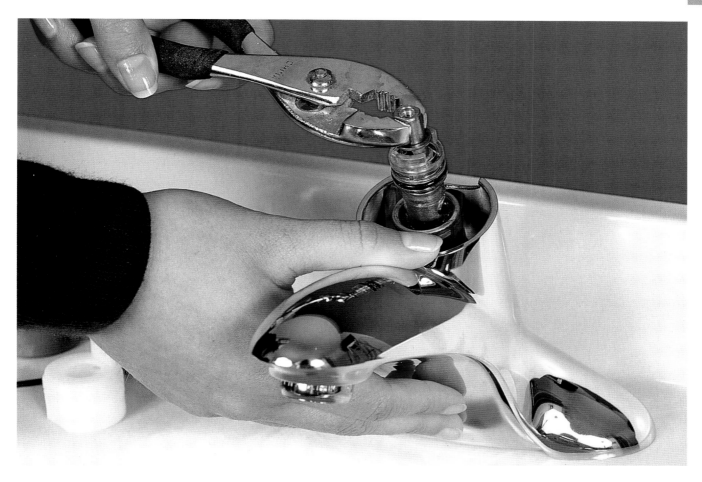

Ditch Witch
1959 W. Fir Ave.
P.O. Box 66
Perry, OK 73077-0066
(800) 266-3255
www.ditchwitch.com
Designs and manufactures trenching products, from compact walk-along models to large riding units.

Fluidmaster Inc.
30800 Rancho Viejo Rd.
San Juan Capistrano, CA 92675
(949) 728-2000
www.fluidmaster.com
Manufactures a full line of toilet repair products.

Gerber Plumbing Fixtures Corp.
2500 Internationale Pkwy.
Woodridge, IL 60517-4073
(630) 679-1420
www.gerberonline.com
Manufactures kitchen and bath plumbing fixtures.

GROHE America
865 Centennial Ave.
Piscataway, NJ 08854
(855) 815-4726
www.grohe.us
Manufactures faucets for kitchens.

International Code Council (ICC)
200 Massachusetts Ave., NW
Washington, DC 20001
(888) 422-7233
www.iccsafe.org
Develops and maintains codes required in the construction of residential and commercial structures.

King Innovation
16250 W. Woods Edge Rd.
New Berlin, WI 53151
(800) 624-4320
www.kinginnovation.com
Designs and manufactures labor-saving solutions for the irrigation industry that speed up installation and prevent problems. (Formerly Blazing Products)

Resource Guide

Kohler
444 Highland Dr.
Kohler, WI 53044
(800) 456-4537
www.kohler.com
Manufactures bath and kitchen fixtures, including a variety of sinks, lavs, faucets, tubs, and accessories.

National Kitchen and Bath Association
1 W. Broad St. Ste. 300
Bethlehem, PA 18018
(800) 843-6522
www.nkba.org
Helps homeowners with everything from finding a designer in their area to information on trends, products, and services.

NSF International – The Public Health and Safety Company
789 N. Dixboro Rd.
Ann Arbor, MI 48105
(800) 673-6275
www.nsf.org
Develops national standards, provides learning opportunities, and provides third-party certification of water-treatment products.

Plumbing-Heating-Cooling Contractors Association
180 S. Washington St.
Falls Church, VA 22046
(800) 533-7694
www.phccweb.org
An organization that helps homeowners find qualified professionals to keep their homes safe and problem free.

Plumbing Manufacturers International (PMI)
1750 Tysons Blvd. Ste. 1500
McLean, VA 22102
(847) 481-5500
www.safeplumbing.org
The national trade association of plumbing products manufacturers.

Rain Bird
970 W. Sierra Madre Ave.
Azusa, CA 91702
(800) 724-6247
www.rainbird.com
Manufactures residential irrigation products and services.

House of Rohl

3 Parker
Irvine, CA 92618
(800) 777-9762
www.houseofrohl.com
Manufactures luxury fixtures and faucets for kitchens and baths.

United States Plastic Corp.

1390 Neubrecht Rd.
Lima, OH 45801
(800) 809-4217
www.usplastic.com
Offers a catalog of plastic products, including pipes, tools, and fittings for all applications, through its Web site.

Viega Manufacturing

585 Interlocken Blvd.
Broomfield, CO 80021
(800) 976-9819
www.viega.us
Manufactures pipe-line systems and pipe connection products.

Water Quality Association (WQA)

2375 Cabot Dr.
Lisle, IL 60532-3696
(630) 505-0160
www.wqa.org
Not-for-profit international trade association that represents the household, commercial, industrial, and small-community water-treatment industry.

World Plumbing Council

P.O. Box 2005, MARMION
Western Australia 6020
www.worldplumbing.org
Seeks to facilitate the exchange of information, ideas, and technology between the plumbing industry and individuals worldwide.

Clamp-type flaring tool For flaring soft copper tubing. It consists of a base clamp, which bites onto the pipe, and a flaring vise with a threaded stem and cone-shaped flare head, which slides over the clamp. As you screw downward, you force the head into the end of the pipe.

Close-quarters tubing cutter A much smaller version of a tubing cutter, it is handy when working in confined spaces.

Closet flange The rim on a closet bend by which that pipe attaches to the floor. See *Toilet flange*.

Compression fittings Replaced most cone and friction washers in the 1960s. These fittings tighten, or compress, over the tapered seats with pressure to lock into place.

Cone washers Also known as friction washers, these transition fittings were used to join permanent pipes to fixture-supply tubes prior to the 1950s.

Corpcock (or corporation stop) The first shutoff in the water line that is tapped directly into an iron or plastic water main and can be reached only by excavating the soil above it.

Coupling A transitional fitting that joins pipes end to end.

CPVC (chlorinated polyvinyl chloride) pipe Plastic water piping.

Curb stop (or stop box) The second shutoff in a water line. It is usually found underground in a public area such as between the street and the sidewalk.

Deglaze (plastic pipe) The process of priming pipe ends and fittings to roughen the surfaces and make it easier for the solvent to achieve a good bond. You can also use an abrasive pad to achieve this.

Dielectric union A special coupling used when connecting piping made of dissimilar metals. It prevents the transfer of electrons, which results in electrolytic corrosion.

Drainpipe Pipe that directs wastewater away from the fixtures and house to the sewer or septic tank.

Drum trap An outdated waste trap that has a juglike shape, with two inlets and one outlet. It was used almost exclusively in bathrooms and can be found in the floor of older houses, near the toilet. It cannot be vented and is no longer allowed by code.

Dual stop A valve with two outlets that splits one water line (usually hot) into two—for example, joining a kitchen's hot-water supply line to the kitchen faucet and a dishwasher.

Elbow A fitting used for making directional changes in pipelines.

Escutcheon A decorative plate that covers the hole in a wall through which a pipe or faucet body passes.

Extension tube   A copper tube with compression fittings at one end. The tube bridges the gap between the supply risers and faucet stubs.

Faucet stem (compression)   The main vertical component of a compression faucet. It has a handle at one end and a seat washer at the other. It is threaded and moves up and down to control water flow.

Female thread   The end of a pipe or fitting with internal threads.

Ferrule   On a compression fitting, the beveled compression ring that is driven into the tapered seats to lock the fitting in place.

Fill valve   The device in a toilet that allows water to refill the tank after each flush.

Fixture   Any of a number of water-using devices hooked up to the main water supply network: sinks, showers, bathtubs, and toilets.

Fixture riser   Any length of vertical pipe that supplies or drains water to or from a fixture. See Riser.

Flapper   A rubber seal that controls the flow of water from a toilet tank.

Flush bag (also called blow bag)   A special bag that attaches to a garden hose and is inserted into a large drainpipe. It uses water pressure in order to clear clogs.

Flush valve   A device at the bottom of a toilet tank for flushing. See *Flapper*.

Freeze-proof sillcock   A sillcock with an extended stem that allows water to be shut off inside rather than outside the house. It prevents the water line from freezing in the cold winter months.

Freeze-proof yard hydrant   The equivalent of a freeze-proof sillcock, but installed in the yard. These faucets supply water to gardens and livestock tanks and drain underground rather than through the spout.

Friction washer   A flat, rubber washer used to make a water seal in chrome P-traps.

Frost-proof flashing   Flashing that has double sheet-metal walls with an air space in between them. This air space acts as an insulator to prevent freezing in colder climates.

Gallons per flush (gpf)   The amount, in gallons of water, used for each flush of a toilet.

Galvanized-steel pipe   Once used for in-house water systems, steel pipe is now used mostly in repair situations only.

Gate valve   A valve that allows an unrestricted flow of water via an internal gate. Gate valves are predominately used at or near the beginning of water supply systems.

Gravity-flow toilet   A toilet designed to use the force of gravity to flush wastewater away, through the drainage system.

Gray water   Wastewater from sinks, baths, and kitchen appliances.

Hammer-type flaring tool   A cone-shaped stud that's hit by a hammer to create a flare in the end of copper tubing for a flared fitting.

Hole strapping   Plastic or copper strapping used to support pipe against studs and joists.

Hootie wench   A special sink-clip wrench used to remove the rim clips that hold a sink in place on the counter.

Hose bibcock   An external-threaded faucet onto which a hose is attached. (See *silcock*.)

Integral trap   A trap molded into a fixture. It sweeps up before it sweeps down, keeping standing water in place.

Laundry box   A recessed connection box for water and drain lines. It is mounted on top of the standpipe.

Lavatory basin   A sink, or wash basin, that is located in a bathroom or powder room.

Line friction   Water pressure loss due to friction caused by pipe walls and fittings.

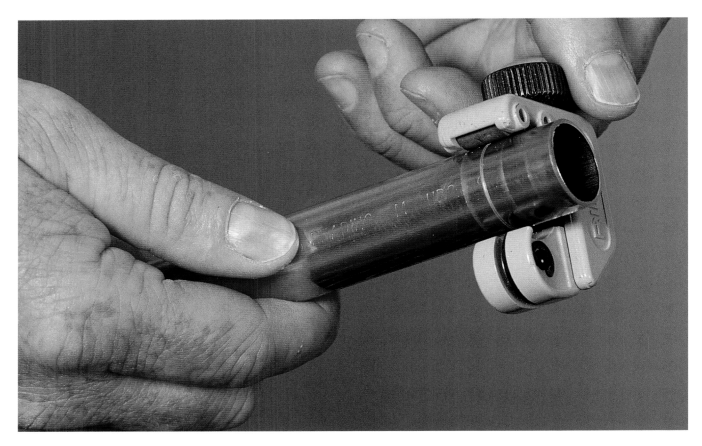

Low-flow toilet A water-saving toilet now mandatory in residential homes since the passing of the Clean Water Act in 1994. These toilets use only 1.6 gallons per flush (gpf).

Male thread The external threads located at the end of a fitting, pipe, or fixture connection.

Meter pit Underground container—usually near the street, on either side of the sidewalk—that houses below grade water meters.

Oakum An oily ropelike material that was commonly used years ago in combination with lead to make watertight seals on joints.

O-ring packing Modern alternative to traditional packing. It consists of a lubricated o-ring that fits over the stem.

PEX Acronym for cross-linked polyethylene plastic tubing. PEX is a strong, flexible tube used for radiant floor heating and water delivery.

P-trap A U-shaped bend and a trap arm, this is the most common type of trap. It may be made of plastic or chrome-plated brass and is commonly used to drain sinks, tubs, and showers.

PVC (polyvinyl chloride) plastic pipe Piping used for drainage and venting systems; it is corrosion-resistant and easy to install.

Packing (compression valve/faucet) Material, often graphite, used to create a seal around the stem in a compression faucet.

Pipe chase A boxed-out cavity, vertical or horizontal, through which pipes may travel from floor to floor or room to room.

Pipe joint compound A pastelike filler material applied to threaded connections to help prevent leaks.

Plumber's grease Heatproof nontoxic grease applied to washers and other moving parts as a lubricant.

Plumber's code The rules and regulations determined and enforced by a state, local, or municipal authority governing all plumbing work done in that jurisdiction.

Pounds per square inch (psi) The measure of water pressure in a system, expressed in pounds of pressure per square inch of surface. Too much pressure causes water-heater relief valves to leak, toilets to keep running, and faucets to pulse and pound when turned off and on.

Potable Water that is free from impurities in amounts sufficient to cause disease or harmful physiological effects.

Pressure-assisted toilet A toilet designed to push water through the system more forcefully with the help of compressed air.

Pressure-reduction valve A valve that reduces excessively high, damaging water pressure and conserves water.

Public parking The area outside a house that lies between the sidewalk and the street.

Reducer A fitting used to join two pipes of different diameters.

Re-vent A pipe installed to vent a fixture trap that connects to a main vent.

Rim holes (toilet) Holes around the underside perimeter of the toilet rim through which water flows into the bowl from the tank.

Riser A water supply or drainage pipe that carries or drains water vertically.

S-trap A sink trap that connects to drainpipes at floor level. These traps are impossible to vent and are no longer allowed by code.

Seat The stationary base in a compression valve onto which the stem closes to shut off water supply. The stem holds a washer that fits into the seat.

Seat-dressing tool A threaded stem with a T-handle and several grinding blades that cut into a pitted seat to make it smooth.

Seat washer The rubber or neoprene washer that covers the lower end of the faucet stem and fits into the seat of the faucet body to form a seal.

Septic system The intricate network of pipes and tanks that collect and dispose of wastewater from the house.

Sewer A public drainline, usually underground, that carries away waste or rainwater.

Shutoff valve A device set into a water line to allow for interruption of the flow of water to a fixture or appliance.

Sillcock An outdoor water faucet; also called a *hose bibcock*.

Siphon jet A hole at the bottom of a toilet's water bowl through which water from the tank flows to aid in creating a flushing siphon.

Siphoned trap A situation in which the flow of water from a sink fills an improperly vented trap and thereafter pulls excess water through the pipe, leaving too little water behind to seal the trap.

Soil pipe A drainpipe that carries wastes to the sewer drain; also, the main drainpipe that receives all the wastewater from a group of plumbing fixtures, including a toilet, or from all the plumbing fixtures in a given installation.

Spigot (cast-iron fitting) The male end of a cast-iron waste pipe that requires a neoprene gasket or packing material to fit into the female end, or bell, of another pipe to make a seal.

Glossary

Spud gasket (flush valve)   Rubber gasket that fits over the flush-valve's shank threads that will connect to the spud nut and washer joining the two sections of toilet.

Spud washer (flush valve)   A large rubber seal that fits over the spud nut connecting the toilet tank to the base and bowl of the toilet.

Stem   The component in a faucet that rises up and down via its threads, when the faucet handle is turned, to control the flow of water.

Stem nut   The nut that connects to and tightens on the faucet stem. It may also serve as the packing nut in some faucet designs.

Step-up   An abrupt vertical offset in a horizontal piping run.

Stop valve   A valve, mandatory under toilets and often under sinks as well, that allows you to shut off a water line completely. There are two basic types: compression stops and ball-valve stops.

Stop-and-waste valve   Usually globe valves with a drain screw on the downstream side of the shutoff. These valves are widely used in the winter months to shut down water running through dedicated outdoor lines.

Stop box   The vertical extension of a curb stop that extends from the water service line up to grade level.

Stub-out   The termination of water-delivery or drainage-network pipe extended into a room through a wall or floor to which a fixture or appliance is to be connected.

Supply tube   The flexible tubing that carries water from the shutoff valves to the fixture above.

Sweat fittings   Copper or brass fittings that are made to be soldered.

Sweep bend   A fitting used for vent and waste piping that has a gradual 90-degree bend. It may have either a short or long vertical bend.

T-fitting   A T-shaped pipe fitting with three points of connection.

Takeoff fitting   A fitting that branches off a drainpipe or vent pipe. The location of a vent takeoff fitting is crucial to proper venting.

Tank ball (toilet)   A hollow ball attached to a lift wire that opens and closes the tank water outlet with each flush.

Temperature-and-pressure (T&P) relief valve   The device installed in a water heater that prevents temperature and pressure from building up inside the tank and causing it to explode.

Thumb cutter   See *Close-quarters tubing cutter.*

Toilet flange   A slotted ring connected to a vertical collar that rests on the floor. Toilets are bolted directly to this fitting.

Trap   The water-filled curved pipe that prevents sewer gas from entering the house through the drainage network.

Trap-preserving vent   A vent added to a drainage line to ensure that a trap receives enough makeup air and functions properly.

Trunk lines   The hot- and cold-water supply lines that usually run side by side along the center beam of the house, branching to serve isolated fixtures or fixture groups along the way.

Tube trap   A conventional P-trap for a sink or other fixture that usually comes as a kit, with all components needed for complete installation.

Union fittings   Threaded fittings that join two lengths of pipe together, allowing you to connect or disconnect pipes repeatedly anywhere they are installed.

Vacuum breaker   A device that prevents back-siphoning of water into a water supply line, usually mounted on the faucet.

Vents   Pipes that pull system makeup air from above the roof. Vents control the water seals for traps. Some may also carry water. See *Wet vent.*

Vent stack   A vertical vent installed for the sole purpose of providing relief air. It does not carry water.

Water hammer   A back shock in water pipes caused by a sudden change in pressure after a faucet or water valve shuts off.

Water spot (toilet)   The amount of water held in the bowl of a toilet. The larger the water spot, the more likely the bowl will remain clean.

Weir   The highest point in any fixture trap or built-in trap of a toilet. It regulates the amount of water held in the trap. In toilets, it determines how much water remains in the bowl.

Wet vent   An extension of a waste pipe that also acts as a vent. It is not permitted to receive waste coming from a toilet or kitchen sink.

Wiped-lead joints   A hand-formed splice between lead and galvanized-steel pipe. Outdated and no longer used.

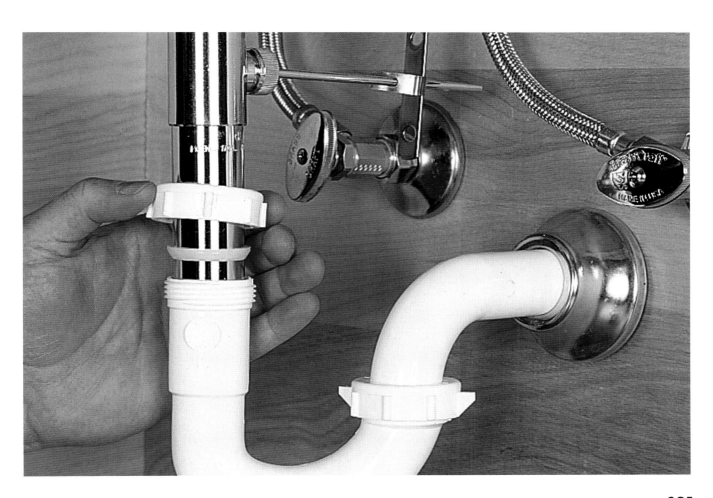

Index

Photo Credits

Back cover: *top* Freeze Frame Studio/CH

page 2: Freeze Frame Studio/CH

page 8: *top* courtesy of Moen; *bottom* courtesy of Kohler

page 14: Freeze Frame Studio/CH

page 38: *top right* Rinku Dua/Shutterstock.com

page 52: John Parsekian/CH

page 59: Brian C. Nieves/CH

page 65: *illustrations* courtesy of American Standard

page 79: *top* courtesy of Kohler; *bottom all* courtesy of TOTO USA

page 109: *all* John Parsekian/CH

page 137: *bottom left* courtesy of GE; *top right* courtesy of Whirlpool; *bottom right* courtesy of Miele

page 147: Freeze Frame Studio/CH

page 154: *left* Freeze Frame Studio/CH

page 156: Freeze Frame Studio/CH

page 162: *top left* & *bottom right* courtesy of Moen; *top right* courtesy of Delta Faucet Company; *bottom left* courtesy of Kohler

page 164: Freeze Frame Studio/CH

page 199: *top* courtesy of Kohler; *bottom illustration by* Ian Warpole

page 200: *all* courtesy of Jason International

page 205: *illustration by* Ian Warpole

page 211: courtesy Marathon Water Heaters

page 212: *top illustrations by* Clarke Barre

page 226: *bottom left* Kharoll Mendoza/Shutterstock.com

page 231: *top illustration* courtesy of Pure Water; *bottom* courtesy of Culligan

pages 238–239: *top illustrations by* George Retseck

page 273: *all* Gary David Gold/CH

pages 292–295: Charles T. Byers

page 300: Artazum/Shutterstock.com

page 303: courtesy of Kohler

page 311: courtesy of Kohler

Metric Equivalents

Length

1 inch	25.4mm
1 foot	0.3048m
1 yard	0.9144m
1 mile	1.61km

Area

1 square inch	645mm²
1 square foot	0.0929m²
1 square yard	0.8361m²
1 acre	4046.86m²
1 square mile	2.59km²

Volume

1 cubic inch	16.3870cm³
1 cubic foot	0.03m³
1 cubic yard	0.77m³

Common Lumber Equivalents

Sizes: Metric cross sections are so close to their U.S. sizes, as noted below, that for most purposes they may be considered equivalents.

Dimensional lumber		
	1 x 2	19 x 38mm
	1 x 4	19 x 89mm
	2 x 2	38 x 38mm
	2 x 4	38 x 89mm
	2 x 6	38 x 140mm
	2 x 8	38 x 184mm
	2 x 10	38 x 235mm
	2 x 12	38 x 286mm

Sheet sizes		
	4 x 8 ft.	1200 x 2400mm
	4 x 10 ft.	1200 x 3000mm

Sheet thicknesses		
	¼ in.	6mm
	⅜ in.	9mm
	½ in.	12mm
	¾ in.	19mm

Stud/joist spacing		
	16 in. o.c.	400mm o.c.
	24 in. o.c.	600mm o.c.

Capacity

1 fluid ounce	29.57mL
1 pint	473.18mL
1 quart	0.95L
1 gallon	3.79L

Weight

1 ounce	28.35g
1 pound	0.45kg

Temperature

Fahrenheit = Celsius x 1.8 + 32

Celsius = Fahrenheit - 32 x ⁵⁄₉

Nail Size and Length

Penny Size	Nail Length
2d	1"
3d	1¼"
4d	1½"
5d	1¾"
6d	2"
7d	2¼"
8d	2½"
9d	2¾"
10d	3"
12d	3¼"
16d	3½"